BIOINFORMATICS

The Impact of Accurate Quantification on Proteomic and Genetic Analysis and Research

BIOINFORMATICS
The Impact of Accurate Quantification on Proteomic and Genetic Analysis and Research

Edited by
Yu Liu, PhD

Apple Academic Press

TORONTO   NEW JERSEY

Apple Academic Press Inc.	Apple Academic Press Inc.
3333 Mistwell Crescent	9 Spinnaker Way
Oakville, ON L6L 0A2	Waretown, NJ 08758
Canada	USA

©2014 by Apple Academic Press, Inc.

First issued in paperback 2021

Exclusive worldwide distribution by CRC Press, a member of Taylor & Francis Group

No claim to original U.S. Government works

ISBN 13: 978-1-77463-340-3 (pbk)
ISBN 13: 978-1-77188-019-0 (hbk)

Library of Congress Control Number: 2013958243

Library and Archives Canada Cataloguing in Publication

Bioinformatics (2014)
Bioinformatics: the impact of accurate quantification on proteomic and genetic analysis and research/edited by Yu Liu, PhD.

Includes bibliographical references and index.
ISBN 978-1-77188-019-0 (bound)
1. Bioinformatics. I. Liu, Yu, 1963-, editor of compilation II. Title.

| QH324.2.B55 2014 | 570.285 | C2014-900045-6 |

Apple Academic Press also publishes its books in a variety of electronic formats. Some content that appears in print may not be available in electronic format. For information about Apple Academic Press products, visit our website at **www.appleacademicpress.com** and the CRC Press website at **www.crcpress.com**

ABOUT THE EDITOR

YU LIU, PhD

As a bioinformatician, Dr. Yu Liu's research has been centered on the development and application of computational tools for the study of complex diseases. He is familiar with data generated from microarray, next generation sequencing, and high-resolution mass spectrometry, and he has extensive experience for developing bioinformatics tools and system biology approaches to study complex diseases such as sleep apnea, neurodegenerative diseases, and cancers. More recently, he has developed systems biology approach that enables the discovery of high-level disease mechanisms and provide testing hypotheses for further research. Currently, he is a senior research associate in the Center for Proteomics and Bioinformatics of Case Western Reserve University, Cleveland, Ohio. He got PhD in Bioinformatics from Montreal University, Montreal, Canada, and had postdoc training in University of Toronto, Canada.

CONTENTS

Part I: RNA-Seq

Part II: Microarray

Part III: GWAS

Part IV: Proteomics

ACKNOWLEDGMENT AND HOW TO CITE

The editor and publisher thank each of the authors who contributed to this book, whether by granting their permission individually or by releasing their research as open source articles. The chapters in this book were previously published in various places in various formats. To cite the work contained in this book and to view the individual permissions, please refer to the citation at the beginning of each chapter. Each chapter was read individually and carefully selected by the editors. The result is a book that provides a nuanced study of the recent advances in bioinformatics.

LIST OF CONTRIBUTORS

Elizabeth A. Ainsworth
USDA ARS Global Change and Photosynthesis Research Unit, 1201 W. Gregory Drive, Urbana, IL 61801, USA and Department of Plant Biology, University of Illinois, Urbana-Champaign, Urbana, IL, 61801, USA

Miguel A. Anton
CEIT and TECNUN, University of Navarra, San Sebastián, Spain

Guillemette Antoni
UMR_S 937, INSERM, Boulevard de l'Hopital, Paris, 75013, France, UMR_S 937, ICAN Institute, Université Pierre et Marie Curie, Boulevard de l'Hopital, 75013, Paris, France and Dalla Lana School of Public Health, University of Toronto, College Street, Toronto, M5T 3M7, Ontario, Canada

M. Michael Barmada
Department of Human Genetics, University of Pittsburgh, Pittsburgh, PA, USA

Pedro Carmona-Saez
Integromics SL, Madrid, Spain

William Cohen
UMR_S 626, INSERM, rue Saint-Pierre, Marseille, 13385, France and UMR_S 626, Université de la Méditerranée, rue Saint-Pierre, Marseille, 13385 France

Colin N. Dewey
Department of Computer Sciences, University of Wisconsin-Madison, Madison, WI, USA and Department of Biostatistics and Medical Informatics, University of Wisconsin-Madison, Madison, WI, USA

Apostolos Dimitromanolakis
Dalla Lana School of Public Health, University of Toronto, College Street, Toronto, M5T 3M7, Ontario, Canada

Eleazar Eskin
Department of Computer Science, University of California Los Angeles, Los Angeles, California, US

France Gagnon
Dalla Lana School of Public Health, University of Toronto, College Street, Toronto, M5T 3M7, Ontario, Canada

Emmanuelle Génin
Inserm U946, F-75010, Paris, France and Institut Universitaire d'Hématologie, Université Paris Diderot, F-75010, Paris, France

Marine Germain
UMR_S 937, INSERM, Boulevard de l'Hopital, Paris, 75013, France and UMR_S 937, ICAN Institute, Université Pierre et Marie Curie, Boulevard de l'Hopital, 75013, Paris, France

Dorleta Gorostiaga
CEIT and TECNUN, University of Navarra, San Sebastián, Spain

Elizabeth Guruceaga
CEIT and TECNUN, University of Navarra, San Sebastián, Spain

Eran Halperin
The Blavatnik School of Computer Science, and the Molecular Microbiology and Biotechnology Department, Tel-Aviv University, Tel-Aviv, 69978, Israel and International Computer Science Institute, 1947 Center St., Berkeley, AC 94704, USA

William E. Haskins
Pediatric Biochemistry Laboratory, University of Texas at San Antonio, TX, 78249, USA, Depts. Biology & Chemistry, University of Texas at San Antonio, TX, 78249, USA, RCMI Proteomics & Protein Biomarkers Cores, University of Texas at San Antonio, San Antonio, TX 78249, USA, and Dept. of Medicine, Division of Hematology & Medical Oncology, Cancer Therapy & Research Center,University of Texas Health Science Center at San Antonio, San Antonio, TX, 78229, USA

Dan He
Program in Medical and Population Genetics, Broad Institute, Cambridge, MA 02142, USA

Farhad Hormozdiari
Department of Computer Science, University of California Los Angeles, Los Angeles, California, US

Alain Hovnanian
INSERM U781, F-75743, Paris, France, Université René Descartes, F-75743, Paris, France and Centre Hospitalier Universitaire Necker-Enfants malades, Departments of Genetics and Dermatology, F-75743, Paris, France

Xia Jiang
Department of Biomedical Informatics, University of Pittsburgh, Pittsburgh, PA, USA

Rémi Kazma
Inserm U946, F-75010, Paris, France and Institut Universitaire d'Hématologie, Université Paris Diderot, F-75010, Paris, France

Mark Lathrop
Institut de Génomique, Centre National de Génotypage, Commissariat à l'Energie Atomique, rue Gaston Crémieux, Evry, 91057, France

Bo Li
Department of Computer Sciences, University of Wisconsin-Madison, Madison, WI, USA

Chun Li
Department of Biostatistics, Vanderbilt University School of Medicine, Nashville, TN 37232, USA and Center for Human Genetics Research, Vanderbilt University School of Medicine, Nashville, TN 37232, USA

Yi-Xue Li
Bioinformatics Center, Key Laboratory of Systems Biology, Shanghai Institutes for Biological Sciences, Chinese Academy of Sciences, 320 Yueyang Road, Shanghai 200031, P. R. China, Shanghai Center for Bioinformation Technology, 100 Qinzhou Road, Shanghai 200235, P. R. China and School of Life Science and Technology, Tongji University, 1239 Siping Road, Shanghai 200092, P.R. China

Yuan-Yuan Li
Bioinformatics Center, Key Laboratory of Systems Biology, Shanghai Institutes for Biological Sciences, Chinese Academy of Sciences, 320 Yueyang Road, Shanghai 200031, P. R. China and Shanghai Center for Bioinformation Technology, 100 Qinzhou Road, Shanghai 200235, P. R. China

Yvonne Liss
Dokumentationszentrum schwerer Hautreaktionen (dZh), Department of Dermatology, D-79095, Freiburg, Germany

Bao-Hong Liu
Shanghai Center for Bioinformation Technology, 100 Qinzhou Road, Shanghai 200235, P. R. China and School of Life Science and Technology, Tongji University, 1239 Siping Road, Shanghai 200092, P.R. China

Douglas W. Mahoney
Division of Biomedical Statistics and Informatics, Department of Health Sciences Research, Mayo Clinic, 200 First Street SW, Rochester, MN 55905, USA

Maja Mockenhaupt
Dokumentationszentrum schwerer Hautreaktionen (dZh), Department of Dermatology, D-79095, Freiburg, Germany

Luis M. Montuenga
Center for Applied Medical Research, University of Navarra, Pamplona, Spain and Department of Histology and Pathology, University of Navarra, Pamplona, Spain

Pierre-Emmanuel Morange
UMR_S 626, INSERM, rue Saint-Pierre, Marseille, 13385, France and UMR_S 626, Université de la Méditerranée, rue Saint-Pierre, Marseille, 13385 France

Ekaterina Myasnikova
Department of Computational Biology, Center for Advanced Studies, St.Petersburg State Polytechnical University, St.Petersburg, 195251, Russia

Luigi Naldi
Department of Dermatology, Azienda Ospedaleria Ospedali Riuniti di Bergamo, Milano University, Bergamo, Italy

Richard E. Neapolitan
Department of Computer Science, Northeastern Illinois University, Chicago, IL, USA

Ann L. Oberg
Division of Biomedical Statistics and Informatics, Department of Health Sciences Research, Mayo Clinic, 200 First Street SW, Rochester, MN 55905, USA

Tiphaine Oudot-Mellakh
UMR_S 937, ICAN Institute, Université Pierre et Marie Curie, Boulevard de l'Hopital, 75013, Paris, France

Nathaniel Parrish
Department of Computer Science, University of California Los Angeles, Los Angeles, California, US

Bogdan Pasaniuc
Department of Epidemiology, Harvard School of Public Health, Boston, Harvard University, MA 02115, USA and Program in Medical and Population Genetics, Broad Institute, Cambridge, MA 02142, USA

Alberto Pascual-Montano
Computer Architecture Department, Facultad de Ciencias Físicas, Universidad Complutense de Madrid, Madrid 28040, Spain

Konstantinos Petritis
Center for Proteomics, Translational Genomics Research Institute, Phoenix, AZ 85004, USA

Ruben Pio
Center for Applied Medical Research, University of Navarra, Pamplona, Spain and Department of Biochemistry, University of Navarra, Pamplona, Spain

Andrei Pisarev
Department of Computational Biology, Center for Advanced Studies, St.Petersburg State Polytechnical University, St.Petersburg, 195251, Russia

Jean-Claude Roujeau
Inserm U448, F-94010, Créteil, France and Service Dermatologie, Hôpital Henri-Mondor, Université Paris-Est, F-94010, Créteil, France

Angel Rubio
CEIT and TECNUN, University of Navarra, San Sebastián, Spain

Maria Samsonova
Department of Computational Biology, Center for Advanced Studies, St.Petersburg State Polytechnical University, St.Petersburg, 195251, Russia

Martin Schumacher
Institute of Medical Biometry and Medical Informatics, University Medical Center, D-79095 Freiburg, Germany

Victor Segura
CEIT and TECNUN, University of Navarra, San Sebastián, Spain

Peggy Sekula
Institute of Medical Biometry and Medical Informatics, University Medical Center, D-79095 Freiburg, Germany

Grigory Stein
Confocal Microscopy and Image Processing Group, Institute of Cytology RAS, St.Petersburg, 194064, Russia

Svetlana Surkova
Department of Computational Biology, Center for Advanced Studies, St.Petersburg State Polytechnical University, St.Petersburg, 195251, Russia

Jyothi Thimmapuram
Roy J. Carver Biotechnology Center, University of Illinois, Urbana-Champaign, Urbana, IL, 61801, USA and Bioinformatics Core, Discovery Park, Purdue University, West Lafayette, IN, 47907, USA

David-Alexandre Tregouet
UMR_S 937, INSERM, Boulevard de l'Hopital, Paris, 75013, France and UMR_S 937, ICAN Institute, Université Pierre et Marie Curie, Boulevard de l'Hopital, 75013, Paris, France

Mark J. van der Laan
Division of Biostatistics, University of California Berkeley, 101 Haviland Hall, Berkeley, CA 94720, USA

Shyam Visweswaran
Department of Biomedical Informatics, University of Pittsburgh, Pittsburgh, PA, USA, Intelligent Systems Program, University of Pittsburgh, Pittsburgh, PA, USA and Clinical and Translational Science Institute, University Pittsburgh, Pittsburgh, PA, USA

Hui Wang
Department of Pediatrics, Stanford University, MSOB X111, Stanford, CA 94305, USA

Philip Wells
Department of Medicine, Ottawa Hopital Research Institute, Carling Avenue, Ottawa, K1Y 4E9, Ontario, Canada

Zhi-Qiang Ye
Bioinformatics Center, Key Laboratory of Systems Biology, Shanghai Institutes for Biological Sciences, Chinese Academy of Sciences, 320 Yueyang Road, Shanghai 200031, P. R. China and Shanghai Center for Bioinformation Technology, 100 Qinzhou Road, Shanghai 200235, P. R. China

Craig R. Yendrek
USDA ARS Global Change and Photosynthesis Research Unit, 1201 W. Gregory Drive, Urbana, IL 61801, USA

Hui Yu
Bioinformatics Center, Key Laboratory of Systems Biology, Shanghai Institutes for Biological Sciences, Chinese Academy of Sciences, 320 Yueyang Road, Shanghai 200031, P. R. China, Graduate University of the Chinese Academy of Sciences, 19A Yuquanlu, Beijing 100049, P. R. China, and Shanghai Center for Bioinformation Technology, 100 Qinzhou Road, Shanghai 200235, P. R. China

Noah Zaitlen
Department of Epidemiology, Harvard School of Public Health, Boston, Harvard University, MA 02115, USA and Program in Medical and Population Genetics, Broad Institute, Cambridge, MA 02142, USA

Jianqiu Zhang
Dept. Electrical and Computer Engineering, University of Texas at San Antonio, TX 78249, USA

INTRODUCTION

This book provides an overview to the field of bioinformatics and statistical methodology. The articles chosen show various approaches to bias correction and error estimation, as well as quantitative methods for genome and proteome analysis. The book is broken in to four parts: Part I focuses on RNA-Seq, Part II is on microarrays, Part III gives several descriptions of various Genome-Wide Association Studies, and Part IV describes proteomics.

RNA sequencing (RNA-Seq) is emerging as a highly accurate method to quantify transcript abundance. However, analyses of the large data sets obtained by sequencing the entire transcriptome of organisms have generally been performed by bioinformatics specialists. Chapter 1, by Yendrek and colleagues, provides a step-by-step guide and outlines a strategy using currently available statistical tools that results in a conservative list of differentially expressed genes. The authors also discuss potential sources of error in RNA-Seq analysis that could alter interpretation of global changes in gene expression. When comparing statistical tools, the negative binomial distribution-based methods, edgeR and DESeq, respectively identified 11,995 and 11,317 differentially expressed genes from an RNA-seq dataset generated from soybean leaf tissue grown in elevated O_3. However, the number of genes in common between these two methods was only 10,535, resulting in 2,242 genes determined to be differentially expressed by only one method. Upon analysis of the non-significant genes, several limitations of these analytic tools were revealed, including evidence for overly stringent parameters for determining statistical significance of differentially expressed genes as well as increased type II error for high abundance transcripts. Because of the high variability between methods for determining differential expression of RNA-Seq data, the authors suggest using several bioinformatics tools, as outlined here, to ensure that a conservative list of differentially expressed genes is obtained. They also conclude that despite these analytical limitations, RNA-Seq provides highly accurate transcript abundance quantification that is comparable to qRT-PCR.

Recent studies in genomics have highlighted the significance of sequence insertions in determining individual variation. Efforts to discover the content of these sequence insertions have been limited to short insertions and long unique insertions. Much of the inserted sequence in the typical human genome, however, is a mixture of repeated and unique sequence. Current methods are designed to assemble only unique sequence insertions, using reads that do not map to the reference. These methods are not able to assemble repeated sequence insertions, as the reads will map to the reference in a different locus. In Chapter 2, Parrish and colleagues present a computational method for discovering the content of sequence insertions that are unique, repeated, or a combination of the two. Their method analyzes the read mappings and depth of coverage of paired-end reads to identify reads that originated from inserted sequence. The authors demonstrate the process of assembling these reads to characterize the insertion content. The method is based on the idea of segment extension, which progressively extends segments of known content using paired-end reads. They apply their method in simulation to discover the content of inserted sequences in a modified mouse chromosome and show that their method produces reliable results at 40x coverage.

RNA-Seq is revolutionizing the way transcript abundances are measured. A key challenge in transcript quantification from RNA-Seq data is the handling of reads that map to multiple genes or isoforms. This issue is particularly important for quantification with *de novo* transcriptome assemblies in the absence of sequenced genomes, as it is difficult to determine which transcripts are isoforms of the same gene. A second significant issue is the design of RNA-Seq experiments, in terms of the number of reads, read length, and whether reads come from one or both ends of cDNA fragments. Chapter 3, by Li and Dewey, presents RSEM, user-friendly software package for quantifying gene and isoform abundances from single-end or paired-end RNA-Seq data. RSEM outputs abundance estimates, 95% credibility intervals, and visualization files and can also simulate RNA-Seq data. In contrast to other existing tools, the software does not require a reference genome. Thus, in combination with a *de novo* transcriptome assembler, RSEM enables accurate transcript quantification for species without sequenced genomes. On simulated and real data sets, RSEM has superior or comparable performance to quantification methods that rely on a reference genome. Taking advantage of RSEM's ability to effectively use ambiguously-mapping reads, the authors show that accurate gene-level abundance estimates are best obtained with large numbers of short single-end reads. On the other hand, estimates of the relative frequencies of isoforms within single genes may be improved through the use of paired-end reads, depending on the number of possible splice forms for each gene. RSEM is an accurate and user-friendly software tool for quantifying transcript abundances from RNA-Seq data. As it does not rely on the existence of a reference genome, it is particularly useful for quantification with *de novo* transcriptome assemblies. In addition, RSEM has enabled valuable guidance for cost-efficient design of quantification experiments with RNA-Seq, which is currently relatively expensive.

Accuracy of the data extracted from two-dimensional confocal images is limited due to experimental errors that arise in course of confocal scanning. The common way to reduce the noise in images is sequential scanning of the same specimen several times with the subsequent averaging of multiple frames. Attempts to increase the dynamical range of an image by setting too high values of microscope PMT parameters may cause clipping of single frames and introduce errors into the data extracted from the averaged images. For the estimation and correction of this kind of errors, a method based on censoring technique (Myasnikova et al., 2009) is used. However, the method requires the availability of all the confocal scans along with the averaged image, which is normally not provided by the standard scanning procedure.

To predict error size in the data extracted from the averaged image, the authors of Chapter 4, Myasnikova and colleagues, developed a regression system. The system is trained on the learning sample composed of images obtained from three different microscopes at different combinations of PMT parameters, and for each image all the scans are saved. The system demonstrates high prediction accuracy and was applied for correction of errors in the data on segmentation gene expression in *Drosophila* blastoderm stored in the FlyEx database (http://urchin.spbcas.ru/flyex/, http://flyex.uchicago.

edu/flyex/). The prediction method is realized as a software tool Correct-Pattern, freely available athttp://urchin.spbcas.ru/asp/2011/emm/. The authors created a regression system and software to predict the magnitude of errors in the data obtained from a confocal image based on information about microscope parameters used for the image acquisition. An important advantage of the developed prediction system is the possibility to accurately correct the errors in data obtained from strongly clipped images, thereby allowing to obtain images of the higher dynamical range and thus to extract more detailed quantitative information from them.

Exon and exon+junction microarrays are promising tools for studying alternative splicing. Current analytical tools applied to these arrays lack two relevant features: the ability to predict unknown spliced forms and the ability to quantify the concentration of known and unknown isoforms. SPACE is an algorithm that has been developed to (1) estimate the number of different transcripts expressed under several conditions, (2) predict the precursor mRNA splicing structure and (3) quantify the transcript concentrations including unknown forms. The results presented in Chapter 5 by Anton and colleagues show its robustness and accuracy for real and simulated data.

Differential coexpression analysis (DCEA) is increasingly used for investigating the global transcriptional mechanisms underlying phenotypic changes. Current DCEA methods mostly adopt a gene connectivity-based strategy to estimate differential coexpression, which is characterized by comparing the numbers of gene neighbors in different coexpression networks. Although it simplifies the calculation, this strategy mixes up the identities of different coexpression neighbors of a gene and fails to differentiate significant differential coexpression changes from those trivial ones. Especially, the correlation-reversal is easily missed although it probably indicates remarkable biological significance. Chapter 6, by Yu and colleagues, developed two link-based quantitative methods, DCp and DCe, to identify differentially coexpressed genes and gene pairs (links). Bearing the uniqueness of exploiting the quantitative coexpression change of each gene pair in the coexpression networks, both methods proved to be superior to currently popular methods in simulation studies. Re-mining of a publicly available type 2 diabetes (T2D) expression dataset from the perspective of differential coexpression analysis led to additional discoveries than those from differential expression analysis. This work pointed out the critical weakness of current popular DCEA methods, and proposed two link-based DCEA algorithms that will make contribution to the development of DCEA and help extend it to a broader spectrum.

When a large number of candidate variables are present, a dimension reduction procedure is usually conducted to reduce the variable space before the subsequent analysis is carried out. The goal of dimension reduction is to find a list of candidate genes with a more operable length ideally including all the relevant genes. Leaving many uninformative genes in the analysis can lead to biased estimates and reduced power. Therefore, dimension reduction is often considered a necessary predecessor of the analysis because it cannot only reduce the cost of handling numerous variables, but also has the potential to improve the performance of the downstream analysis algorithms. In Chapter 7, Wang and van der Laan propose a TMLE-VIM dimension reduction procedure based on the variable importance measurement (VIM) in the frame work of targeted maximum likelihood estimation (TMLE). TMLE is an extension of maximum likelihood estimation targeting the parameter of interest. TMLE-VIM is a two-stage procedure. The first stage resorts to a machine learning algorithm, and the second step improves the first stage estimation with respect to the parameter of interest. The authors demonstrate with simulations and data analyses that their approach not only enjoys the prediction power of machine learning algorithms, but also accounts for the correlation structures among variables and therefore produces better variable rankings. When utilized in dimension reduction, TMLE-VIM can help to obtain the shortest possible list with the most truly associated variables.

Stevens-Johnson syndrome (SJS) and Toxic Epidermal Necrolysis (TEN) are rare but extremely severe cutaneous adverse drug reactions in which drug-specific associations with HLA-B alleles were described. Chapter 8, by Génin and colleagues, seeks to investigate genetic association at a genome-wide level on a large sample of SJS/TEN patients and they performed a genome wide association study on a sample of 424 European cases and 1,881 controls selected from a Reference Control Panel. Six SNPs located in the HLA region showed significant evidence for association (OR range: 1.53-1.74). The haplotype formed by their risk allele was more associated with the disease than any of the single SNPs and was even much stronger in patients exposed to allopurinol (ORallopurinol = 7.77, 95%CI = [4.66; 12.98]). The associated haplotype is in linkage disequilibrium with the HLA-B*5801 allele known to be associated with allopurinol induced SJS/TEN in Asian populations. The involvement of genetic variants located in the HLA region in SJS/TEN is confirmed in European samples, but no other locus reaches genome-wide statistical significance in this sample that is also the largest one collected so far. If some loci outside HLA play a role in SJS/TEN, their effect is thus likely to be very small.

Recent advances in sequencing technologies set the stage for large, population-based studies, in which the ANA or RNA of thousands of individuals will be sequenced. Currently, however, such studies are still infeasible using a straightforward sequencing approach; as a result, recently a few multiplexing schemes have been suggested, in which a small number of ANA pools are sequenced, and the results are then deconvoluted using compressed sensing or similar approaches. These methods, however, are limited to the detection of rare variants. In Chapter 9, He and colleagues provide a new algorithm for the deconvolution of DNA pools multiplexing schemes. The presented algorithm utilizes a likelihood model and linear programming. The approach allows for the addition of external data, particularly imputation data, resulting in a flexible environment that is suitable for different applications. Particularly, the authors demonstrate that both low and high allele frequency SNPs can be accurately genotyped when the DNA pooling scheme is performed in conjunction with microarray genotyping and imputation. Additionally, they demonstrate the use of our framework for the detection of cancer fusion genes from RNA sequences.

Gene-gene epistatic interactions likely play an important role in the genetic basis of many common diseases. Recently, machine-learning and data mining methods have been developed for learning epistatic relationships from data. A well-known combinatorial method that has been successfully applied for detecting epistasis is Multifactor Dimensionality Reduction (MDR). Jiang et al. created a combinatorial epistasis learning method called BNMBL to learn Bayesian network (BN) epistatic models.

They compared BNMBL to MDR using simulated data sets. Each of these data sets was generated from a model that associates two SNPs with a disease and includes 18 unrelated SNPs. For each data set, BNMBL and MDR were used to score all 2-SNP models, and BNMBL learned significantly more correct models. In real data sets, we ordinarily do not know the number of SNPs that influence phenotype. BNMBL may not perform as well if we also scored models containing more than two SNPs. Furthermore, a number of other BN scoring criteria have been developed. They may detect epistatic interactions even better than BNMBL. Although BNs are a promising tool for learning epistatic relationships from data, we cannot confidently use them in this domain until we determine which scoring criteria work best or even well when we try learning the correct model without knowledge of the number of SNPs in that model. In Chapter 10, Jiang and colleagues evaluated the performance of 22 BN scoring criteria using 28,000 simulated data sets and a real Alzheimer's GWAS data set. Their results were surprising in that the Bayesian scoring criterion with large values of a hyperparameter called α performed best. This score performed better than other BN scoring criteria and MDR at recall using simulated data sets, at detecting the hardest-to-detect models using simulated data sets, and at substantiating previous results using the real Alzheimer's data set. The authors conclude that representing epistatic interactions using BN models and scoring them using a BN scoring criterion holds promise for identifying epistatic genetic variants in data. In particular, the Bayesian scoring criterion with large values of a hyperparameter α appears more promising than a number of alternatives.

Elevated levels of factor VIII (FVIII) and von Willebrand Factor (vWF) are well-established risk factors for cardiovascular diseases, in particular venous thrombosis. Although high, the heritability of these traits is poorly explained by the genetic factors known so far. The aim of this work was to identify novel single nucleotide polymorphisms (SNPs) that could influence the variability of these traits. Antoni and colleagues conduct three independent genome-wide association studies for vWF plasma levels and FVIII activity in Chapter 11, and their results were combined into a meta-analysis totalling 1,624 subjects. No single nucleotide polymorphism (SNP) reached the study-wide significance level of 1.12×10^{-7} that corresponds to the Bonferroni correction for the number of tested SNPs. Neverthe-less, the recently discovered association of *STXBP5, STX2, TC2N* and *CLEC4M* genes with vWF levels and that of *SCARA5* and *STAB2* genes with FVIII levels were confirmed in this meta-analysis. Besides, among the fifteen novel SNPs showing promising association at $p < 10^{-5}$ with either vWF or FVIII levels in the meta-analysis, one located in *ACCN1* gene also showed weak association ($P = 0.0056$) with venous thrombosis in a sample of 1,946 cases and 1,228 controls. This study has generated new knowledge on genomic regions deserving further investigations in the search for genetic factors influencing vWF and FVIII plasma levels, some potentially implicated in VT, as well as providing some supporting evidence of previously identified genes.

Mass spectrometry utilizing labeling allows multiple specimens to be subjected to mass spectrometry simultaneously. As a result, between-ex-periment variability is reduced. In Chapter 12, Oberg and Mahoney de-scribe use of fundamental concepts of statistical experimental design in the labeling framework in order to minimize variability and avoid biases. They demonstrate how to export data in the format that is most efficient for statistical analysis. The authors demonstrate how to assess the need for normalization, perform normalization, and check whether it worked. They describe how to build a model explaining the observed values and test for differential protein abundance along with descriptive statistics and mea-sures of reliability of the findings. Concepts are illustrated through the use of three case studies utilizing the iTRAQ 4-plex labeling protocol.

Relative isotope abundance quantification, which can be used for pep-tide identification and differential peptide quantification, plays an impor-tant role in liquid chromatography-mass spectrometry (LC-MS)-based proteomics. However, several major issues exist in the relative isotopic quantification of peptides on time-of-flight (TOF) instruments: LC peak boundary detection, thermal noise suppression, interference removal and mass drift correction. The authors of Chapter 13, Haskins and col-leagues, propose to use the Maximum Ratio Combining (MRC) method to extract MS signal templates for interference detection/removal and LC peak boundary detection. In their method, MRCQuant, MS templates are extracted directly from experimental values, and the mass drift in each LC-MS run is automatically captured and compensated. They compared the quantification accuracy of MRCQuant to that of another representative LC-MS quantification algorithm (msInspect) using datasets downloaded from a public data repository. MRCQuant showed significant improvement in the number of accurately quantified peptides. MRCQuant effectively addresses major issues in the relative quantification of LC-MS-based proteomics data, and it provides improved performance in the quantification of low abundance peptides.

PART I

RNA-SEQ

CHAPTER 1

THE BENCH SCIENTIST'S GUIDE TO STATISTICAL ANALYSIS OF RNA-SEQ DATA

CRAIG R. YENDREK, ELIZABETH A. AINSWORTH, AND JYOTHI THIMMAPURAM

1.1 BACKGROUND

As a method for characterizing global changes in transcription, RNA-Seq is an attractive option because of the ability to quantify differences in mRNA abundance in response to various treatments and diseases, as well as to detect alternative splice variants and novel transcripts [1]. Compared to microarray techniques, RNA-Seq eliminates the need for prior species-specific sequence information and overcomes the limitation of detecting low abundance transcripts. In addition, early studies have demonstrated that RNA-Seq is very reliable in terms of technical reproducibility [2]. As a result, biologists studying an array of model and non-model organisms are beginning to utilize RNA-Seq analysis with ever growing frequency [3-7]. However, without experience using bioinformatics methods, the large number of choices available to analyze differential expression can be overwhelming for the bench scientist (see Table one in [8]).

This chapter was originally published under the Creative Commons Attribution License. Yendrick CR, Ainsworth EA, and Thimmapuram J. The Bench Scientist's Guide to Statistical Analysis of RNA-Seq Data. BMC Research Notes 5,506 (2012), doi:10.1186/1756-0500-5-506.

Essentially, RNA-Seq consists of five distinct phases, 1) RNA isolation, 2) library preparation, 3) sequencing-by-synthesis, 4) mapping of raw reads to a reference transcriptome or genome and 5) determining significance for differential gene expression (for review see [1]). In an effort to familiarize the bench scientist with the post-sequencing analysis of RNA-Seq data (phase 5), we have developed an analysis strategy based on currently available bioinformatics tools. Here, we compare three statistical tools used to analyze differential gene expression: edgeR, DESeq and Limma [9-11]. Based on their performance, we present an analysis strategy that combines these tools in order to generate an optimized list of genes that are differentially expressed. Finally, we highlight several aspects of RNA-Seq analysis that have the potential to lead to misleading conclusions and discuss options to minimize these pitfalls.

1.2 RESULTS

1.2.1 GENERATING HIGH QUALITY READS IS DEPENDENT ON INITIAL RNA QUALITY

Prior to library construction and sequencing-by-synthesis, the quality of the isolated RNA was assessed by gel electrophoresis to ensure purity (Additional file 1). Three replicate samples were isolated from soybean leaves that had been grown in either chronic O_3 (150 parts per billion) or ambient O_3 for six weeks. No degradation was observed in any of the samples and staining of the 26S rRNA band was more intense compared to the 18S rRNA band, indicating that high quality RNA had been isolated. In addition, there was no evidence that genomic DNA was co-purified during RNA extraction. Following library preparation and sequencing-by-synthesis, analysis of the raw reads determined that all six samples had a median quality score (QS) of 34 (Table 1). As a result, averages of ~28 million high quality reads were obtained for each sample.

1.2.2 UTILIZING STATISTICAL TOOLS THAT ARE COMPATIBLE WITH RNA-SEQ DATA

The raw reads described in Table 1 were aligned to the soybean reference transcriptome [12] using the mapping tool Novoalign, a short read aligner demonstrated to be highly accurate [13,14]. When differential expression was analysed subsequently, the total number of genes with significantly altered transcript abundance in plants exposed to elevated ozone was 11,995 for edgeR, 11,317 for DESeq and 9,131 for Limma. Since RNA-Seq generates count data, it is more appropriate to use a discrete probability distribution to analyze differential gene expression [15]. Therefore, edgeR and DESeq, which are based on the negative binomial distribution, are compatible with the data generated by RNA-Seq [9,10]. In contrast, Limma [16] was adapted to analyze RPKM values using a method previously developed for continuous data from microarray studies (fluorescence values) and is based on the t-distribution [11]. The Limma method was clearly very different from the two negative binomial distribution methods, but even between edgeR and DEseq there were 678 additional genes identified by edgeR as differentially expressed, representing approximately 6% of the significant genes.

TABLE 1: Post sequencing analysis of raw reads

Sample	Treatment	Flowcell Lane	Number of Reads	Q.S. (median)	Q.S. (interquartile range)
1	Ambient	4	36,408,402	34	26–36
2	Elevated O_3	4	28,554,551	34	26–36
3	Ambient	5	16,862,414	34	29–37
4	Elevated O_3	5	17,575,844	34	29–37
5	Ambient	6	31,889,531	34	28–37
6	Elevated O_3	6	37,605,167	34	28–37

For each sample, the total number of reads and read quality score (QS) is listed. A QS of 34 indicates one sequencing error per 4,000 base pairs. Generally, a QS over 20 (1% error rate) is considered acceptable for RNA-Seq. One control (−) and one elevated ozone (+) replicate were pooled and run on a single lane of the flow-cell.

1.2.3 WORKFLOW FOR RNA-SEQ DATA OPTIMIZATION

In response to the differences described above, we developed a strategy to integrate the results analyzed separately by edgeR and DESeq into one optimized dataset. As a first step, any gene that had zero mapped reads for all six samples was removed, resulting in 40,537 genes mapped by Novoalign out of the 46,367 genes comprising the soybean reference transcriptome (Figure 1, Step A) [12]. Software code to carry out this preliminary step as well as the subsequent analyses using edgeR and DESeq (Figure 1, Step B) using the R statistical package [17] is provided. These analyses are performed independently using the same mapping file and result in two excel files containing $\log_2$ fold change values and p-values that have been adjusted for multiple testing for each gene that was mapped by Novoalign.

In order to identify the common genes determined to be differentially expressed by both DESeq and edgeR, we intersected the two lists of significant genes (Figure 1, Step C). As a result, the genes that were determined to be significantly regulated by only one statistical method were eliminated. A comparison of the 2,242 eliminated genes revealed that the non-significant p-value responsible for the gene's removal was generally close to, but above $p = 0.05$ (Figure 2). Therefore, we classified these genes as marginally significant. The optimized list after these filtering and merge steps totalled 10,535 differentially expressed genes. Many of these genes had very low read counts for all samples, potentially making conclusions related to biological relevance misleading. To deal with this issue, we removed any gene with a control and treatment RPKM value of < 1.0 (Figure 1, Step D), reducing the total number of differentially expressed genes to 8,927. However, this step is optional and should be performed only after careful consideration.

1.2.4 COMPARING THE ACCURACY OF RNA-SEQ DATA WITH QRT-PCR

Several genes known to be regulated by elevated ozone were chosen to analyze via qRT-PCR. The targets chosen include genes involved with photosynthesis, carbohydrate metabolism and oxidative stress, all bio-

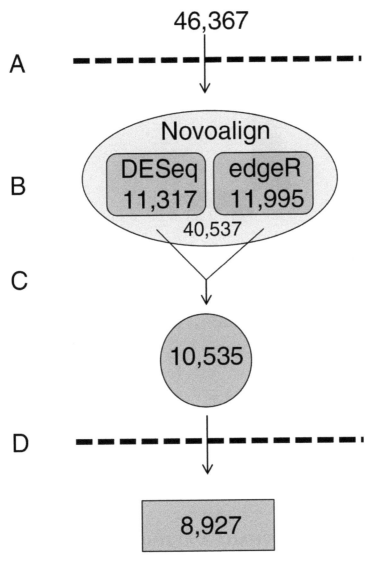

FIGURE 1: RNA-Seq data optimization strategy. The flowchart outlines the strategy for identifying soybean leaf transcripts significantly changing in response to elevated ozone. All genes mapping zero reads for all samples were removed (A) after aligning raw reads to the reference transcriptome, consisting of 46,367 genes. Differential expression was then separately determined using DESeq and edgeR (B). The two lists of significant genes were intersected to obtain a single list of differentially expressed genes (C). Finally, low expression genes (RPKM < 1.0) were removed (D).

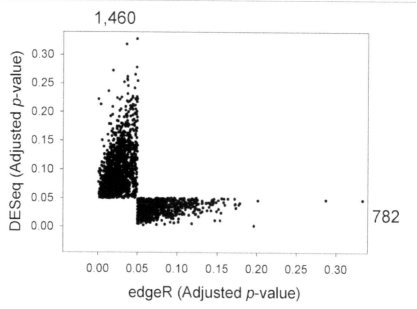

FIGURE 2: p-value comparison between edgeR and DESeq. The edgeR and DESeq p-values of the 2,242 marginally significant genes eliminated in Step C of Figure 1 are compared.

logical processes that have been well characterized to be responsive to elevated ozone at the level of transcription [18]. The response of each of the targets was consistent with the documented effects of elevated ozone. In addition, the expression ratios for both methods were similar (Figure 3), thus validating the previously reported accuracy of RNA-Seq data.

1.2.5 POTENTIAL PITFALLS AND LIMITATIONS OF RNA-SEQ ANALYSIS

A first potential limitation of this approach is that it may be too conservative, as evidenced by the 2,242 marginally significant genes that were removed from the final optimized list (Figure 1, Step C). The behavior

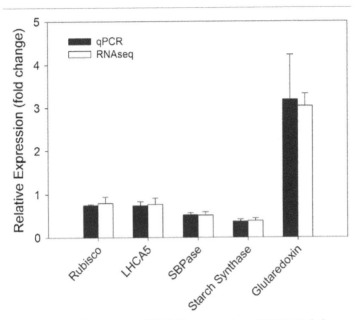

FIGURE 3: Comparing the accuracy of RNA-Seq data using qRT-PCR. Relative expression ratios determined by qRT-PCR were compared to RNAseq results for several genes known to be regulated by elevated ozone.

of these genes was analysed in the context of changes to transcripts with broadly similar functions, using the MapMan expression tool [19] to analyze functional category significance for each of the lists of marginally significant genes (Table 2). This tool first identified 11 functional categories from the optimized list of differentially expressed genes consisting of a subset of genes that collectively responded to elevated ozone in a similar manner; i.e., the expression profile of each significant functional category was different from the expression profile of all other categories. When the lists of marginally significant genes were analyzed subsequently, most of these categories were found not to be significantly different, indicating that the eliminated genes did not respond in a manner similar to the optimized list of genes. However, statistical significance was achieved for several categories. Despite having an expression profile consistent with the remaining genes included in the optimized list, 320 RNA, 70 stress, 36 hormone metabolism, 19 DNA, and 10 mitochondrial electron transport-related genes were eliminated based on a non-significant determination by one of the two statistical tools.

TABLE 2: Functional category significance of optimized and marginally significant genes

Functional Category	Optimized		DESeq Marginal		edgeR Marginal	
	# of genes	p-valie	# of genes	p-value	# of genes	p-value
Stress	497	0*	70	2.20E-03*	19	0.17
Signaling	909	0*	102	0.43	40	0.70
Cell wall	263	8.51E-29*	28	0.14	4	0.50
Photosynthesis	117	3.29E-05*	22	0.76	4	0.23
RNA	1132	6.04E-05*	222	0.01*	98	4.40E-03*
Hormone metabolism	321	3.08E-04*	36	0.01*	19	0.51
DNA	133	0.002*	34	0.70	19	0.03*
Major CHO metabolism	76	0.003*	5	0.72	7	0.42
Lipid metabolism	223	0.023*	23	0.34	16	0.88
Mitochondrial electron transport / ATP synthesis	71	0.042*	2	0.17	10	0.04*
TCA cycle	44	0.049*	—	—	7	0.53

The genes eliminated from Step C in Figure 1 are grouped into functional categories and compared with the final optimized list of significant genes. p-value indicates the significance that transcript abundance of all the genes within a specified category are changing in a similar manner compared to all other categories. Asterisks signify p-value below p = 0.05.

An additional limitation was uncovered by further investigation of the final list of optimized genes. After a cursory examination of several genes that were previously characterized to be regulated by growth in elevated ozone, we identified a potential issue with the statistical analysis that preferentially impacted the high abundance genes. It is well-documented that plants grown in elevated ozone exhibit reduced photosynthesis, increased antioxidant capacity and increased protein turnover [18]. Four high abun-

Table 3: Statistical limitations are revealed by independent analysis of ozone-responsive genes

Functional Annotation	Locus ID	Transcript Length	Control RPKM	Treatment RPKM	Fold Change	DESeq	edgeR	Individual t-test
Light-harvesting complex II CAB protein	Glyma05g25810	1100	7733.20 ± 783.0	4575.86 ± 429.3	0.59	0.063	2.34E-29*	0.004*
Ubiquitin	Glyma20g27950	1540	1504.18 ± 149.8	2425.15 ± 206.1	1.61	0.061	0.007*	0.003*
Thioredoxin	Glyma17g37280	1134	265.49 ± 11.1	220.79 ± 17.8	0.83	0.14	0.213	0.021*
L-ascorbate peroxidase	Glyma11g11460	1278	84.73 ± 2.6	107.98 ± 4.4	1.27	0.214	0.272	0.001*
Polysaccharide catabolism	Glyma06g45700	1831	70.47 ± 8.6	18.91 ± 9.2	0.27	0.214	4.48E-19*	0.002*
Glutaredoxin	Glyma13g30770	747	11.31 ± 1.4	34.68 ± 6.8	3.07	8.39E-14*	1.56E-11*	0.004*
Protein degradation DER1 like	Glyma04g14250	1088	5.23 ± 0.2	44.52 ± 11.8	8.51	7.06E-49*	8.30E-38*	0.005*
Lipoxygenase	Glyma03g42500	2833	2.90 ± 0.4	5.64 ± 1.3	1.95	1.96E-04*	2.26E-04*	0.027*
Starch synthase catalytic domain	Glyma20g36040	1954	2.84 ± 0.8	0.12 ± 0.1	0.04	1.14E-23*	1.60E-36*	0.005*
WRKY trascription factor	Glyma10g27860	1468	1.69 ± 0.4	70.92 ± 20.80	41.97	9.23E-121*	2.76E-92*	0.005*

Genes known to be regulated by elevated ozone that had a range of transcript abundances were selected from the optimized list of differentially regulated genes. In addition to p-values from DESeq and edgeR, an ANOVA was performed on RPKM values. Asterisks signify p-value below p = 0.05.

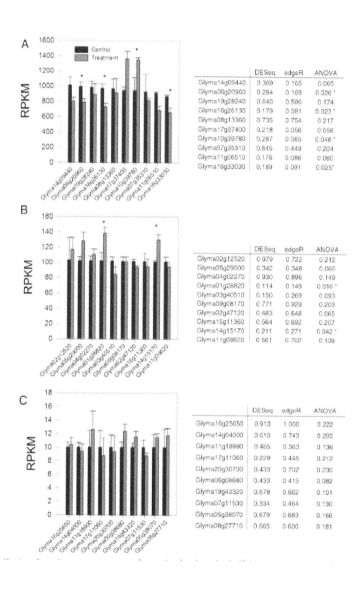

FIGURE 4: Identification of type II error across a range of transcript abundance levels. RPKM values were compared between control and treatment for 10 randomly selected genes, ranging from high (A), moderate (B) and low (C) abundance transcripts. Also included are the p-values from DESeq, edgeR and an ANOVA performed using RPKM data. Asterisks signify p-value below p = 0.05.

dance genes (Glyma05g25810, Glyma20g27950, Glyma17g37280 and Glyma11g11460) involved with these processes were not found to be differentially expressed by at least one of the statistical tools used in this analysis, despite RPKM values with obvious differences and analysis of variance (ANOVA) results that indicated significance (Table 3). A more detailed examination across a range of RPKM values support the finding of an increase in type II error for high abundance genes. Four out of 10 randomly selected genes with RPKM values near 1000 that were determined not to be differentially regulated by both edgeR and DESeq did, in fact, have significantly altered transcript abundance when analyzed using ANOVA (Figure 4A). In contrast, none of the genes with RPKM values near 10 were identified as false negatives (Figure 4C).

1.3 DISCUSSION

While the aim of this paper is to familiarize the molecular biologist interested in undertaking an RNA-Seq project with the methods and issues related to post-sequencing analysis, emphasis still needs to be placed on proper handling of RNA samples. Here, we isolated high quality RNA (Additional file 1) using a well-established protocol for soybean leaf tissue [20]. In addition, care was taken during the library construction and sequencing-by-synthesis phases, as evidenced by the high quality scores for each sample (Table 1). As a result, the average number of usable reads per sample was >20 million, which is the recommended depth required to quantify differential expression in a species with a referenced genome [21].

It is also important to utilize a valid experimental design for RNA-Seq projects, which includes the use of biological replicates. Reports demonstrating highly reproducible RNA-Seq results [2,22] make it tempting to reduce sequencing costs by only using one replicate per treatment group. However, without replication it is impossible to estimate error, without which there is no basis for statistical inference [23]. Therefore, it is recommended that RNA-Seq experiments include at least three biological replicates per treatment group [24], as was done in the experiment presented here.

Along these lines, it is important to understand the nature of RNA-Seq data and why it is necessary to use a compatible statistical method, such as a negative binomial distribution [9,10]. For discrete variables such as count data, it is possible to associate all observed values with a non-zero probability. In contrast, there is zero probability that a specific fluorescence value (continuous variable) will be obtained from microarray hybridization. This distinction is important in the context of the varying number of total reads obtained for individual RNA-Seq samples. For example, the probability of mapping 100 reads out of 16.86 million (Table 1; Sample3) for a particular gene is different than mapping 100 reads out of 36.41 million (Table 1; Sample1). To deal with this issue, both edgeR [9] and DESeq [10] normalize the read data based on the total number of reads per sample prior to differential expression analysis.

The main goal of this work was to compare the accuracy of two statistical tools, edgeR and DEseq. At first glance, it appears that both tools perform equally well (Figure 1, Step B). However, when the differentially expressed genes from edgeR and DEseq were intersected (Figure 1, Step C), quite a few genes from each list were eliminated (2,242 total genes). Because of this, we adopted a strategy to identify genes that were determined to be differentially expressed by both edgeR and DESeq. In other words, greater confidence was achieved if a gene was determined significant by each of the statistical tools.

This strategy made it possible to follow the genes that were eliminated and to identify aspects of the analysis that have the potential to lead to erroneous conclusions. One aspect to consider is how each of the different statistical tools is designed to handle and report 'zero reads' or transcripts that are not expressed in a given treatment. For example, DESeq will output 'Inf' or '-Inf' to excel as the log2 fold change value for genes that fail to align any reads for all control or treatment samples (Table 4). In contrast, edgeR outputs log2 fold changes values that are unrealistically large. It is possible that some of these genes could reveal important aspects of global transcription that were altered (i.e., genes that were turned on or off by the treatment) and should not be inadvertently removed. In many cases, however, these genes had very few reads for each replicate as well as for each treatment (Table 4). Transcript abundance this low, while determined

TABLE 4: Expression data for low abundance genes

	DESeq		edgeR		Raw Counts						RPKM	
	$\log_2$ FC	Padj	$\log_2$ FC	Padj	con1	con2	con3	trt1	trt2	trt3	Control	Treatment
Genes turned on			27.39									0.08 ± 0.032
Glyma18g02680	Inf	0.0224	27.26	0.0187	0	0	0	5	1	4	0	0.30 ± 0.206
Glyma01g41980	Inf	0.0331	27.25	0.0187	0	0	0	3	4	2	0	0.17 ± 0.054
Glyma11g04880	Inf	0.0324	27.24	0.0187	0	0	0	5	2	2	0	0.14 ± 0.057
Glyma16g06500	Inf	0.0320	27.24	0.0326	0	0	0	5	1	3	0	0.06 ± 0.030
Glyma12g05780	Inf	0.0488	27.08	0.0326	0	0	0	3	1	4		
Genes turned off												
Glyma07g02590	-Inf	0.0011	-28.11	0.0004	7	4	5	0	0	0	0.20 ± 0.050	0
Glyma17g17930	-Inf	0.0011	-28.12	0.0004	3	5	8	0	0	0	0.17 ± 0.084	0
Glyma17g34230	-Inf	0.0016	-28.02	0.0006	9	3	3	0	0	0	0.54 ± 0.292	0
Glyma12g14620	-Inf	0.0052	-27.71	0.0035	3	5	4	0	0	0	0.57 ± 0.372	0
Glyma03g37640	-Inf	0.0075	-27.56	0.0061	4	2	5	0	0	0	0.14 ± 0.013	0

Log$_2$ fold change, p-value, raw count data and RPKM vaules for representative samples from gene clusters turned on or off by elevated ozone. DESeq outputs an 'Inf' or '-Inf' log$_2$ fold change value to excel when all control or treatment replicates map zero reads.

to be significantly different, is unlikely to be biologically relevant and should be removed from the analysis. Care should be taken when choosing an arbitrary cutoff, however, to prevent the elimination of genes that may play a transcriptional role in response to the treatment being investigated. In this case, we used a conservative RPKM value <1.0 that resulted in the removal of 1,608 low abundance genes (Figure 1, Step D).

Another aspect that has the potential to confound RNA-Seq analysis deals with the issue of statistical stringency. In Table 2, we demonstrated that for several functional categories, the marginally significant genes eliminated from the optimized list did, in fact, respond to elevated ozone in a manner similar to the genes ultimately determined to be differentially expressed. Therefore, it may be more appropriate to perform network analysis for individual metabolic or signal transduction pathways using the entire RNA-Seq dataset, not just the genes determined to be differentially expressed [25]. However, this strategy is limited by pathways that have been previously characterized and would fail to uncover new connections, especially unknown signalling relationships.

One final issue revealed by this analysis was the increase in type II error for high abundance genes (Table 3 and Figure 4). Several of the genes determined not to be differentially regulated by one or both of the statistical tools are involved with processes that have been well characterized to be regulated to elevated ozone, including decreased photosynthesis (Glyma05g25810 and Glyma17g37280) [16], increased antioxidant capacity (Glyma11g11460) [26] and increased protein turnover (Glyma20g27950) [27]. However, these genes were determined to be differentially expressed based on statistical analysis of RPKM values. This problem undermines the effectiveness of performing RNA-Seq analysis to uncover novel relationships because it will fail to identify all of the high abundance genes that are differentially regulated in response to elevated ozone; genes that are more likely to impact biological processes, especially metabolic functions.

1.4 CONCLUSIONS

There are many new challenges facing the bench scientist when undertaking an RNA-Seq project, especially regarding the large number of bioinformatics tools that have been developed to analyze the post-sequencing dataset [28-32]. Here, we provide a step-by-step guide for analyzing RNA-seq data. In addition, we identified limitations that exist for widely used methods to determine differential expression of RNA-seq data. Therefore, we suggest that our strategy to merge the common genes identified by multiple tools and examine the eliminated genes is an improvement that better ensures confidence in generating a list of differentially expressed genes. We also demonstrate that the results obtained from a select set of genes using qRT-PCR closely agree with the RNA-Seq data. Because of this high accuracy, we envision RNA-Seq replacing microarrays as the new standard for global transcript quantification.

1.5 METHODS

1.5.1 BACKGROUND

Soybean plants (Glycine max cv. Be Sweet 292) were grown in environmentally controlled growth chambers for six weeks in either ambient or elevated ozone conditions (150 ppb for 8 h d-1). Tissue was collected from mature leaves and ground to a fine powder in a liquid nitrogen cooled mortar and pestle. Total RNA was isolated following the protocol of Bilgin et al. [20] and DNase treated using the TURBO DNA-free kit (Life Technologies, Grand Island, NY). Each sample (5 µg) was resolved on a 1% agarose gel containing 40 mM MOPS (pH 7.0), 2 mM EDTA (pH 8.0) and 5 mM iodoacetamide. Before loading the gel, each sample was diluted to 10μL with nuclease free water and heated at 70°C for 5 min along with 7.5μL MOPS/EDTA buffer and 5μL formaldehyde (37% wt.).

1.5.2 LIBRARY PREPARATION AND SEQUENCING-BY-SYNTHESIS

The DNase-treated RNA (1 μg) was used to prepare individually barcoded RNA-Seq libraries with the TruSeq RNA Sample Prep kit (Illumina, San Diego, CA). Pools of two samples per lane were sequenced on a HiSeq2000 for 100 cycles using version 2 chemistry and analysis pipeline 1.7 according to the manufacturer's protocols (Illumina, San Diego, CA). All raw data has been submitted to the NCBI [GenBank:SRP009826].

1.5.3 ALIGNING RAW READS TO THE SOYBEAN TRANSCRIPTOME

Illumina sequences from each of the samples from three biological replicates of control and treatment (elevated ozone) were cleaned using the FASTX toolkit, with a minimum quality score of 20 and minimum length of 75 nt. Soybean genome (Gmax_109) and gff file (Gmax_109.gff3) were downloaded from phytozome (http://www.phytozome.net/soybean). Soybean transcripts were extracted from the genome sequences based on the. gff file. These soybean transcripts (46,367 transcripts) were considered as reference transcriptome for RNA-Seq analysis.

Mapping of Illumina sequences with Novoalign was done with –H (for hard clipping the reads), –l 65, -rA10 (to allow 10 multiple alignments). With these parameters at least 90% of the each read's length should map to the reference to consider it as a mapped read. After mapping with Novoalign, read counts for each gene were generated using PERL scripts. These reads counts were used for statistical analysis using DESeq and edgeR packages of 'R' to determine differential expression at the gene level. Since approximately 92% of the mapped reads aligned to the transcriptome uniquely, multireads were not considered. All biological replicates demonstrated a >0.93 correlation when RPKM values were compared, indicating high reproducibility of replicates. See online user guides for more information about performing alignments with Novoalign (http://www. novocraft.com/wiki/tiki-index.php).

1.5.4 STATISTICAL ANALYSES

Gene lengths and count data for the three independent control and ozone-treated replicates were used to analyze differential expression using R software (Version 2.13.0) [33]. The Limma-RPKM method is based on a two-group Affymetrix dataset design included as part of the Limma package [11,17]. For the edgeR analysis, the trimmed mean of the M values method (TMM; where M = log2 fold change) was used to calculate the normalization factor and quantile-adjusted conditional maximum likelihood (qCML) method for estimating dispersions was used to calculate expression differences using an exact test with a negative binomial distribution [9,15,34]. For the DESeq analysis, differential expression testing was performed using the negative binomial test on variance estimated and size factor normalized data [10]. All p-values presented were adjusted for false discovery rate to control for type I error due to multiple hypothesis testing. The programming code for each of the specific packages can be found by viewing the vignette details in R using the 'openVignette()' command.

Log2 fold change values were loaded into the MapMan expression tool to link gene identifiers with functional annotations using the Gmax_109_ peptide mapping file. This tool automatically analyzes functional category significance base on the Wilcoxon rank sum test [19].

Differential expression of RPKM normalized data was tested by ANOVO and corrected for multiple comparisons following the methods of Benjamini and Hochberg (1995) [35] with a false discovery rate of 0.25 using SAS (Version 9.2, Cary, NC; Table 4).

1.5.5 QRT-PCR

First-strand cDNA synthesis was performed using 1 μg of DNase treated RNA and was reverse transcribed in a 20 μl reaction with Superscript II

(Life Technologies, Grand Island, NY) and oligo(dT) primers according to the manufacturer's instructions. Quantitative PCR was performed on an Applied Biosystems 7900HT Fast Real-Time PCR System (Life Technologies, Grand Island, NY) using Power SYBR Green PCR master mix (Life Technologies, Grand Island, NY) and 400nM of each primer in a 10 μl reaction. Primers were aliquoted onto a 384-well PCR plate using a JANUS automated liquid handling system (Perkin Elmer, Waltham, MA). The following are the primer sequences for each of the target genes: Rubisco (Glyma19g06340), primer A- GCACAATTGGCAAAGGAAGT, primer B- GAGAAGCATCAGTGCAACCA; LHCA5 (Glyma06g04280), primer A- GTGGAGCATCTTTCCAATCC, primer B- TGGATAAGCTCAAGCCCAAG; SBPase (Glyma11g34900), primer A- ATAAGTTGACCGGCATCACC, primer B- GGGTTGTCAGATGTGGCTCT; starch synthase (Glyma13g27480), primer A- GACCCTCTCGATGTTCAAGC, primer B- ATTCTCTGAGGTGGCAATGG; glutaredoxin (Glyma13g30770), primer A- AATCCAATGGCACCTATCCA, primer B- AGGGTTCACTCCCAGACCTT. Target gene expression was normalized to cons14 [36]. Each PCR amplification curve was analyzed with LinRegPCR software [37] to calculate the PCR efficiency and threshold value from the baseline-corrected delta-Rn values in the log-linear phase. The normalized expression level for each gene was determined as reported in [38].

REFERENCES

1.   Wang Z, Gerstein M, Snyder M. RNA-seq: a revolutionary tool for transcriptomics. Nat Rev Genet. 2009;10:57–63. doi: 10.1038/nrg2484.
2.   Marioni JC, Mason CE, Mane SM, Stephens M, Gilad Y. RNA-seq: An assessment of technical reproducibility and comparison with gene expression arrays. Genome Res. 2008;18:1509–1517. doi: 10.1101/gr.079558.108.
3.   Brautigam A, Gowik U. What can next generation sequencing do for you? Next generation sequencing as a valuable tool in plant research. Plant Biology. 2010;12:831–841. doi: 10.1111/j.1438-8677.2010.00373.x.
4.   Nowrousian M. Next-generation sequencing techniques for eukaryotic microorganisms: sequencing-based solutions to biological problems. Eukaryot Cell. 2010;9:1300–131015. doi: 10.1128/EC.00123-10.

5. Perez-Enciso M, Feretti L. Massive parallel sequencing in animal genetics: where-froms and wheretos. Anim Genet. 2010;41:561–56913. doi: 10.1111/j.1365-2052.2010.02057.x.

6. Croucher NJ, Thomson NR. Studying bacterial transcriptomes using RNA-seq. Curr Opin Microbiol. 2010;13:619–624. doi: 10.1016/j.mib.2010.09.009.

7. Sutherland GT, Janitz M, Kril JJ. Understanding the pathogenesis of Alzheimer's disease: will RNA-Seq realize the promise of transcriptomics? J Neurochem. 2011;166:937–946.

8. Garber M, Grabher MG, Guttman M, Trapnell. Computational methods for tran-scriptome annotation and quantification using RNA-seq. Nat Methods. 2011;8:469–477. doi: 10.1038/nmeth.1613.

9. Robinson MD, McCarthy DJ, Smyth GK. edgeR: a Bioconductor package for differential expression analysis of digital gene expression data. Bioinformatics. 2009;26:139–140.

10. Anders S, Huber W. Differential expression analysis for sequence count data. Ge-nome Biol. 2010;11:R106. doi: 10.1186/gb-2010-11-10-r106.

11. Smyth GK. Linear models and empirical Bayes methods for assessing differential expression in microarray experiments. Stat Appl Genet Mol Biol. 2004;3:Article 3.

12. Schmutz. et al. Genome sequence of the palaeopolyploid soybean. Nature. 2010;463:178–183. doi: 10.1038/nature08670.

13. Ruffalo M, LaFramboise T, Koyuturk M. Comparative analysis of algorthms for next-generation sequencing read alignment. Bioinformatics. 2011;27:2790–2796. doi: 10.1093/bioinformatics/btr477.

14. Li H, Homer N. A survey of sequence alignment algorithms for next-generation sequencing. Brief Bioinform. 2010;11:473–483. doi: 10.1093/bib/bbq015.

15. Robinson MD, Smyth GK. Moderated statistical tests for assessing differences in tag abundance. Bioinformatics. 2007;23:2881–2887. doi: 10.1093/bioinformatics/btm453.

16. Cloonan. et al. Stem cell transcriptome profiling via massive-scale mRNA sequenc-ing. Nat Methods. 2008;5:613–619. doi: 10.1038/nmeth.1223.

17. Smyth GK. In: Bioinformatics and Computational Biology Solutions using R and Bioconductor. Gentleman R, Carey V, Dudoit S, Irizarry R, Huber W, editor. Spring-er, New York; 2005. Limma: linear models for microarray data; pp. 397–420.

18. Ainsworth EA, Yendrek CR, Sitch S, Collins WJ, Emberson LD. The effects of tropo-spheric ozone on net primary production and implications for climate change. Annu Rev Plant Biol. 2012;63:637–661. doi: 10.1146/annurev-arplant-042110-103829.

19. Thimm O, Blaesing O, Gibon Y, Nagel A, Meyer S, Krüger P, Selbig J, Müller LA, Rhee SY, Stitt M. MAPMAN: a user-driven tool to display genomics data sets onto diagrams of metabolic pathways and other biological processes. Plant J. 2004;37:914–939. doi: 10.1111/j.1365-313X.2004.02016.x.

20. Bilgin DD, DeLucia EH, Clough SJ. A robust plant RNA isolation method suitable for Affymetrix GeneChip analysis and quantitative real-time RT-PCR. Nat Protoc. 2009;4:333–340. doi: 10.1038/nprot.2008.249.

21. Li H, Lovci MT, Kwon YS, Rosenfeld MG, Fu XD, Yeo GW. Determination of tag density required for digital transcriptome analysis: Application to an androgen-

sensitive prostate cancer model. PNAS. 2008;105:20179–20184. doi: 10.1073/pnas.0807121105.

22. Mortazavi A, Williams BA, McCue K, Schaeffer L, Wold B. Mapping and quantifying mammalian transcriptomes by RNA-seq. Nat Methods. 2008;5:621–628. doi: 10.1038/nmeth.1226.

23. Fisher RA. The design of experiments. 6. Edinburgh, Oliver and Boyd Ltd; 1951.

24. Auer P, Doerge RW. Statistical design and analysis of RNA sequencing data. Genetics. 2010;185:405–416. doi: 10.1534/genetics.110.114983.

25. Leakey ADB, Xu F, Gillespie KM, McGrath JM, Ainsworth EA, Ort DR. Genomic basis for stimulated respiration by plants growing under elevated carbon dioxide. PNAS. 2009;106:3597–3602. doi: 10.1073/pnas.0810955106.

26. Conklin PL, Barth C. Ascorbic acid, a familiar small molecule intertwined in the response of plants to ozone, pathogens, and the onset of senescence. Plant Cell Environ. 2004;27:959–970. doi: 10.1111/j.1365-3040.2004.01203.x.

27. Pell EJ, Schlagnhaufer CD, Arteca RN. Ozone-induced oxidative stress: Mechanisms of action and reaction. Physiol Plant. 1997;100:264–273. doi: 10.1111/j.1399-3054.1997.tb04782.x.

28. Howe EA, Sinha R, Schlauch D, Quackenbush J. RNA-Seq analysis in MeV. Bioinformatics. 2011;27:3209–3210. doi: 10.1093/bioinformatics/btr490.

29. Cumbie. et al. GENE-Counter: a computational pipeline for the analysis of RNA-Seq data for gene expression differences. PLoS One. 2011;6:e25279. doi: 10.1371/journal.pone.0025279.

30. Zhao WM. et al. wapRNA: a web-based application for the processing of RNA sequences. Bioinformatics. 2011;27:3076–3077. doi: 10.1093/bioinformatics/btr504.

31. Wang L, Si YQ, Dedow LK, Shao Y, Liu P, Brutnell TP. A low-cost library construction protocol and data analysis pipeline for Illumina-based strand-specific multiplex RNA-Seq. PLoS One. 2011;6:e26426. doi: 10.1371/journal.pone.0026426.

32. Zytnicki M, Quesneville H. S-MART, a software toolbox to aid RNA-seq data analysis. PLoS One. 2011;6:e25988. doi: 10.1371/journal.pone.0025988.

33. R Development Core Team. R: A language and environment for statistical computing. R Foundation for Statistical Computing, Vienna, Austria; 2011. ISBN 3-900051-07-0, URL http://www.R-project.org/

34. Robinson MD, Smyth GK. Small-sample estimation of negative binomial dispersion, with applications to SAGE data. Biostatistics. 2008;9:321–332.

35. Benjamini Y, Hochberg Y. Controlling the false discovery rate: a practical and powerful approach to multiple testing. J R Stat Soc B. 1995;57:289–300.

36. Libault M, Thibivilliers S, Bilgin DD, Radwan O, Benitez M, Clough SJ, Stacey G. Identification of four soybean reference genes for gene expression normalization. Plant Genome. 2008;1:44–54. doi: 10.3835/plantgenome2008.02.0091.

37. Ruijter JM, Ramakers C, Hoogaars WM, Karlen Y, Bakker O, van den Hoff MJ, Moorman AF. Amplification efficiency: linking baseline and bias in the analysis of quantitative PCR data. Nucleic Acids Res. 2009;37:e45. doi: 10.1093/nar/gkp045.

38. Gillespie KM, Rogers A, Ainsworth EA. Growth at elevated ozone or elevated carbon dioxide concentration alters antioxidant capacity and response to acute oxidative stress in soybean (Glycine max) J Exp Bot. 2011;62:2667–2678. doi: 10.1093/jxb/erq435.

CHAPTER 2

ASSEMBLY OF NON-UNIQUE INSERTION CONTENT USING NEXT-GENERATION SEQUENCING

NATHANIEL PARRISH, FARHAD HORMOZDIARI, AND ELEAZAR ESKIN

2.1 INTRODUCTION

The genetic variation between two individuals may total as much as 8 Mb of sequence content [1]. These variations can vary in size, from single nucleotides up to entire Mb-sized segments of the genome. Variations at the nucleotide level are referred to as single-nucleotide polymorphisms (SNPs), while larger differences spanning an entire segment of the genome are called structural variations (SVs). Structural variations may include instances where a segment of genome is inserted, deleted or inverted in an individual genome. Identifying the variation between two individuals is an essential part of genetic studies. Knowing the content of these variations can help us answer questions such as whether an individual is susceptible to a disease, or why a drug may affect individuals differently. Numerous studies have shown a high correlation between SV and genetic disorders among individuals [2-4]. The variation between one individual (the donor) and another (the reference) is computed by collecting sequence data from the donor, then comparing this sequence to that of the reference. In practice, the reference is typically the NCBI human reference genome (hg17, hg18).

This chapter was originally published under the Creative Commons Attribution License. Parrish N, Hormozdiari F, and Eskin E. Assembly of Non-Unique Insertion Content Using Next-Generation Sequencing. BMC Bioinformatics 12(Supple 6),S3 (2011), doi:10.1186/1471-2105-12-S6-S3.

One decade after the emergence of high throughput sequencing (HTS) technology, thousands of genomes have been sequenced using Illumina, ABSOLiD, Solexa, and 454 technology. These technologies are able to sequence a mammalian-size genome in a matter of days, at a cost on the order of a few thousand dollars. This has attracted much attention from both research and industry. HTS has revolutionized the sequencing process, but it has its own drawbacks. Although the technology can generate a very large number of reads in a short amount of time, the length of each read is significantly shorter than is achieved using Sanger sequencing. This limitation has raised a number of challenging computational problems.

Two orthogonal methods have been introduced to detect the variation between an individual diploid donor genome and a haploid reference genome. The first method, known as resequencing, maps all donor reads to the reference [5,6] and uses this mapping information to predict the variation between the donor and the reference [7-11]. In the second method, called assembly, a de novo assembler [12-15] is used to assemble the sequence of the donor genome and then detect the differences between the donor and the reference.

One type of structural variation is an insertion of a segment in the donor genome compared to the reference genome. Insertions can be classified as either a unique inserted segment of genome in the donor that does not align to the reference genome, or a copied insertion, where the inserted segment exists in the reference at a different locus.

Kidd et al. 2008 was the first study to tackle the unique insertion problem, and did so by using traditional Sanger sequencing of entire unmapped fosmid clones [16,17]. Unfortunately, this method is costly to apply to HTS data. Many studies in recent years have tried to solve the general SV problem using HTS data [7-11], though these methods were not designed to detect novel insertions. De novo assembly [12-15] can be used to detect the unique and copied insertions, however the high computational cost and memory requirements have made them difficult to use in practice. Moreover, as it is shown by Alkan et.al 2010, de novo assemblers have limitations in how accurately they can construct the genome [18]. The only efficient method to assemble unique insertions was introduced by Hajirasouliha et.al [19], which uses paired-end mappings and

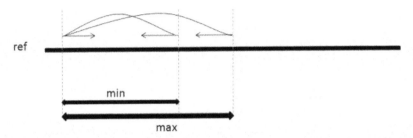

FIGURE 1: The sampling of paired-end reads from a genome The sampling of paired-end reads from a genome in which the distance between sampling positions is at least min, but not more than max

the unmapped reads to construct the unique insertions using a de novo assembler.

In this study we attempt to solve both the unique and copied insertion problems. We will use a hybrid method similar to the method mentioned in [19] using both the reference and a specialized assembler to solve the problem. Our study differs from that carried out by Hajirasouliha et.al [19] in that we are able to successfully assemble insertions comprised of both copied and unique content.

When a paired-end read is sampled from a genome, the distance between the two mates can be modeled as a normal distribution with a mean distance μ and standard deviation σ. We will further assume for simplicity that the separation distance lies within some well-defined interval, as shown in Figure 1. The first mate is mapped in the forward direction (+) and the corresponding mate is mapped in the reverse direction (−). Given a set of paired-end reads originating from insertion sequences, we will use this fact to "anchor" one mate in the pair to a known segment of the genome, then find the most likely mapping position of the opposite mate by aligning it with the endpoint of that segment. By finding all such reads that can be anchored and aligned in this way, we are able to discover the content of the insertion sequence beyond the endpoint of the known segment. Iteratively repeating this process allows us to extend these known segments and assemble the insertions. Figures 2 and 3 illustrate this approach.

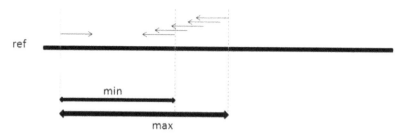

FIGURE 2: Estimating the mate position It is shown that when a mate in a paired-end read is aligned to the reference, one can estimate the position where the second mate aligns

2.2 METHODS

2.2.1 NOTATION AND DEFINITIONS

The set of paired-end reads from the donor genome is represented by $R = \{r_1, r_2, r_3, ...r_n\}$ where r_i^+ and r_i^- are the forward and reverse strands, respectively, of read r_i. r_i^+loc is the set of positions read r_i^+ maps to in the reference. $r_i.loc_k$ is the k-th position among all possible mappings of read i. Δ_{min} and Δ_{max} are the minimum and maximum insert sizes, where $\Delta_{min} = \mu - 3\sigma$ and $\Delta_{max} = \mu + 3\sigma$ (three standard deviations from the mean). The set of insertion locations is represented by $Loc = \{loc_1, loc_2, ...loc_m\}$, with Segs $= \{seg_{0,1}, seg_{1,2}, ...seg_{m-1,m}\}$ representing the segments of the donor genome between those insertions. The entire donor genome sequence is Donor.

We classify reads into the follow five categories:

- One-end anchored (OEA): Read-pairs in which one mate maps to the reference genome, and one does not.
- Orphan: Read-pairs in which neither mate maps to the reference.
- Concordant: Read-pairs in which both reads map to the reference, and the distance between their mapped locations is within the range [Δmin, Δmax]. Furthermore, one mate should map in the forward direction, and one in the reverse.
- Discordant: Read-pairs in which both reads map to the reference, but are not concordant.

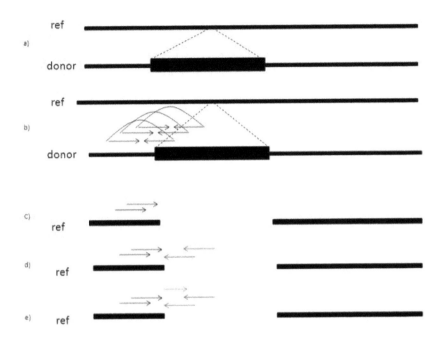

FIGURE 3: Illustrating our approach Each step of our approach is illustrated in this figure. (a) indicates the donor and reference genome, where a segment is inserted in the donor genome. In (b) we show how the reads are sampled from the donor genome. Reads preceding the insertion in (c) map to the reference while their corresponding mate fails to map to the reference. In (d) we show how to use the information from (c) to predict the position of unmapped mates. In the end (e) displays how we can use the (c) and (d) mapping to align additional unmapped reads.

- Over-coverage: Reads which are concordant, but which map to a region with a higher depth-of-coverage than is expected, where the depth-of-coverage in a region is simply the number of reads that map to that region divided by the length of the region.

We will define the following notation for working with slices of strings and arrays: Given an array A, A[i : j] denotes a contiguous segment of A of length j − i beginning at position i.

We use the ⊕ operation to represent whether or not two strings have an alignment score above some threshold. $s_1 \oplus s_2$ = true iff $\text{Align}(s_1, s_2) > \tau$. The exact value of the threshold τ varies depending on the context and in all cases is user-configurable, so we leave it as an implicit parameter and omit it from the notation.

The depth-of-coverage for a particular position in the reference is defined as the number of reads that cover that position. We will define μDOC as the mean depth-of-coverage across the entire reference genome, and $\mu_{DOC[r]}$ as the mean depth-of-coverage across the positions in the reference that are covered by read r.

2.2.2 ASSEMBLING THE INSERTION

Given the set of insertion locations hoc and the set of reads R, our goal is to identify the subset of reads that were sampled from a particular insertion, determine the correct layout of those reads, and finally to decide the consensus value for each position in the insertion. We aim to solve this problem using an iterative approach based on the notion of segment extension, which is analogous to building and traversing a path through the string graph [20] simultaneously. We will first present the mathematical foundations of our approach, then describe the optimizations that make this approach practical on common desktop computing hardware.

We begin the insertion assembly process by partitioning the donor genome according to the insert loc_i Loc = $\{loc_1, loc_2, ...loc_m\}$. This results in a set of segments Segs = $\{seg_{0,1}, seg_{1,2}, ... seg_{m-1,m}\}$, where $seg_{i,i+1}$ represents the segment of the donor genome between insertion loci i and i + 1. For each segment $seg_{i,i+1}$, we attempt to assemble insertions i and i +1 by extending the segment at each endpoint using an iterative process. For the sake of simplicity, we will formulate only extension in the forward direction.

The segment extension method is based on identifying a set of reads which have a high likelihood of covering a particular position pos in the donor genome, where pos lies at the edge of some segment seg. Identification of this set occurs in two passes. The first pass is performed only once, and selects reads from R which are likely to have been sampled from any insertion in the donor genome. We refer to the result of this first pass as the insertion read set. The second pass is performed for every position pos, and further filters the insertion read set to select reads that are likely to cover position pos. We refer to the result of the second pass as the covering set for position pos. Once we identified the covering set, we decide the value of Donor[pos] by finding the consensus among all reads in the set. We then move to position pos + 1 and repeat the process.

2.2.3 INSERTION READ SET

Consider a paired-end read r in which one mate covers the insertion and the other mate does not. In the case that the insertion sequence is unique, it follows that $r^-.loc = \varnothing$ or $r^+.loc = \varnothing$ ($\varnothing$ being the empty set), categorizing the read as OEA. In the case where the insertion sequence is copied, then both mates will map somewhere in the reference, however the distance between them is unlikely to be consistent with the expected insert size ($|r^+.loc - r^-.loc| < \Delta_{min}$ or $|r^+.loc - r^-.loc| > \Delta_{max}$). In this case the read will be categorized as Discordant.

Now consider a paired-end read r in which both reads cover the insertion. If the insertion sequence is unique, then neither mate mate will map to the reference and the read will be categorized as an Orphan read. On the other hand, if the insertion sequence is copied, then both mates will map to some region in the reference, and the distance between them will be consistent with the expected insert size. However, if we calculate the depth of coverage in this region, we will find it to be higher than the sequencing coverage. These reads will be categorized as over-coverage.

Based on this analysis, we define the following four functions:

$$\text{IsOEA}(r) = \begin{cases} \text{true} & \text{if}(r^-.loc = \varnothing \text{ or } r^+.loc = \varnothing) \text{and } (r^-loc \cup r^+.loc = \varnothing) \\ \text{false} & \text{otherwise} \end{cases} \quad (1)$$

$$\text{IsOrphan}(r) = \begin{cases} \text{true} & \text{if } r^-.loc = \varnothing \text{ and } r^+.loc = \varnothing \\ \text{false} & \text{otherwise} \end{cases} \quad (2)$$

$$\text{IsDiscordant(r)} = \begin{cases} \text{true} & \text{if } |r^+.loc_k - r^-.loc_k| < \Delta_{min} \text{ or } |r^+.loc - r^-.loc| < \Delta_{maxforallk} \\ \text{false} & \text{otherwise} \end{cases} \quad (3)$$

$$\text{IsOverCoverage(r)} = \begin{cases} \text{true} & \text{with probability } Pr = \dfrac{\max\left(0, \mu_{DOC[r]} - \mu_{DOC}\right)}{\mu_{DOC[r]}} \\ \text{false} & \text{with probability } 1 - Pr \end{cases} \quad (4)$$

We now construct a subset IRS = {r ∈ R : IsOEA(r)| IsOrphan(r)| IsDiscordant(r)| IsOverCoverage(r)}, representing the insertion read set. Note that the IsOverCoverage function is designed such that we select the appropriate fraction of reads from an over-coverage region. For example, if a region has mean read-depth $2_{\mu DOC}$, we are only interested in 50% of the reads from that region. This is captured by the probabilistic function.

2.2.4 COVERING SET

Consider a position in the donor genome Donor[pos] belonging to an insertion and for which the correct nucleotide assignment is unknown. Our goal is to determine the exact set of reads CS_{pos} that cover Donor[pos], which we will refer to as the covering set. Assume that Donor[j] is known for all j ∈ [(pos – 2l – Δ_{max}),pos] and consider a paired-end read r = (r+, r–). We assert that if r– covers Donor[pos], in other words r ∈ CS_{pos}, then the following conditions must hold:

1. r^+ covers some set of positions in Donor[pos – I – Δ_{max} : pos – Δ_{min}].
2. If r⁻[ext] covers Donor[pos], then Donor[pos – ext – 1 : pos – 1] ⊕ r⁻[0: ext – 1].

The region of the donor genome denoted by Donor[pos – I – Δ_{max} : pos – Δ_{min}] is referred to as the anchor window, and so reads that meet condition 1 are considered anchored. Reads that meet condition 2 are referred to as extending reads. We will capture these conditions formally in two functions ΦA and ΦE (anchors and extends, respectively), defined as follows:

$$\Phi_A(r,pos)=\begin{cases}\text{true} & \text{if }\exists\,anchor:Donor[anchor:anchor+1]\oplus r^+,\\ & (pos-1-\Delta_{max}\ )\leq anchor\leq(pos-1-\Delta_{min}\\ \text{false} & \text{otherwise}\end{cases} \tag{5}$$

$$\Phi_E(r,pos)=\begin{cases}\text{true} & \text{if }\exists\,ext:Donor[pos-ext-1:pos-1]\oplus r^-[0:ext-1],ext>k\\ \text{false} & \text{otherwise}\end{cases} \tag{6}$$

Refer to Figures 4, 5, and 6 for an illustration of the process identifying the covering set. Note that for our purposes, small values of ext are not informative, as there is a relatively high probability, given two short strings s_1 and s_2, that $s_1 \oplus s_2$. Therefore we will further require that ext $>\kappa$, where κ is user-configurable. In practice, given a paired-end read it is not known a priori which mate is the forward strand and which is the reverse. During construction of the covering set we therefore test both orientations and settle on one should it be found to meet the two conditions.

We can now compute an approximation of CS_{pos} as follows: $CS'_{pos} = \{r \in IRS: \Phi_A(r, pos) = true$ and $\Phi_E(r, pos) = true$. We note that this is only an approximation, as the repetitive nature of genomic sequence dictates that there will be reads in CS'_{pos} that do not truly cover Donor[pos]. Furthermore, our choice of κ as a lower threshold means there will be reads in CS_{pos} that are not in CS_{pos}'.

Using the covering set we can now decide the value of Donor[pos] as follows, where ext_s is the value of ext computed for read s. Note that this is merely a formal statement of the standard consensus problem. Refer to Figure 7 for an illustration of this.

Anchor window   Extension window

FIGURE 4: Defining the anchor and extension windows

Anchor window   Extension window

FIGURE 5: Identifying anchoring reads

$$Donor[pos] = \text{argmax}(|s \in CS'_{pos} : s[ext_s] = c|) \tag{7}$$

Once the value of Donor[pos] is known, we can iteratively repeat this process for pos + 1, pos + 2,.... Recall that we initially assumed that the value of Donor [j] is known for all $j : i-21-\Delta_{max} <j <i$. This will always be the case at the boundaries of the insertion, providing a base from which we can iterate. The iteration ter minates when there is insufficient consensus to decide the value of Donor[pos]. That is, $|s \in CS'_{pos} : s[ext_s]=c|< \varepsilon$, where ε is a user-defined threshold.

ALGORITHM 1: ASSEMBLING THE INSERTION

1:	Segs ← Segements(Donor, Loc)
2:	IRS ← InsertionReadSet (R)
3:	**for** seg ∈ Segs **do**
4:	i ← |seg| + 1
5:	CS'$_i$ ← CoveringSet (IRS, i)
6:	**while** |CS'$_i$| > ∈ **do**
7:	seg[i] ← Consensus (CS'$_i$, i)
8:	i ← i +1
9:	CS'$_i$ ← CoveringSet (IRS, i)
10:	**end while**
11:	**end for**

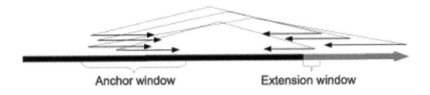

Anchor window   Extension window

FIGURE 6: Identifying extending reads

Algorithm 1 illustrates the high-level algorithm. Once this iterative process has been applied to each segment, we are left with a set of extended segments $Seg' = \{seg'_{0,1}, seg'_{1,2}, \ldots seg'_{m-1,m}\}$. In order to assemble the complete donor genome, we compute the overlap of each adjacent pair of segments. Pairs with a low overlap score indicate that only a portion of the insertion between them was assembled, in which case it may be possible to revisit the insertion using a more relaxed set of alignment functions. Pairs with high overlap scores indicate a successful insertion assembly and may be merged into a single contig.

Note that in algorithm described above, at each iteration we selected the value that is the consensus of all reads in the covering set. In general, this approach works well when applied to non-repetitive insertion sequences. For insertion sequences that are repetitive, however, there will be multiple values of c for which $|r \in CS'_1 : r[ext_s] = c| > \varepsilon$, where ε is suitably large to eliminate the effect of read errors. In this case we say that our iterative algorithm has encountered a divergence, and we proceed to branch and explore each supported value of c. Ultimately, each branch will return a set of hypothetical sequences. In attempting to select the most probable sequence from this set, we reason that an ideal assembly of all insertions would account for every read in IRS. Therefore when assembling each individual insertion, we select the hypothesis that accounts for the greatest number of reads.

We assume that the locations of the insertions are provided to us as input to the assembly method. There are two main methods for determining these locations that have been presented in previous work:

1. Using existing SVs frameworks: Many efficient tools have been developed in past few years to detect the SVs efficiently and accurately [8,10]. We can use the output of their methods as the input to our algorithm.

2. Clustering the OEA Reads: OEA reads are indicator of unique insertion, we will cluster the OEA reads and pick the cluster set which has the most number of OEA reads. Clustering the OEA reads will increase our confidence level if an insertion has occurred in the donor genome. Furthermore, it will reduce the estimated number of unique insertions in the donor genome, which follows the maximum parsimony [19].

2.2.5 OPTIMIZATION

Given a donor genome containing a total length L_1 of all insert content, a naïve implementation has a running time that is dominated by the insertions assembly step, with a running time of $O(L_1 |IRS| \max_{pos}(|CS'_{pos}|))$. For every position pos in each insertion, we must search through $|IRS|$, computing Φ_A and Φ_E for each read, to identify the approximate covering set. Note that as $|IRS|$ will be dictated by L_1, and will be fairly constant, this can be roughly simplified to .

In order to reduce the computational complexity of this search problem, we make use of recent methods developed for read-mapping applications using the Burrows-Wheeler Transform (BWT) [21]. While we will not discuss the implementation details, the advantages of using a BWT can be summarized as follows. Given two strings x and y, we would like to find all instances of string y in x, allowing for d mismatches. While a naïve search algorithm would require $O(|x||y|)$ operations, using a BWT we can achieve this in only $O(\ell^d|y|)$ operations, irrespective of $|x|$, where ℓ is the size of the language (4 in our case). Furthermore, unlike the common suffix tree-based approaches, the BWT can be represented compactly, making this approach feasible on standard desktop hardware.

In read-mapping applications such as BWA and Bowtie[22,23], x is the reference genome, and y is an individual read. We instead set , the concatenation of all forward-end reads, and search for substrings s of the anchor window. Given a function BWTSearch that returns the set of matching indices, we can now use the BWT to locate all anchored reads:

$$\{r \in R': \Phi_A(r,p) = \text{true}\} = \{r_j \in R': \frac{j}{\epsilon} \cup BWTSearch(s,x)\}$$

That is, if read r_j is anchored, then one of the calls to BWTSearch should return the index jl. The key difference here is that computing the set on the left requires computing Φ_A for all reads in R', while the set on the right can be computed using only $\Delta_{max} - \Delta_{min}$ (the number of substrings of length l in the anchor window) calls to BWTSearch.

We also note here that as each insertion is assembled independently, it is straightforward to parallelize our approach on multiple processors. Once the insertion read set IRS has been generated, it can be read by all processes on a single machine or cloned on each machine in a cluster. Each process is then assigned a single segment to extend. Furthermore, the construction of IRS itself can be parallelized simply by dividing up the set of reads among multiple processes.

2.3 RESULTS

In this part of the paper we will report the accuracy of our method in assembling the insertions. We designed a simulated framework in which the reference genome is the C57BL/6J (NCBI m37) chromosome 17 and the donor genome is simulated by inserting sequence segments into the reference genome. Unique insertions were generated using a uniform distribution over the four bases. Copied insertions were generated by choosing a uniformly random position in the genome and duplicating the content at that position. The mean size of the inserted segments is 2kbps, with a standard deviation of 200bp. We generate a set of reads from the donor genome using MetaSim [24], using a read length of 36 and a mean insert size of 200bp with a standard deviation of 25bp. We generate reads at 40X coverage. Moreover, we vary the number of inserted segments from 10-1000. We calculate the accuracy of our method by counting the number of insertions that were assembled correctly within some small margin of error (an edit distance of 10bp was used in the results shown). Table 1, shows the results of this calculation, confirming that our method maintains high reliability as the number of insertions grows. In these results, each insertion contains equal parts unique and non-unique content, generated by copying a segment of the reference genome and inserting a unique segment. The decrease in accuracy as the number of insertions grows can be attributed to the increase in the number of reads contained in the Insertion Read Set. As the size of this set grows, the probability of selecting the wrong reads during segment extension also increases. Our results demonstrate, however, that this effect is fairly small.

TABLE 1: Accuracy of our method at varying numbers of insertions, from 10 to 1000

#Insertion	Accuracy(%)	Standard Deviation(%)
10	98.00%	4.47%
50	92.80%	1.79%
100	94.20%	2.49%
500	91.64%	1.04%
1000	89.92%	0.77%

In Figure 8 we show that the running time of the algorithm increases quadratically if we apply the naive indexing algorithm. Using the Burrows-Wheeler Transformation discussed in the Optimization section results in a running time that grows linearly.

In order to test our method in different insertion categories, we run our method on three different cases as shown in Figure 9. Case 1 is where the insertion is unique and the sequence is inserted in a unique region in the reference. In Case 2 the insertion is copied but it is inserted in a unique region. Case 3 is similar to case 2, where the insertion is copied but contains a unique segment as well. Table 2 indicates the assembly accuracy for 1000 insertions in the 3 different categories. In the first case we are testing how accurately our method can assemble the unique insertions. This is the simplest case among the three. In the second case, not only we are testing our assembly accuracy, but our success is also an indication of how well we can detect the set of over-coverage reads. High accuracy in the second case is not only important for insertion assembly, but it can also be widely used in the CNV detections. In case 3, in addition to the complexity in

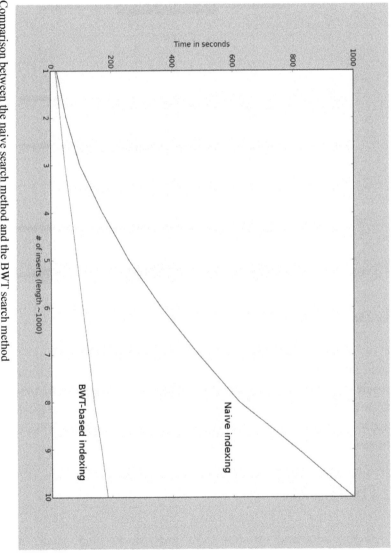

FIGURE 8: Comparison between the naive search method and the BWT search method

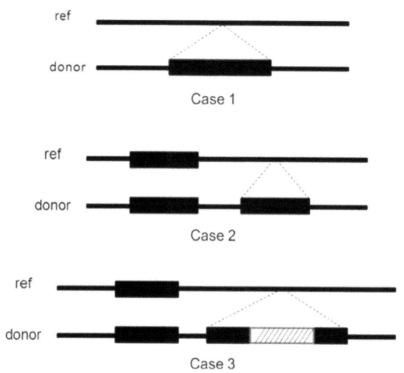

FIGURE 9: Three different categories of insertions

case 2 and case 1, we have to deal with the case where there is an insertion inside another insertion. As the results indicate our method maintains high accuracy as the complexity of the insertions grows, which suggest we can apply our method to any insertion assembly problem, without any assumptions as to the type of the insertions.

2.4 DISCUSSION

Detecting structural variation (SVs) between two individuals has been studied widely in the past few years. Although detecting the presence of SVs is an important problem, assembling the actual sequence of the SV accurately is invaluable. While high throughput sequencing (HTS) has revolutionized genomics by giving us the opportunity to cost effectively sequence many individuals, this technology limits the extent to which reference-based assemblers can discover the content of inserted sequences. In this study we addressed the insertion assembly problem using paired-end HTS data. While previous methods were focused on assembling the content of unique insertion sequences, and thus are not able to assemble insertions containing copied regions, our method is able to assemble both copied and unique insertions. Furthermore, it is independent of any de novo assemblers, and as such it can be used as a stand alone tool to assemble insertion sequences. We have shown that at 40X coverage we can assemble the insertions with very high accuracy. Finally, we have demonstrated the practicality of our approach by presenting both algorithmic optimizations and parallelization opportunities that make this method feasible even for mammalian-size genomes.

TABLE 2: Accuracy of our method in 3 different categories

Category	Accuracy (%)	Standard Deviation
Case 1	98.40%	0.23%
Case 2	92.06%	0.98%
Case 3	89.92%	0.77%

Accuracy of our method in 3 different categories. In Case 1 both the inserted sequence and the region in the reference is unique. In Case 2 the copied sequence is inserted in a unique region in the reference, while in Case 3 the copied insertion contains a unique insertion as well. For each case, 5 simulations were performed.

REFERENCES

1. Li R, et al.: Building the sequence map of the human pan-genome. Nat Biotechnol. 2010, 28(1):57-63. P
2. Korbel JO, et al.: The genetic architecture of Down syndrome phenotypes revealed by high-resolution analysis of human segmental trisomies. Proceedings of the National Academy of Sciences of the United States of America 2009, 106(29):12031-12036.

3. Sharp AJ, et al.: Discovery of previously unidentified genomic disorders from the duplication architecture of the human genome. Nature Genetics 2006, 38(9):1038-1042.

4. McCarroll SA, et al.: Deletion polymorphism upstream of IRGM associated with altered IRGM expression and Crohn's disease. Nature Genetics 2008, 40(9):1107-1112.

5. Alkan C, et al.: Personalized copy number and segmental duplication maps using next-generation sequencing. Nature Genetics 2009, 41(10):1061-1067.

6. Hach F, et al.: mrsFAST: a cache-oblivious algorithm for short-read mapping. Nat Methods. 2010, 7(8):576-577.

7. Korbel JO, et al.: Paired-End Mapping Reveals Extensive Structural Variation in the Human Genome. Science 2007, 318(5849):420-426.

8. Hormozdiari F, et al.: Combinatorial algorithms for structural variation detection in high-throughput sequenced genomes. Genome Research 2009, 19(7):1270-1278.

9. Chen K, et al.: BreakDancer: an algorithm for high-resolution mapping of genomic structural variation. Nature Methods 2009, 6(9):677-681.

10. Medvedev P, et al.: Computational methods for discovering structural variation with next-generation sequencing. Nature Methods 2009, 6(11 Suppl):S13-S20.

11. Lee S, et al.: A robust framework for detecting structural variations in a genome. Bioinformatics 2008, 24(13):i59-i67.

12. Pevzner PA, et al.: A new approach to fragment assembly in DNA sequencing. Proceedings of the fifth annual international conference on Computational biology RECOMB 01 2001, :256-267.

13. Simpson JT, et al.: ABySS: a parallel assembler for short read sequence data. Genome Research 2009, 19(6):1117-1123.

14. Butler J, et al.: ALLPATHS: De novo assembly of whole-genome shotgun microreads. Genome Research 2008, 18(5):810-820.

15. Zerbino DR, Birney E: Velvet: Algorithms for de novo short read assembly using de Bruijn graphs. Genome Research 2008, 18(5):821-829.

16. Kidd JMo: Mapping and sequencing of structural variation from eight human genomes. Nature 2008, 453(7191):56-64.

17. Kidd JM, et al.: Characterization of missing human genome sequences and copy-number polymorphic insertions. Nature Methods 2010, 7(5):365-372.

18. Alkan C, et al.: Limitations of next-generation genome sequence assembly. Nat Methods 2011, 8(1):61-65. advance on(november), http://www.ncbi.nlm.nih.gov/pubmed/21102452

19. Hajirasouliha I, et al.: Detection and characterization of novel sequence insertions using paired-end next-generation sequencing. Bioinformatics 2010, 26(10):1277-1283.

20. Myers EW: The fragment assembly string graph. Bioinformatics 2005, 21 Suppl 2(Suppl 2):ii79-85.

21. Burrows M, Wheeler DJ: A block-sorting lossless data compression algorithm. [http://citeseerx.ist.psu.edu/viewdoc/summary?doi=10.1.1.121.6177] 1994.

22. Langmead B, et al.: Ultrafast and memory-efficient alignment of short DNA sequences to the human genome. Genome Biology 2009, 10(3):R25.

23. Li H, Durbin R: Fast and accurate short read alignment with Burrows–Wheeler transform. Bioinformatics 2009, 25(14):1754-1760.
24. Richter DC, et al.: MetaSim: a sequencing simulator for genomics and metagenomics. PLoS ONE 2008, 3(10):e3373.

RSEM: ACCURATE TRANSCRIPT QUANTIFICATION FROM RNA-SEQ DATA WITH OR WITHOUT A REFERENCE GENOME

BO LI AND COLIN N. DEWEY

3.1 BACKGROUND

RNA-Seq is a powerful technology for analyzing transcriptomes that is predicted to replace microarrays [1]. Leveraging recent advances in sequencing technology, RNA-Seq experiments produce millions of relatively short reads from the ends of cDNAs derived from fragments of sample RNA. The reads produced can be used for a number of transcriptome analyses, including transcript quantification [2-7], differential expression testing [8,9], reference-based gene annotation [6,10], and de novo transcript assembly [11,12]. In this paper we focus on the task of transcript quantification, which is the estimation of relative abundances, at both the gene and isoform levels. After sequencing, the quantification task typically involves two steps: (1) the mapping of reads to a reference genome or transcript set, and (2) the estimation of gene and isoform abundances based on the read mappings.

A major complication in quantification is the fact that RNA-Seq reads do not always map uniquely to a single gene or isoform. Previously, we have shown that properly taking read mapping uncertainty into account with a statistical model is critical for achieving the most accurate abundance estimates [7]. In this paper, we present a user-friendly software package, RSEM (RNA-Seq by Expectation Maximization), which implements our quantification method and provides extensions to our original model. A key feature unique to RSEM is the lack of the requirement of a reference genome. Instead, it only requires the user to provide a set of reference transcript sequences, such as one produced by a de novo transcriptome assembler [11,12]. Extensions to our original methodology include the modeling of paired-end (PE) and variable-length reads, fragment length distributions, and quality scores. In addition, a 95% credibility interval (CI) and posterior mean estimate (PME) are now computed for the abundance of each gene and isoform, along with a maximum likelihood (ML) estimate. Lastly, RSEM now enables visualization of its output through probabilistically-weighted read alignments and read depth plots.

Through experiments with simulated and real RNA-Seq data, we find that RSEM has superior or comparable quantification accuracy to other related methods. With additional experiments, we obtained two surprising results regarding the value of PE data and quality score information for estimating transcript abundances. Although a PE read provides more information than a single-end (SE) read, our experiments indicate that for the same sequencing throughput (in terms of the number of bases sequenced), short SE reads allow for the best quantification accuracy at the gene-level. And while one would assume that quality scores provide valuable information for the proper mapping of reads, we find that for RNA-Seq reads with Illumina-like error profiles, a model that takes into account quality scores does not significantly improve quantification accuracy over a model that only uses read sequences.

3.1.1 RELATED WORK

A simple quantification method that was used in some initial RNA-Seq papers [13,14] and that is still used today is to count the number of reads that map uniquely to each gene, possibly correcting a gene's count by the "mappability" of its sequence [15] and its length. The major problems with this type of method are that it: (1) throws away data and produces biased estimates if "mappability" is not taken into account, (2) produces incorrect estimates for alternatively-spliced genes [16], and (3) does not extend well to the task of estimating isoform abundances. A couple of methods were later developed that addressed the first problem by "rescuing" reads that mapped to multiple genes ("multireads") [17,18]. Some other methods addressed the latter two problems, but not the first, by modeling RNA-Seq data at the isoform level [5]. Later, we developed the methodology behind RSEM, which addressed all of these issues by using a generative model of RNA-Seq reads and the EM algorithm to estimate abundances at both the isoform and gene levels [7]. Since the publication of the RSEM methodology, a number of methods utilizing similar statistical methods have been developed [3,4,6,19-22].

Of the methods developed, only RSEM and IsoEM are capable of fully handling reads that map ambiguously between both isoforms and genes, which the authors of both methods have shown is important for achieving the best estimation accuracies [4,7]. In contrast with IsoEM, RSEM is capable of modeling non-uniform sequence-independent read start position distributions (RSPDs), such as 3'-biased distributions that are produced by some RNA-Seq protocols [1]. In addition, RSEM can compute PME and 95% CIs, whereas IsoEM only produces ML estimates. Lastly, RSEM is the only statistical method that we are aware of that is designed to work without a whole genome sequence, which allows for RNA-Seq analysis of species for which only transcript sequences are available.

3.2 IMPLEMENTATION

A typical run of RSEM consists of just two steps. First, a set of reference transcript sequences are generated and preprocessed for use by later RSEM steps. Second, a set of RNA-Seq reads are aligned to the reference transcripts and the resulting alignments are used to estimate abundances and their credibility intervals. The two steps are carried out by the user-friendly scripts rsem-prepare-reference and rsem-calculate-expression. The steps of the RSEM workflow are diagrammed in Figure 1 and described in more detail in the following sections.

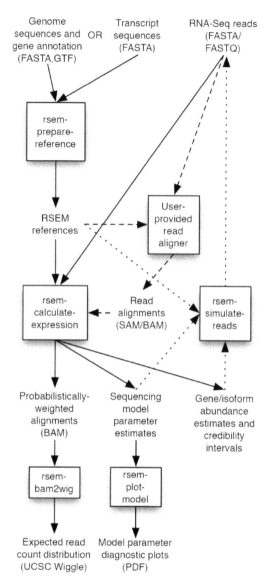

FIGURE 1: The RSEM software workflow. The standard RSEM workflow (indicated by the solid arrows) consists of running just two programs (rsem-prepare-reference and rsem-calculate-expression), which automate the use of Bowtie for read alignment. Workflows with an alternative alignment program additionally use the steps connected by the dashed arrows. Two additional programs, rsem-bam2wig and rsem-plot-model, allow for visualizing the output of RSEM. RNA-Seq data can also be simulated with RSEM via the workflow indicated by the dotted arrows.

3.2.1 REFERENCE SEQUENCE PREPARATION

RSEM is designed to work with reads aligned to transcript sequences, as opposed to whole genome sequences. There are several advantages to using transcript-level alignments. First, for eukaryotic samples, alignment of RNA-Seq reads to a genome is made complicated by splicing and poly-adenylation. Reads that span splice junctions or that extend into poly(A) tails are challenging to align at the genome level, although there are tools available for aligning splice junction reads [23-25]. Second, using transcript-level alignments easily allows for analyses of samples from species without sequenced genomes but with a decently characterized transcriptome (perhaps via RNA-Seq transcriptome assembly [11,12]). Lastly, the total length of all possible transcripts is often much smaller than the length of the genome, allowing for faster alignment at the transcript-level.

A set of transcripts may be specified to rsem-prepare-reference in one of two ways. The simplest approach is to supply a FASTA-formatted file of transcript sequences. For example, such a file could be obtained from a reference genome database, a de novo transcriptome assembler, or an EST database. Alternatively, using the --gtf option, a gene annotation file (in GTF format) and the full genome sequence (in FASTA format) may be supplied. For commonly-studied species, these files may be easily downloaded from databases such as the UCSC Genome Browser Database [26] and Ensembl [27]. If the quality of existing gene annotations is in question, one can use a reference-based RNA-Seq transcriptome assembler, such as Cufflinks [28], to provide an improved set of gene predictions in GTF format. When gene-level abundance estimates are desired, an additional file specifying which transcripts are from the same gene may be specified (via the --transcript-to-gene-map option), or, if a GTF file is provided, the "gene_id" attribute of each transcript may be used to determine gene membership. With either method of specifying transcripts, RSEM generates its own set of preprocessed transcript sequences for use by later steps. For poly(A) mRNA analysis, RSEM will append poly(A) tail sequences to reference transcripts to allow for more accurate read alignment (disabled with --no-polyA). The scripts for preparing the reference sequences need only be run once per reference transcriptome as the transcript sequences are preprocessed in a sample-independent manner.

3.2.2 READ MAPPING AND ABUNDANCE ESTIMATION

The rsem-calculate-expression script handles both the alignment of reads against reference transcript sequences and the calculation of relative abundances. By default, RSEM uses the Bowtie alignment program [29] to align reads, with parameters specifically chosen for RNA-Seq quantification. Alternatively, users may manually run a different alignment program and provide alignments in SAM format [30] to rsem-calculate-expression.

When using an alternative aligner, care must be taken to set the aligner parameters appropriately so that RSEM may provide the best abundance estimates. First, and most critically, aligners must be configured to report all valid alignments of a read, and not just a single "best" alignment. Second, we recommend that aligners be configured so that only matches and mismatches within a short prefix (a "seed") of each read be considered when determining valid alignments. For example, by default, RSEM runs Bowtie to find all alignments of a read with at most two mismatches in its first 25 bases. The idea is to allow RSEM to decide which alignments are most likely to be correct, rather than giving the aligner this responsibility. Since RSEM uses a more detailed model of the RNA-Seq read generation process than those used by read aligners, this results in more accurate estimation. Lastly, in order to reduce RSEM's running time and memory usage, it is useful to configure aligners to suppress the reporting of alignments for reads with a large number (e.g., > 200) of valid alignments.

While the original RSEM package only supported fixed-length SE RNA-Seq reads without quality score information, the new package supports a wide variety of input data types. RSEM now supports both SE and PE reads and reads of variable lengths. Reads may be given in either FASTA or FASTQ format. If reads are given in FASTQ format, RSEM will use quality score data as part of its statistical model. If quality scores are not provided, RSEM uses a position-dependent error model that we described previously [7].

After the alignment of reads, RSEM computes ML abundance estimates using the Expectation-Maximization (EM) algorithm for its statistical model (see Methods). A number of options are available to specify the model that is used by RSEM, which should be customized according to the RNA-Seq protocol that produced the input reads. For example, if a strand-specific protocol is used, the --strand-specific option should be specified. Otherwise, it is assumed that a read has equal probability of coming from the sense or antisense directions. The fragment length distribution is controlled by the --fragment-length- family of options, which are particularly important for SE analysis. For PE analysis, RSEM learns the fragment length distribution from the data. If the protocol produces read position distributions that are highly 5' or 3' biased, then the --estimate-rspd option should be specified so that RSEM can estimate a read start position distribution (RSPD), which may allow for more accurate abundance estimates [7].

In addition to computing ML abundance estimates, RSEM can also use a Bayesian version of its model to produce a PME and 95% CI for the abundance of each gene and isoform. These values are computed by Gibbs sampling (see Methods) and can be obtained by specifying the --calc-ci option. The 95% CIs are valuable for assessing differential expression across samples, particularly for repetitive genes or isoforms because the CIs capture uncertainty due to both random sampling effects and read mapping ambiguity. We recommend using the CIs in combination with the results of differential expression tools, which currently do not take into account variance from multiread allocation. The PME values may be used in lieu of the ML estimates as they are very similar, but have the convenient property of generally being contained within the 95% CIs, which is sometimes not the case for small ML estimates.

The primary output of RSEM consists of two files, one for isoform-level estimates, and the other for gene-level estimates. Abundance estimates are given in terms of two measures. The first is an estimate of the number of fragments that are derived from a given isoform or gene. We can only estimate this quantity because reads often do not map uniquely to a single transcript. This count is generally a non-integer value and is the expectation of the number of alignable and unfiltered fragments that are derived from a isoform or gene given the ML abundances. These (possibly rounded) counts may be used by a differential expression method such as edgeR [9] or DESeq [8]. The second measure of abundance is the estimated fraction of transcripts made up by a given isoform or gene. This measure can be used directly as a value between zero and one or can be multiplied by 106 to obtain a measure in terms of transcripts per million (TPM). The transcript fraction measure is preferred over the popular RPKM [18] and FPKM [6] measures because it is independent of the mean expressed transcript length and is thus more comparable across samples and species [7].

3.2.3 VISUALIZATION

RSEM can produce output for two different visualizations of RNA-Seq data as tracks in genome browsers, such as the UCSC Genome Browser [31]. When the --out-bam option is specified, RSEM maps read alignments from transcript to genomic coordinates and outputs the resulting alignments in BAM format [30]. Each alignment in the BAM file is weighted (using the MAPQ field) by the probability that it is the true alignment, given the ML parameters learned by RSEM. Visualization of the BAM file in a genome browser enables a user to see all of the read alignments and the posterior probabilities assigned to them by RSEM. The BAM file can be further processed by the rsem-bam2wig program to produce a UCSC WIG-formatted file that gives the expected number of reads overlapping each genomic position, given the ML parameters. Wiggle visualizations are useful for looking at the distributions of reads across transcripts. An example of the BAM and WIG visualizations within the UCSC Genome Browser is shown in Figure 2. To produce either visualization, one must have provided a GTF-formatted annotation file to the reference preparation script so that read alignments can be mapped back to genomic coordinates.

To help with diagnosing potential issues in RNA-Seq data generation or quantification, RSEM additionally allows for visualization of the sequencing model it learns from a given sample. This is accomplished by running the rsem-plot-model program on the output of rsem-calculate-expression. A number of plots are produced by rsem-plot-model, including the learned fragment and read length distributions, RSPD, and sequencing error parameters. Three of the plots generated for the RNA-Seq data set from SRA experiment SRX018974 [25] are shown in Additional file 1.

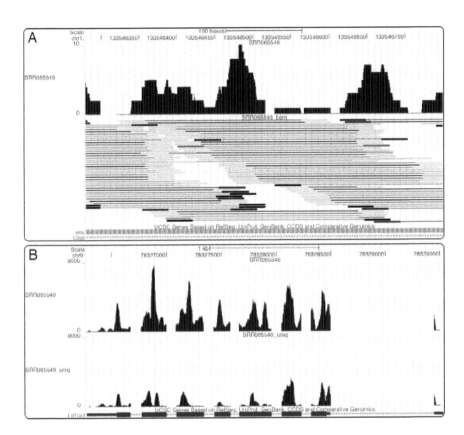

FIGURE 2: RSEM visualizations in the UCSC Genome Browser. Example visualizations of RSEM output from mouse RNA-Seq data set SRR065546 in the UCSC Genome Browser. (A) Simultaneous visualization of the wiggle output, which gives the expected read depth at each position in the genome, and the BAM output, which gives probabilistically-weighted read alignments. In the BAM track, paired reads are connected by a thin black line and the darkness of the read indicates the posterior probability of its alignment (black meaning high probability). (B) An example gene for which the expected read depth (top track) differs greatly from the read depth computed from uniquely-mapping reads only (bottom track).

3.2.4 SIMULATION

RSEM additionally allows for the simulation of RNA-Seq data sets according to the generative model on which it is based (see Methods). Simulation is performed by the rsem-simulate-reads program, which takes as input abundance estimates, sequencing model parameters, and reference transcripts (as prepared by rsem-prepare-reference). Typically, the abundance estimates and sequencing model are obtained by running RSEM on a real data set, but they may also be set manually.

3.3 RESULTS AND DISCUSSION

3.3.1 COMPARISON TO RELATED TOOLS

To evaluate RSEM, we compared its performance to a number of related quantification methods. We compared with IsoEM (v1.0.5) [4], Cufflinks (v1.0.1) [6], rQuant (v1.0) [2], and the original implementation of RSEM (v0.6) [32]. MISO [3], which uses a similar probabilistic model as RSEM, IsoEM, and Cufflinks, was not included in the comparison because it currently only computes the relative frequencies of alternative splice forms for each gene, not global transcript fractions. To make the comparisons fair, we ran Cufflinks only in its quantification mode. That is, it was configured to compute abundance estimates for the set of gene annotations that we provided to all methods and was not allowed to predict novel transcripts. Cufflinks and rQuant both require alignments of reads to a genome sequence and we used TopHat [24] for this purpose. TopHat was provided with the gene annotations and mean fragment length and was not allowed to predict novel splice junctions. For RSEM and IsoEM, which require alignments to transcript sequences, we used Bowtie [29]. As there are limited "gold-standard" data with which to evaluate the accuracy of RNA-Seq quantification methods, we tested the methods on both simulated and real data. On the simulated data, we additionally measured the computational performance (in terms of time and memory) of the methods.

3.3.2 SIMULATED DATA

As there are no published RNA-Seq data simulators, we performed experiments with the simulator included in the RSEM software package. This simulator uses the simple and widely-used model of RNA-Seq fragments being sampled uniformly and independently across all possible start sites from transcripts in a sample. The model used for the simulation is identical to that explicitly assumed by Cufflinks and IsoEM, and implicitly used by rQuant. Therefore, our simulation experiment is a test of how well the various methods perform when the data is generated from the model that they assume. We initially attempted to use an unpublished external simulation software package, Flux Simulator [33], but several bugs in the software prevented us from using it for the purposes of this paper.

We used the simulator to generate a set of 20 million RNA-Seq fragments in a non-strand-specific manner from the mouse transcriptome. Paired-end reads were simulated from these fragments, and a single-end read set was constructed by simply throwing out the second read of each pair. Two mouse reference transcript sets were used: the RefSeq annotation [34] and the Ensembl annotation [27] (see Methods). The RefSeq set is conservative with 20,852 genes and 1.2 isoforms per gene on average. In contrast, the Ensembl set has 22,329 genes and 3.4 isoforms per gene on average. We have made the simulation data for this experiment available on the RSEM website.

For each simulation set, we computed abundance estimates with the tested methods and measured the accuracy of the transcript fraction estimates using the median percent error (MPE), error fraction (EF), and false positive (FP) statistics that we used previously [7]. The MPE is the median of the percent errors of the estimated values from the true values. The 10% EF is the fraction of transcripts for which the percent error of the

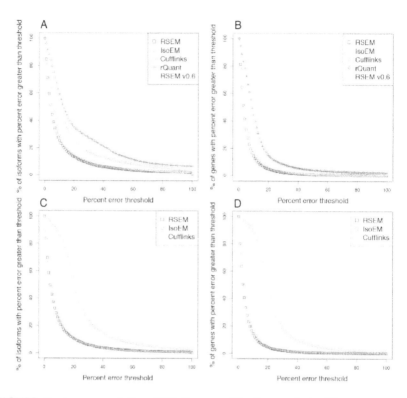

FIGURE 3: Accuracy of four RNA-Seq quantification methods. The percent error distributions of estimates from RSEM, IsoEM, Cufflinks, and rQuant on simulated RNA-Seq data. The error distributions of global isoform and gene estimates from PE data are shown in (A) and (B), respectively. Global isoform and gene estimate error distributions for SE data are shown in (C) and (D), respectively.

abundance estimate is greater than 10%. Lastly, the FP statistic is the fraction of transcripts with true abundance less than 1 TPM that are predicted to have abundance of at least 1 TPM. These statistics were calculated for three levels of estimates: (1) gene relative abundances, (2) global isoform relative abundances, and (3) within-gene isoform relative abundances.

Figure 3 gives the distributions of the errors of the abundance estimates from the five methods on the RefSeq simulated sets, using a style of plot introduced by [4]. Table 1 gives the MPE, 10% EF, and FP rates for the methods. The results for the Ensembl simulated sets are shown in Additional file 2. RSEM v0.6 and rQuant were only run on the SE data, as they do not handle PE data.

For both PE and SE reads, RSEM and IsoEM outperform Cufflinks and rQuant. There are likely two major reasons for the gap in performance between these two pairs of methods. First, Cufflinks and rQuant do not fully handle reads that map to multiple genes ("gene multireads"). Cufflinks uses a "rescue"-like strategy for an initial fractional allocation of multireads, which is roughly equivalent to one iteration of the EM algorithm used by RSEM and IsoEM. As for rQuant, it is not clear from [2] if and how this method handles gene multireads. A second reason for the performance gap is the fact that Cufflinks and rQuant require alignment of reads to the genome, not to a transcript set. As we discussed in the Implementation section, alignment of RNA-Seq reads to a genome sequence is challenging for eukaryotic species, whose RNA transcripts are spliced and polyadenylated.

The relative performance of the methods was similar across the RefSeq and Ensembl sets, although Cufflinks had surprisingly poor accuracy on the Ensembl set. A closer examination of the Cufflinks results revealed that this method was producing abnormally high abundance estimates on a subset of transcripts. This subset consisted of transcripts that were shorter (excluding poly(A) tails) than the mean fragment length (280 bases), indicating that the current implementation of Cufflinks does not properly handle short transcripts.

RSEM and IsoEM are comparable for PE data, but for SE data, RSEM is slightly more accurate. This relatively small improvement of RSEM over IsoEM is likely due to a more detailed implementation of poly(A) tail handling, which was not present in the original version of IsoEM and has only been recently introduced into its software. The improvement of the current version of RSEM over RSEM v0.6 is due to the modeling of fragment lengths for SE data, which was originally shown by [4] to improve accuracy.

TABLE 1: Accuracy measures for quantification methods applied to simulated data

Method	SE			PE		
	MPE	SE 10% EF	FP	MPE	PE 10% EF	FP
RSEM	3.1/4.1/7.1	14.1/25.1/44.1	0.9/1.7/12.0	3.1/4.0/5.2	14.4/23.3/35.7	1.0/1.8/11.3
IsoEM	4.6/5.6/8.0	18.1/29.0/45.4	1.7/3.2/31.3	4.0/4.8/5.1	16.9/25.5/35.3	1.2/1.7/11.1
Cufflinks	9.5/10.7/18.1	46.6/53.3/65.3	2.5/11.1/98.5	20.2/20.4/10.5	89.3/86.4/51.1	2.7/8.6/98.4
rQuant	8.9/12.6/43.9	44.4/58.9/89.0	7.0/16.9/99.1			
RSEM v0.6	10.1/11.2/10.4	50.5/56.0/50.7	1.4/2.9/39.5			

Accuracy of five RNA-Seq quantification methods on simulated SE and PE data. Values are given as gene/global isoform/within-gene isoform.

3.3.3 MAQC DATA

It is challenging to benchmark RNA-Seq quantification methods on real data as we rarely know the "true" transcript abundances in a sample. Currently, qRT-PCR appears to be the most popular technology for producing "gold standard" abundance measurements, although without careful experimental design and data analysis it can give inaccurate results [35]. While RNA-Seq is generally accepted as being a more accurate quantification technology than microarrays [1], it remains to be seen whether it is also superior to qRT-PCR.

For our tests we used data generated from samples used in the Microarray Quality Control (MAQC) Project [36], as has been done in a number of other studies of RNA-Seq quantification accuracy [37,38]. The MAQC project evaluated a variety of microarray platforms and technologies, including TaqMan qRT-PCR, on two human RNA samples, one from brain tissue (HBR) and another from a mixture of tissue types (UHR). The TaqMan qRT-PCR measurements from this project consist of abundance values for a small subset (1,000) of genes, with four technical replicates on each of the two samples. Recently, three groups have generated RNA-Seq data on the two MAQC samples [25,37,39].

We applied the quantification methods on each of the MAQC RNA-Seq data sets and compared their abundance predictions to the qRT-PCR values. All methods were provided with the human RefSeq gene annotation. As for the simulation experiments, Cufflinks was only run in quantification mode and TopHat was only allowed to map to splice junctions present in the annotation. Cufflinks and IsoEM were run with and without their sequence-specific bias correction modes, which can improve quantification accuracy for RNA-Seq libraries generated with a random hexamer priming protocol, which was used for all of the MAQC RNA-Seq data. We did not run RSEM with its position-specific bias correction (RSPD) as this is only appropriate for oligo-dT primed RNA-Seq libraries, which generally have a bias towards reads originating from the 3' end of transcripts.

To assess the similarity of the RNA-Seq abundance predictions with the qRT-PCR measurements, we calculated the Pearson correlation of the logarithm of the abundance values. We used a log transformation to prevent the correlation values from being dominated by the most abundant transcripts. To avoid problems with zeros, correlation values were calculated for only those genes that were predicted to have non-zero abundance by qRT-PCR and all methods. We additionally computed the false positive (FP), true positive (TP), false negative (FN), and true negative (TN) counts for each method, where "positive" means non-zero predicted abundance and truth is determined by the qRT-PCR measurements.

The correlation values for the tested methods on each of the MAQC RNA-Seq samples are shown in Table 2. In general, the methods gave comparable correlation values for each sample. Confirming the results of [38], the bias correction mode of Cufflinks gave predictions with higher correlation than the other methods, particularly on the HBR samples. Unlike Cufflinks, the bias correction mode of IsoEM did not have a significant effect on its correlation with the qRT-PCR values for these samples. Spearman and Pearson correlation values computed without log-transformed abundances yielded similar results (Additional file 3). The TP, FP, TN, and FN counts for the methods were also comparable (Additional file 3).

TABLE 2: Correlation of quantification method predictions with MAQC qRT-PCR values

SRA ID	Read type	Sample	RSEM	IsoEM	IsoEM (C)	Cufflinks	Cufflinks (C)	rQuant
SRX016366	SE	HBR	0.69	0.68	0.68	0.71	0.79	0.72
SRX003926	SE	HBR	0.68	0.67	0.67	0.7	0.73	0.71
SRX018974	PE	HBR	0.69	0.69	0.69	0.69	0.78	NA
SRX016368	SE	UHR	0.71	0.71	0.72	0.72	0.77	0.72
SRX016369	SE	UHR	0.73	0.74	0.74	0.73	0.76	0.74
SRX016370	SE	UHR	0.74	0.75	0.75	0.74	0.77	0.75
SRX016371	SE	UHR	0.74	0.75	0.75	0.74	0.77	0.75
SRX016372	SE	UHR	0.75	0.75	0.75	0.74	0.77	0.76
SRX003927	SE	UHR	0.72	0.71	0.72	0.71	0.74	0.72

Correlation values (Pearson r^2 of log-transformed abundance values) were computed between the predictions of four methods and "gold-standard" values from qRT-PCR for nine different RNA-Seq data sets. IsoEM and Cufflinks were run with (C) and without their bias correction modes.

The lack of a clear distinction between the methods (except for Cufflinks with bias correction enabled) on these data sets can be explained by a number of factors. First, qRT-PCR measurements are only available for 1,000 (5%) out of a total of 19,005 genes in the RefSeq set. After filtering for the qRT-PCR genes that were consistent in their annotation with RefSeq and had non-zero abundance (see Methods), only 716 could be used for correlation analysis. Second, this set of genes is biased towards single-isoform genes and genes that have relatively unique sequences, reducing the ability of these data to distinguish those methods that are better at isoform quantification or multiread handling. The mean number of isoforms per gene in this set is 1.1, compared to 1.7 for all genes (p $< 10^{-115}$, Mood's median test). Similarly, the mean "mappability" (see Methods) of genes in the set is 0.96, compared to 0.91 for all genes (p $< 10^{-6}$). Lastly, biases in the qRT-PCR values, perhaps due to variable amplification efficiencies [35], may have resulted in an inaccurate gold standard.

3.3.4 RUNNING TIME AND MEMORY

In addition to comparing the accuracies of the quantification methods, we also measured their running times and memory usage. For this purpose, we used our simulated mouse RefSeq data set of 20 million fragments, which is comparable in size to data produced by a single lane of the Illumina Genome Analyzer IIx. Table 3 lists the running times and peak memory usage for each method, on both SE and PE data. Additional file 4 gives the corresponding values for the simulated mouse Ensembl data set. All methods were run on an 8-core 2.93 GHz Linux server with 32 GB of RAM and hyper-threading enabled. Alignment with Bowtie against a transcript sequence set and quantification with RSEM uses the least amount of memory, at around 1.1 GB. The peak memory usage for Cufflinks and rQuant is due to running TopHat for aligning reads to the genome. The quantification programs for these two methods required 0.4 and 1.6 GB of memory, respectively, on the RefSeq data set. IsoEM is the fastest method, but has the largest memory requirement, up to 14 GB. It should be noted that the running times of the methods are not completely comparable, as RSEM and Cufflinks compute CIs in addition to ML estimates, whereas the other methods only compute ML estimates.

TABLE 3: Running time and memory usage of quantification methods on SE and PE data

| Method | SE | | | | PE | | | |
	Alignment time	Quant. time	Total time	Peak Memory	Alignment time	Quant. time	Total time	Peak Memory
RSEM	24	50	74	1.1	15	50	66	1.1
IsoEM	5	6	12	12	14	10	24	14
Cufflinks	33	3	36	2.0	60	6	66	2.0
rQuant	33	183	216	2.0				
RSEM v0.6	9	22	31	1.1				

The alignment time, quantification time, total time, and peak memory usage of the tested quantification methods on SE and PE simulated data sets with 20 million fragments. Times are in minutes and memory is specified in GB.

The running time and memory required by RSEM scales linearly with the number of read alignments, which is generally proportional to the number of reads. Although the current version of RSEM has a parallelized EM algorithm, it is not faster than the original version for two reasons. First, the current version runs the EM algorithm for many more iterations to improve accuracy. On this data set, the current version ran for 4,802 iterations, compared to 643 for the older version. Second, the running time for the current version includes the time for computing 95% credibility intervals, which requires significant computation and was not a feature of the original version.

TABLE 4: Accuracies obtained from RNA-Seq data sets with various properties

Species	Seq. Error	Read type	Read length	Read number (×10⁶)	Through-put (MB)	MPE	10% EF	FP
M	N	SE	35	20	700	3.1/4.2/7.1	13.7/25.0/44.0	1.0/1.8/10.9
M	N	SE	70	20	1400	3.1/4.1/6.0	13.6/23.9/40.4	0.8/1.5/8.2
M	N	PE	35	20	1400	3.0/3.9/4.9	13.3/21.9/**34.2**	1.1/1.8/13.4
M	N	SE	35	40	1400	**2.3/3.1/4.8**	**8.4/18.5**/35.8	**0.7/1.3/10.5**
M	Y	SE	35	20	700	3.1/4.2/6.9	14.1/25.2/43.0	1.1/1.9/13.2
M	Y	SE	70	20	1400	3.0/4.0/6.0	14.2/24.4/40.6	1.1/1.6/9.5
M	Y	PE	35	20	1400	3.0/3.9/5.1	14.0/22.9/**35.5**	1.3/1.9/12.1
M	Y	SE	35	40	1400	**2.2/3.0/5.0**	**8.5/18.4**/35.7	**0.9/1.5/11.3**
H	N	SE	35	20	700	4.0/7.8/14.3	20.2/43.5/58.6	3.6/8.0/21.6
H	N	SE	70	20	1400	3.9/7.3/11.9	19.3/41.0/54.4	3.5/7.0/17.3
H	N	PE	35	20	1400	3.7/6.2/**9.0**	17.0/36.4/**47.3**	3.7/6.4/**14.0**
H	N	SE	35	40	1400	**2.9/5.7**/10.1	**13.4/35.4**/50.3	**2.7/6.3**/20.3
H	Y	SE	35	20	700	3.9/7.7/14.8	19.5/43.2/59.3	3.9/8.1/20.8
H	Y	SE	70	20	1400	3.8/7.2/12.4	19.2/40.8/55.2	3.9/7.1/17.7
H	Y	PE	35	20	1400	3.8/6.5/**9.2**	19.0/37.8/**48.2**	4.2/6.4/**13.9**
H	Y	SE	35	40	1400	**2.9/5.6**/10.3	**12.8/35.5**/50.8	**3.0/6.3**/18.8

Accuracy of abundance estimates from RNA-Seq data sets varying in species (H = human, M = mouse), sequencing error, type (SE or PE) of reads, number of reads, and length of reads. Values are given as gene/global isoform/within-gene isoform.

3.3.5 EXPERIMENTAL RESULTS

With RSEM extended to model PE data and reads with quality score information, we set out to determine whether these more complex data types allow for improvement in abundance estimation accuracy. To this end, we performed two sets of simulation experiments. With the first set of experiments we compared the performance of PE reads against that of SE reads. With the second, we tested whether quality scores provide information that improves estimation accuracy.

3.3.6 PAIRED VS. SINGLE END READS

We previously showed that for SE RNA-Seq protocols, the number of reads is more important than the length of reads for increasing the accuracy of gene-level abundance estimates [7]. Given fixed sequencing throughput (in terms of the total number of bases), we found that the optimal read length was around 25 bases for SE RNA-Seq analysis in both mouse and maize. This result was confirmed by a later study [4]. Recent studies have reached the conclusion that PE reads can offer improved estimation accuracy over SE reads, particularly for isoforms of alternatively-spliced genes [3,4]. With RSEM now extended to model PE data, we decided to test these results with our own simulations.

We simulated RNA-Seq data with four different configurations: (1) 20 million, 35 base SE reads, (2) 20 million, 70 base SE reads, (3) 20 million, 35 base PE reads, and (4) 40 million 35 base SE reads. The latter three configurations give the same throughput in terms of the number of bases sequenced, and thus are the most comparable in terms of cost, given a simple economic model in which one pays per sequenced base. We simulated for both human and mouse, and with both RefSeq and Ensembl annotations, to determine if the species or annotation set is a factor. In addition to simulating with different species and annotation sets for each configuration, we also simulated with and without sequencing error to assess whether variable read alignment sensitivity had an impact.

Table 4 gives the MPE, 10% EF, and FP of the RSEM estimates computed from the RefSeq simulated data sets (Additional file 5 gives the corresponding values for the Ensembl sets). As expected, with the number of reads fixed, the 70 base reads gave better estimation accuracy than the 35 base reads. Confirming previous results [3,4], with the number of reads and total throughput fixed, PE reads improved estimation accuracy over SE reads (compare the PE accuracies with those of the SE 70 base accuracies). However, with the same sequencing throughput, short SE reads offered the highest estimation accuracy at the gene level. This result held across both species and regardless of whether reads contained sequencing errors. These results suggest that if the primary goal is the accurate estimation of gene abundances, then the sequencing of a large number of short SE reads is best. For example, given a choice between one Illumina lane of PE 35 base reads and two Illumina lanes of SE 35 base reads, our simulations show that the latter will provide the best overall quantification results for gene-level estimates. An additional advantage of using SE reads in this scenario is that two lanes of SE reads can be run in parallel whereas the two ends of a PE lane are currently generated one after the other. Thus, using short SE reads can save sequencing time. This result depends on the SE estimation procedure being provided with a fragment length distribution, as SE data is not easily used to automatically determine this distribution. However, this distribution can usually be obtained by other means ahead of time.

TABLE 5: The effect of quality score modeling on quantification accuracy

Simulation model	Estimation model	MPE	10% EF	FP
theoretical	quality	3.1/4.1/7.2	13.8/25.2/43.5	1.0/1.8/11.6
theoretical	profile	3.1/4.1/7.2	13.9/25.3/43.6	1.0/1.8/11.7
empirical	quality	3.1/4.0/7.0	14.2/25.3/43.0	1.2/2.0/11.4
empirical	profile	3.1/4.1/7.0	14.3/25.4/43.2	1.1/2.0/11.2

Accuracy of abundance estimates from RNA-Seq data sets with different combinations of sequencing error models for simulation and estimation. Values are given as gene/global isoform/within-gene isoform.

On the other hand, if the primary interest is in the relative frequencies of alternative splicing events within single genes, then PE data can provide more accurate estimates, depending on the transcript set. The result that the PE data show a larger accuracy improvement over SE data for the human RefSeq simulations is explained by the fact that the human RefSeq annotation has more isoforms per gene on average (1.6) than the mouse RefSeq annotation (1.2). This is further supported by the results of the simulations using the Ensembl annotations, which have significantly more isoforms per gene on average (6.3 for human and 3.4 for mouse). Thus, for species with genes that undergo a large number of alternative splicing events, PE data will likely be better for inferring the relative frequencies of these events. Although the results for gene-level and within-gene isoform-level estimates are clear, those for global isoform-level estimates are mixed. In some simulation sets, SE data performs better than PE data (with the same throughput), and in others, the opposite is true. This is explained by the fact that the global abundance of an isoform is the product of its gene's abundance and its within-gene abundance. Thus, one can improve global isoform abundance accuracy by producing better abundance estimates at either of the other two levels. Global isoform-level estimates are improved by SE data through more accurate gene-level estimates and by PE data through more accurate within-gene isoform estimates.

Overall, we suggest that researchers carefully consider the objectives of their RNA-Seq experiments before deciding on sequencing parameters, such as read length and number of reads. While one may be inclined to produce long and PE reads, it may be more cost efficient to use a larger number of SE reads if the only goal is quantification of gene abundances. If the goal is instead to analyze within-gene isoform frequencies or to perform non-quantification tasks such as transcriptome assembly, then PE reads should be preferred. To determine the optimal sequencing strategy for quantification with a particular transcript set, the RSEM simulation tool can be used.

3.3.7 THE VALUE OF QUALITY SCORES FOR RNA-SEQ QUANTIFICATION

We performed simulation experiments to determine if the use of quality scores (rather than just the read sequences themselves) improves the accuracy of quantification with RNA-Seq data. Two SE simulations were performed, each with a different sequencing error model. The simulations used the mouse RefSeq transcript set as a reference. In the first simulation, an error was introduced at a given read position according to the theoretical probability of an error given the quality score at that position. That is, the probability that an error was introduced at a position with Phred quality score q was $10^{-q/10}$. In the second simulation, the probability of a sequencing error given a quality score q was determined from the training data (we call this the "empirical" model). For the two simulated data sets, we estimated abundances with RSEM using two different models: one that takes the quality scores into account (the "quality score" model), and a second that uses our original error model, which does not take into account quality scores and instead estimates a sequencing error model that is position and base-dependent (the "profile" model). The MPE, 10% EF, and FP statistics were calculated for the abundance estimates of the two RSEM models on the two simulated data sets (Table 5). We found that even when sequencing errors followed the theoretical probabilities given by the quality scores, the accuracy of the quality score model was practically indistinguishable from that of the profile model. Simulations with the Ensembl transcript set gave similar results (Additional file 6). This indicates that for the purposes of quantification from RNA-Seq data, quality scores from Illumina-generated reads provide little additional information. This does not suggest that sequencing errors do not need to be modeled, however. Instead, these results suggest that an effective sequencing error model can be learned from the read sequences alone. We stress that these results are only for the task of quantification. Applications such as SNP detection will certainly need to take quality score information into account.

3.4 CONCLUSIONS

We have presented RSEM, a software package for performing gene and isoform level quantification from RNA-Seq data. Through simulations and evaluations with real data, we have shown that RSEM has superior or comparable performance to other quantification methods. Unlike other tools, RSEM does not require a reference genome and thus should be useful for quantification with de novo transcriptome assemblies. The software package has a number of other useful features for RNA-Seq researchers including visualization outputs and CI estimates. In addition, the software is user-friendly, typically requiring at most two commands to estimate abundances from raw RNA-Seq reads and uses reference transcript files in standard formats. Lastly, RSEM's simulation module is valuable for determining optimal sequencing strategies for quantification experiments. Taking advantage of this module, we have determined that a large

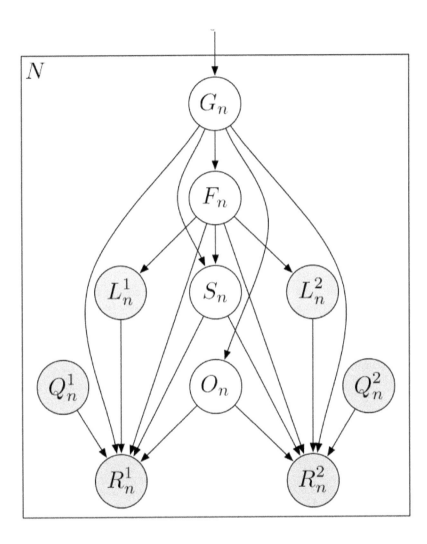

FIGURE 4: The directed graphical model used by RSEM. The model consists of N sets of random variables, one per sequenced RNA-Seq fragment. For fragment n, its parent transcript, length, start position, and orientation are represented by the latent variables G_n, F_n, S_n and O_n respectively. For PE data, the observed variables (shaded circles), are the read lengths (L_n^1 and L_n^2), quality scores (Q_n^1 and Q_n^2), and sequences (R_n^1 and R_n^2). For SE data, L_n^2, Q_n^2, and R_n^2 are unobserved. The primary parameters of the model are given by the vector θ, which represents the prior probabilities of a fragment being derived from each transcript.

number of short SE reads is best for gene-level quantification, while PE reads may improve within-gene isoform frequencies for the mouse and human transcript sets.

RSEM will continue to be developed to remain up to date with the latest sequencing technologies and research about details of the RNA-Seq protocol. Future work will include incorporating additional biases into the model, such as sequence-specific read position preferences [38,40] and transcript-specific read distributions [41]. We also intend to add support for color-space reads generated by ABI SOLiD sequencers and indels within read alignments.

3.5 AVAILABILITY AND REQUIREMENTS

- Project name: RSEM
- Project home page: http://deweylab.biostat.wisc.edu/rsem webcite
- Operating systems: Any POSIX-compatible platform (e.g., Linux, Mac OS X, Cygwin)
- Programming languages: C++, Perl
- Other requirements: Pthreads; Bowtie [29] for the default alignment mode of rsem-calculate-expression; R for rsem-plot-model.
- License: GNU GPL.

3.6 METHODS

3.6.1 STATISTICAL MODEL

The statistical model used by RSEM can be represented by the directed graphical model shown in Figure 4. Compared to our original statistical model [7], this model has been extended in four ways. First, PE reads are now modeled, using a pair of observed random variables, R^1 and R^2. For the case of SE reads, R^2 is treated as a latent random variable. Second, the length of the fragment from which a read or pair of reads is derived is now modeled and is represented by the latent random variable F. The distribution of F is specified using a global fragment length distribution λ_F, which is truncated and normalized given that a fragment is derived from a specific transcript of finite length. That is,

$$P(F = x|G = i) = \lambda_F(x) \left(\sum_{x'=1}^{\ell_i} \lambda_F(x') \right)^{-1}$$

where ℓ_i is the length of transcript i. The use of a fragment length distribution for RNA-Seq quantification was first introduced by [6] for paired-end data and later described by [4] for single-end data.

A third extension allows the lengths of reads to vary (such as for 454 data). The length of a read is represented by the observed random variable L (or L^1 and L^2 for PE reads). Similar to the fragment length model, the distribution of L is specified using a global read length distribution λ_R, which is truncated and normalized given a specific fragment length. In symbols:

$$P(L = y|F = x) = \lambda_R(y) \left(\sum_{x'=1}^{x} \lambda_R(y') \right)^{-1}$$

Lastly, the quality scores for a read are now used to model the probability of that read's sequence. The quality score string for a read is represented by the random variable Q. For the purposes of quantification, we do not specify a distribution for the Q random variables, as they are observed and not dependent on any of the other random variables (i.e., we are only interested in the conditional likelihood of the reads given their quality scores). Rather than rely on the theoretical probabilities of errors implied by the quality scores, we use an empirical error function, ε. Given that read position i has quality score q_i and is derived from the reference character c, the conditional probability of the read character r_i is $P(r_i|q_i, c) = \varepsilon(r_i, q_i, c)$. If quality scores are not available or reliable, then our position and reference character-dependent error model [7] may be used.

3.6.2 EXPECTATION-MAXIMATION

Given a set of RNA-Seq data, RSEM's primary goal is to compute the ML values of the parameters, θ, of the model presented in the previous section, where θ_i represents the probability that a fragment is derived from transcript i (with θ_0 representing the "noise" transcript from which reads that have no alignments may be derived). Once estimated, the θ values are converted to transcript fractions (which we denote by τ) using the equation

$$\tau_i = \frac{\theta_i / \ell'_i}{\sum_{j \neq 0} \theta_j / \ell'_j}$$

where ℓ'_i is the effective length of transcript i [6], given by

$$\sum_{x \leq \ell_i} \lambda_F(x) \, (\ell_i - x + 1)$$

for poly(A)- transcripts and

$$\sum_{x \leq \ell_i + \ell_A} \lambda_F(x) \min(\ell_i + \ell_A - x + 1, \ell_i)$$

for poly(A)+ transcripts, where ℓ_A is the length of a poly(A) tail. The effective length can be thought of as the mean number of positions from which a fragment may start within the sequence of transcript i.

RSEM computes approximate ML estimates for θ using the EM algorithm (for details, see [7]). The estimates are approximate because alignments are used to restrict the possible positions from which reads may be derived. During the first 20 iterations (and every 100 iterations) of EM, the parameters of the fragment length, RSPD, and sequencing error distributions are updated along with θ. During all other iterations, only the θ parameters are updated. This estimation strategy is an improvement over the original implementation of RSEM, which estimated all parameters other than θ before EM using uniquely-mapping reads. The algorithm is stopped when all θ_i with value $\geq 10^{-7}$ have a relative change of less than 10^{-3}. After convergence, RSEM outputs the ML τ values, as well as the expected value of the number of RNA-Seq fragments derived from each transcript, given the ML parameters.

To speed up inference, reads with a large number (at least 200, by default) of alignments are filtered out. We additionally filter out reads that are likely to be derived from poly(A) tails, as aligners may not always detect that these reads have many alignments. Due to the alignment approximation and this filtering strategy, a straightforward application of the EM procedure described will lead to biased abundance estimates for transcripts that contain highly-repetitive sequences (including poly(A) tails). Therefore, we apply a slight modification to our ML estimator to adjust for this bias. For transcript i, we calculate a value m_i, which is the probability that a read (fragment) generated from transcript i will not have a large number of alignments. In general, the value of m_i depends on the fragment length distribution, the read length distribution, the RSPD, the strand-specificity of the protocol, and the length of a poly(A) tail. During the maximization step of EM, our modification is to set θ_i to be proportional to $c_i/(Nm_i)$, where c_i is the expected number of fragments derived from transcript i and N is the total number of unfiltered fragments.

3.6.2 GIBBS SAMPLING

In addition to computing ML estimates, RSEM uses a Bayesian version of its model to compute PME and 95% CIs of abundances. In the Bayesian model, the θ parameters are treated as latent random variables with a Dirichlet prior distribution. The parameters of the Dirichlet distribution (α) are set to one, which makes the prior equivalent to a uniform distribution and the maximum a posteriori estimates of θ equal to the ML estimates.

RSEM computes PMEs and 95% CIs with a two-stage sampling process. First, a standard application of the collapsed Gibbs sampling algorithm [42] is used to obtain a sampled set of count vectors, where each vector represents the number of fragments that are mapped to each transcript. During each round of the Gibbs sampling algorithm, the true mapping of each fragment is resampled given the current mappings of all other fragments. The initial mapping of each fragment is sampled according to the ML parameters computed by the EM algorithm. The algorithm is run to sample 1000 count vectors.

The second stage of the sampling process involves sampling values of θ given each count vector sampled from the first stage. Given a count vector, c, a θ vector is sampled from its posterior distribution, which is simply a Dirichlet distribution with $\alpha_i = c_i + 1$. For each count vector, 50 θ vectors are sampled, resulting in 50,000 total samples for θ. The θ samples are converted to transcript fractions (τ) and then summarized to produce a PME and 95% CI for the abundance of each transcript.

To validate the CIs generated by RSEM, we simulated an RNA-Seq data set with the mouse RefSeq annotation and estimated CIs with RSEM from 50% credibility up to 95% credibility. We then computed the fraction of transcripts for which the true abundances fell within the credibility intervals, out of all transcripts with abundance at least 1 TPM (Table 6). The results indicate that the 95% credibility intervals are reasonably accurate and that these intervals are tight (since the fraction of correctly predicted transcript levels goes down in step with the credibility level). CIs estimated from data simulated with the mouse Ensembl annotation were less accurate (Additional file 7). We investigated why the CIs were less accurate on this set and found that many of the CIs were biased downward due to the Dirichlet prior and the larger number of transcripts in the Ensembl set. Although the CIs for the Ensembl set did not perform as well as those for the RefSeq set, we expect that they are still useful for comparing abundances across samples, as the biases in the CIs should be consistent. However, these results suggest that further work is needed to develop prior distributions that can better handle the large numbers of transcripts with zero abundance that are typical of RNA-Seq data sets.

3.6.3 REFERENCE SEQUENCES

Two sources were used for reference transcript set annotations: the RefSeq gene annotations from the UCSC Genome Browser Database [26] and the Ensembl release 63 annotations [27]. The genome versions used for the RefSeq annotations of human and mouse were build 36.1 (UCSC hg18) and build 37 (UCSC mm9), respectively. For the Ensembl human annotation, build 37 (UCSC hg19) was used instead. Both the RefSeq and Ensembl annotations were filtered to remove non-coding genes and genes located on non-standard chromosomes (e.g., chr1_random and chr5_h2_hap1). In addition, we identified a small fraction of RefSeq genes that were located at multiple, non-overlapping positions and renamed them so that each gene originated from a unique locus.

TABLE 6: Accuracy of RSEM's credibility interval estimates

Credibility level	Isoforms with true abundance within estimated CI (%)	Genes with true abundance within estimated CI (%)
95	93.1	94.3
90	87.7	89.0
85	82.6	83.9
80	77.3	78.6
75	72.1	73.4
70	67.0	68.4
65	62.0	63.5
60	57.0	58.5
55	52.0	53.4
50	46.9	48.4

Accuracies of credibility intervals computed by RSEM for credibility levels ranging from 50% to 95%.

3.6.4 SIMULATION

The generative statistical model used by RSEM is easily used to simulate RNA-Seq data. In addition to the primary parameters of the model (e.g., abundances, fragment and read length distributions, and sequencing error model parameters), quality score information must be provided to simulate reads. For the purposes of the simulations in this paper, we used a first-order Markov chain model of quality scores to generate quality score strings for each read. The parameters of the simulation model were learned from real RNA-Seq data sets from the Sequence Read Archive (SRA). The mouse simulation parameters were learned from SRA accession SRX026632, which consists of ~ 4.2 million PE 35 base reads sequenced from a library of poly(A)+ RNA from C2C12 mouse myoblasts [3]. For the human simulations, we learned parameters from SRA accession SRX016368, which consists of ~ 93 million SE 35 base reads sequenced from a MAQC UHR sample [37]. As the human data were SE reads, RSEM was provided with a fragment length distribution with μ = 200 and σ = 29 in order to learn the other model parameters. However, for the simulations, both human and mouse data were generated with a fragment length distribution with μ = 280 and σ = 17, which was used in [3] for similar simulations. Lastly, to model the fact that the mRNAs have poly(A) tails, we appended 125 As to the end of each transcript.

3.6.5 MAQC VALIDATION

TaqMan qRT-PCR measurements were downloaded from the Gene Expression Omnibus (GEO) (Platform GPL4097). For each sample, the abundance of a gene was taken as the mean of the values that passed the detection threshold for all probes assigned to the gene across all technical replicates. Following [37], a gene was considered expressed if 75% of its probes passed the detection threshold. The RefSeq transcript accessions listed for each gene in the GEO record were compared to the RefSeq accessions for each gene in the genome annotation. Only those genes for which the GEO accessions were a superset of the annotation accessions were kept. This was done to ensure that the RNA-Seq estimates were comparable to the values for the qRT-PCR probes, which are only guaranteed to correspond to the accessions given in the GEO record. This filtering resulted in a set of 716 genes, 656 and 618 of which were detected in UHR and HBR, respectively.

    To analyze how representative the filtered qRT-PCR genes were of the entire human RefSeq gene set, we computed the "mappability" of each gene. For each isoform we generated all possible 35 base reads from its sequences and aligned them to the entire transcript set with Bowtie, allowing at most two mismatches. The mappability of an isoform was computed as the fraction of reads derived from it that only aligned with isoforms of its gene. The mappability of a gene was then computed as the mean of its isoform mappabilities.

REFERENCES

1. Wang Z, Gerstein M, Snyder M: RNA-Seq: a revolutionary tool for transcriptomics. Nature Reviews Genetics 2009, 10:57-63.
2. Bohnert R, Rätsch G: rQuant.web: a tool for RNA-Seq-based transcript quantitation. Nucleic Acids Research 2010, (38 Web Server):W348-51.
3. Katz Y, Wang ET, Airoldi EM, Burge CB: Analysis and design of RNA sequencing experiments for identifying isoform regulation. Nature Methods 2010, 7(12):1009-15.
4. Nicolae M, Mangul S, Măndoiu I, Zelikovsky A: Estimation of alternative splicing isoform frequencies from RNA-Seq data. In Algorithms in Bioinformatics, Lecture Notes in Computer Science. Edited by Moulton V, Singh M. Liverpool, UK: Springer Berlin/Heidelberg; 2010::202-214.
5. Jiang H, Wong WH: Statistical inferences for isoform expression in RNA-Seq. Bioinformatics 2009, 25(8):1026-1032.
6. Trapnell C, Williams B, Pertea G, Mortazavi A, Kwan G, van Baren M, Salzberg S, Wold B, Pachter L: Transcript assembly and quantification by RNA-Seq reveals unannotated transcripts and isoform switching during cell differentiation. Nature Biotechnology 2010, 28(5):511-515.
7. Li B, Ruotti V, Stewart RM, Thomson JA, Dewey CN: RNA-Seq gene expression estimation with read mapping uncertainty. Bioinformatics 2010, 26(4):493-500.
8. Anders S, Huber W: Differential expression analysis for sequence count data. Genome Biology 2010, 11(10):R106.
9. Robinson MD, McCarthy DJ, Smyth GK: edgeR: a Bioconductor package for differential expression analysis of digital gene expression data. Bioinformatics 2010, 26:139-40.
10. Guttman M, Garber M, Levin JZ, Donaghey J, Robinson J, Adiconis X, Fan L, Koziol MJ, Gnirke A, Nusbaum C, Rinn JL, Lander ES, Regev A: Ab initio reconstruction of cell type-specific transcriptomes in mouse reveals the conserved multi-exonic structure of lincRNAs. Nature Biotechnology 2010, 28(5):503-510.
11. Robertson G, Schein J, Chiu R, Corbett R, Field M, Jackman SD, Mungall K, Lee S, Okada HM, Qian JQ, Griffith M, Raymond A, Thiessen N, Cezard T, Butterfield YS, Newsome R, Chan SK, She R, Varhol R, Kamoh B, Prabhu AL, Tam A, Zhao Y, Moore RA, Hirst M, Marra MA, Jones SJM, Hoodless PA, Birol I: De novo assembly and analysis of RNA-seq data. Nature Methods 2010, 7(11):909-12.
12. Grabherr MG, Haas BJ, Yassour M, Levin JZ, Thompson Da, Amit I, Adiconis X, Fan L, Raychowdhury R, Zeng Q, Chen Z, Mauceli E, Hacohen N, Gnirke A, Rhind N, di Palma F, Birren BW, Nusbaum C, Lindblad-Toh K, Friedman N, Regev A: Full-length transcriptome assembly from RNA-Seq data without a reference genome. Nature Biotechnology 2011, 29(7):644-52.

13. Nagalakshmi U, Wang Z, Waern K, Shou C, Raha D, Gerstein M, Snyder M: The Transcriptional Landscape of the Yeast Genome Defined by RNA Sequencing. Science 2008, 320(5881):1344-1349.

14. Marioni JC, Mason CE, Mane SM, Stephens M, Gilad Y: RNA-seq: an assessment of technical reproducibility and comparison with gene expression arrays. Genome Research 2008, 18(9):1509-17.

15. Morin R, Bainbridge M, Fejes A, Hirst M, Krzywinski M, Pugh T, McDonald H, Varhol R, Jones S, Marra M: Profiling the HeLa S3 transcriptome using randomly primed cDNA and massively parallel short-read sequencing. BioTechniques 2008, 45:81-94.

16. Wang X, Wu Z, Zhang X: Isoform abundance inference provides a more accurate estimation of gene expression levels in RNA-seq. Journal of Bioinformatics and Computational Biology 2010, 8(Suppl 1):177-92.

17. Faulkner GJ, Forrest ARR, Chalk AM, Schroder K, Hayashizaki Y, Carninci P, Hume DA, Grimmond SM: A rescue strategy for multimapping short sequence tags refines surveys of transcriptional activity by CAGE. Genomics 2008, 91(3):281-8.

18. Mortazavi A, Williams BA, McCue K, Schaeffer L, Wold B: Mapping and quantifying mammalian transcriptomes by RNA-Seq. Nature Methods 2008, 5(7):621-8.

19. Feng J, Li W, Jiang T: Inference of isoforms from short sequence reads. Journal of Computational Biology 2011, 18(3):305-21.

20. Paşaniuc B, Zaitlen N, Halperin E: Accurate Estimation of Expression Levels of Homologous Genes in RNA-seq Experiments. Journal of Computational Biology 2011, 18(3):459-68.

21. Richard H, Schulz MH, Sultan M, Nürnberger A, Schrinner S, Balzereit D, Dagand E, Rasche A, Lehrach H, Vingron M, Haas SA, Yaspo ML: Prediction of alternative isoforms from exon expression levels in RNA-Seq experiments. Nucleic Acids Research 2010, 38(10):e112..

22. Taub M, Lipson D, Speed TP: Methods for allocating ambiguous short-reads. Communications in Information and Systems 2010, 10(2):69-82.

23. De Bona F, Ossowski S, Schneeberger K, Ratsch G: Optimal spliced alignments of short sequence reads. Bioinformatics 2008, 24(16):i174-180.

24. Trapnell C, Pachter L, Salzberg SL: TopHat: discovering splice junctions with RNA-Seq. Bioinformatics 2009, 25(9):1105-11.

25. Au KF, Jiang H, Lin L, Xing Y, Wong WH: Detection of splice junctions from paired-end RNA-seq data by SpliceMap. Nucleic Acids Research 2010, 38(14):4570-8.

26. Fujita PA, Rhead B, Zweig AS, Hinrichs AS, Karolchik D, Cline MS, Goldman M, Barber GP, Clawson H, Coelho A, Diekhans M, Dreszer TR, Giardine BM, Harte RA, Hillman-Jackson J, Hsu F, Kirkup V, Kuhn RM, Learned K, Li CH, Meyer LR, Pohl A, Raney BJ, Rosenbloom KR, Smith KE, Haussler D, Kent WJ: The UCSC Genome Browser database: update 2011. Nucleic Acids Research 2011, (39 Database):D876-82.

27. Flicek P, Amode MR, Barrell D, Beal K, Brent S, Chen Y, Clapham P, Coates G, Fairley S, Fitzgerald S, Gordon L, Hendrix M, Hourlier T, Johnson N, Kähäri A, Keefe D, Keenan S, Kinsella R, Kokocinski F, Kulesha E, Larsson P, Longden I, McLaren W, Overduin B, Pritchard B, Riat HS, Rios D, Ritchie GRS, Ruffier M, Schuster M, Sobral D, Spudich G, Tang YA, Trevanion S, Vandrovcova J, Vilella

AJ, White S, Wilder SP, Zadissa A, Zamora J, Aken BL, Birney E, Cunningham F, Dunham I, Durbin R, Fernández-Suarez XM, Herrero J, Hubbard TJP, Parker A, Proctor G, Vogel J, Searle SMJ: Ensembl 2011. Nucleic Acids Research 2011, (39 Database):D800-6.

28. Roberts A, Pimentel H, Trapnell C, Pachter L: Identification of novel transcripts in annotated genomes using RNA-Seq. Bioinformatics 2011. first published online June 21, 2011

29. Langmead B, Trapnell C, Pop M, Salzberg SL: Ultrafast and memory-efficient alignment of short DNA sequences to the human genome. Genome Biology 2009, 10(3):R25..

30. Li H, Handsaker B, Wysoker A, Fennell T, Ruan J, Homer N, Marth G, Abecasis G, Durbin R: The Sequence Alignment/Map format and SAMtools. Bioinformatics 2009, 25(16):2078-9.

31. Kent WJ, Sugnet CW, Furey TS, Roskin KM, Pringle TH, Zahler AM, Haussler , David : The Human Genome Browser at UCSC. Genome Research 2002, 12(6):996-1006.

32. Li J, Jiang H, Wong WH: Modeling non-uniformity in short-read rates in RNA-Seq data. Genome Biology 2010, 11(5):R50..

33. Flux Simulator [http://flux.sammeth.net/simulator.html]

34. Pruitt KD, Tatusova T, Klimke W, Maglott DR: NCBI Reference Sequences: current status, policy and new initiatives. Nucleic Acids Research 2009, (37 Database):D32-6.

35. Bustin SA: Why the need for qPCR publication guidelines?-The case for MIQE. Methods 2010, 50(4):217-26.

36. Shi L, Reid LH, Jones WD, Shippy R, Warrington JA, Baker SC, Collins PJ, de Longueville F, Kawasaki ES, Lee KY, Luo Y, Sun YA, Willey JC, Setterquist RA, Fischer GM, Tong W, Dragan YP, Dix DJ, Frueh FW, Goodsaid FM, Herman D, Jensen RV, Johnson CD, Lobenhofer EK, Puri RK, Schrf U, Thierry-Mieg J, Wang C, Wilson M, Wolber PK, Zhang L, Amur S, Bao W, Barbacioru CC, Lucas AB, Bertholet V, Boysen C, Bromley B, Brown D, Brunner A, Canales R, Cao XM, Cebula TA, Chen JJ, Cheng J, Chu TM, Chudin E, Corson J, Corton JC, Croner LJ, Davies C, Davison TS, Delenstarr G, Deng X, Dorris D, Eklund AC, Fan Xh, Fang H, Fulmer-Smentek S, Fuscoe JC, Gallagher K, Ge W, Guo L, Guo X, Hager J, Haje PK, Han J, Han T, Harbottle HC, Harris SC, Hatchwell E, Hauser CA, Hester S, Hong H, Hurban P, Jackson SA, Ji H, Knight CR, Kuo WP, LeClerc JE, Levy S, Li QZ, Liu C, Liu Y, Lombardi MJ, Ma Y, Magnuson SR, Maqsodi B, McDaniel T, Mei N, Myklebost O, Ning B, Novoradovskaya N, Orr MS, Osborn TW, Papallo A, Patterson T: The MicroArray Quality Control (MAQC) project shows inter- and intraplatform reproducibility of gene expression measurements. Nature Biotechnology 2006, 24(9):1151-61.

37. Bullard JH, Purdom E, Hansen KD, Dudoit S: Evaluation of statistical methods for normalization and differential expression in mRNA-Seq experiments. BMC Bioinformatics 2010, 11:94.

38. Roberts A, Trapnell C, Donaghey J, Rinn JL, Pachter L: Improving RNA-Seq expression estimates by correcting for fragment bias. Genome Biology 2011, 12(3):R22..

39. Wang ET, Sandberg R, Luo S, Khrebtukova I, Zhang L, Mayr C, Kingsmore SF, Schroth GP, Burge CB: Alternative isoform regulation in human tissue transcriptomes. Nature 2008, 456(7221):470-6.

40. Hansen KD, Brenner SE, Dudoit S: Biases in Illumina transcriptome sequencing caused by random hexamer priming. Nucleic Acids Research 2010, 38(12):e131.

41. Wu Z, Wang X, Zhang X: Using non-uniform read distribution models to improve isoform expression inference in RNA-Seq. Bioinformatics 2011, 27(4):502-8.

42. Liu JS: The Collapsed Gibbs Sampler in Bayesian Computations with Applications to a Gene Regulation Problem. Journal of the American Statistical Association 1994, 89(427):958-966.

There are several supplemental files that are not available in this version of the article. To view this additional information, please use the citation information cited on the first page of this chapter.

PART II

MICROARRAY

CHAPTER 4

A REGRESSION SYSTEM FOR ESTIMATION OF ERRORS INTRODUCED BY CONFOCAL IMAGING INTO GENE EXPRESSION DATA *IN SITU*

EKATERINA MYASNIKOVA, SVETLANA SURKOVA, GRIGORY STEIN, ANDREI PISAREV, AND MARIA SAMSONOVA

4.1 BACKGROUND

Confocal scanning microscopy is a commonly used method for acquisition of high-quality digital two- and three-dimensional images of molecular biological objects. The high quality of confocal images makes it possible to extract quantitative data at a single cell resolution, the availability of which is a necessary prerequisite for successful systems biology studies. However the data accuracy is limited due to errors that arise in the course of confocal scanning. In our recent papers [1,2] we analyzed the sources of errors introduced by two-dimensional confocal imaging into the data on gene expression in situ and described algorithms for estimation and correction of these errors. For example, confocal images are inevitably

This chapter was originally published under the Creative Commons Attribution License. Myasnikova E, Surkova S, Stein G, Pisarev A, and Samsonova M. A Regression System for Estimation of Errors Introduced by Confocal Imaging into Gene Expression Data In Situ. BMC Bioinformatics *12*,320 (2011), doi:10.1186/1471-2105-12-320.

contaminated by photon shot noise [3] and a common way to reduce the noise is the averaging of multiple separate scans. However, the information about the averaged image will be lost if pixels with high or/and low intensities are clipped in single scans. Image clipping is a form of signal distortion related to the limited grayscale range of an image. Pixel values that exceed an upper threshold of the grayscale range (e.g., 255 for an 8-bit format) are cut-off at the threshold value, all the pixels with negative intensities are set to zero. Such pixels are referred to as over- and under-saturated, respectively. Averaging of clipped scans results in errors in the data extracted from the averaged image. In our previous work we developed a method [1] based on censoring technique for estimation and correction of this kind of errors, however the method implementation requires not only the averaged image but also all the confocal scans which are not provided by the standard procedure of image acquisition.

The degree of image distortion and hence the size of data error caused by clipping depends on microscope parameters, most of all on the values of gain and offset of the photomultiplier tube (PMT), the detection device to measure photons. These parameters are adjusted to control the dynamical range of an image: the PMT gain (voltage) exponentially amplifies a weak signal, while offset defines the background level of intensities subtracted from the image to increase its brightness. Although the PMT parameters are chosen to ensure that in the averaged image pixels take their value inside the grayscale range and do not look clipped, some of the pixels in single scans may be saturated due to photon noise and clipped off. The adjustment of PMT gain affects signal-to-noise ratio (SNR) in the image amplifying the noise level exponentially. The severity of clipping increases with the increase of gain and offset values, the distortions being the largest when the photomultiplier is adjusted to the limits of its sensitivity. Besides the PMT adjustment SNR can be improved by increase of laser power, however, this approach leads to fluorophore saturation and photobleaching. In practice the laser power is kept at a constant high level and the amount of light admitted to the specimen is reduced through AOTF control, which does not amplify the noise. We have conducted experiments to estimate to what extent other microscope parameters, besides the PMT gain and offset, influence the size of data error.

In the present work we introduce a regression system for prediction of error magnitude in the data extracted from the averaged image. The learning samples are composed of images obtained at different combinations of gain and offset values of three different microscopes. The experiments were designed in a way that for each learning image all the scans were saved as separate image files. The linear regression model involves the values of gain and offset as independent variables while the error value estimated for the given mean intensity level is a dependent variable.

Obviously the magnitude of error may vary among the data obtained with different microscopes and under different experimental conditions, and thus application of the prediction system requires a representative learning sample obtained by the same confocal system and the same scanning experiment as the image subject to error correction. To apply the developed regression system for predicting errors in new data we standardize all our training data obtained with three microscopes; we combine them in one sample and train the system on the combined sample.

The error prediction system was applied to correct errors in the data on expression of segmentation genes in *Drosophila* that are stored in the FlyEx database http://urchin.spbcas.ru/flyex/. This data are widely used in research labs. Our aim was to corroborate the high-precision of the data that was used for construction of the integrated atlas of segmentation gene expression.

The proposed method has important applications. Usually it is recommended to adjust the parameters of microscope to almost avoid pixel saturation in single frames; this approach limits the brightness and contrast of averaged images. The newly developed system provides an opportunity to obtain images in a higher dynamical range and thereby to extract more detailed quantitative information from microscope experiments.

The regression system is implemented as a software tool CorrectPattern freely available at http://urchin.spbcas.ru/asp/2011/emm/.

4.2 METHODS

4.2.1 ALGORITHM

4.2.1.1 ESTIMATION OF BETWEEN-SCAN NOISE

The photon shot noise is an inevitable consequence of the basic properties of confocal microscopy. Among the main advantages of this imaging technology over conventional optical microscopy is the presence of a confocal pinhole, which let only light from the focus plane to reach the detector. Pinhole removes "out of focus" light from the image, thereby decreasing the number of photons reaching detectors. The photon noise arises from a discrete nature and small number of detected photons and in a properly aligned microscope is the major source of errors [3]. This noise is signal dependent and follows the Poisson distribution.

The noise level in the averaged image may be characterized by between-scan variance defined for each pixel in the image as a variance of values of the same pixel in all the scans. To illustrate how the between-scan variance depends on the PMT parameters the mean variances are plotted in Figure 1 against the mean pixel values for different combinations of gain and offset. Although the offset adjustment does not directly affect the PMT noise, subtraction of the background from an image decreases mean intensities leaving the noise unchanged and thereby decreasing the signal-to-noise ratio. For example, noise in the image obtained at the gain 1000V and offset -4% is noticeably higher than in the image from the same microscope obtained at the same gain and zero offset. As it is predicted by the optical theory the noise increases exponentially with gain and linearly with offset. It is clearly seen from the figure that in accordance with the properties of the Poisson distribution the variance linearly depends on the mean pixel value at low and intermediate intensities, while at high intensities the variance values dramatically fall as a result of image clipping.

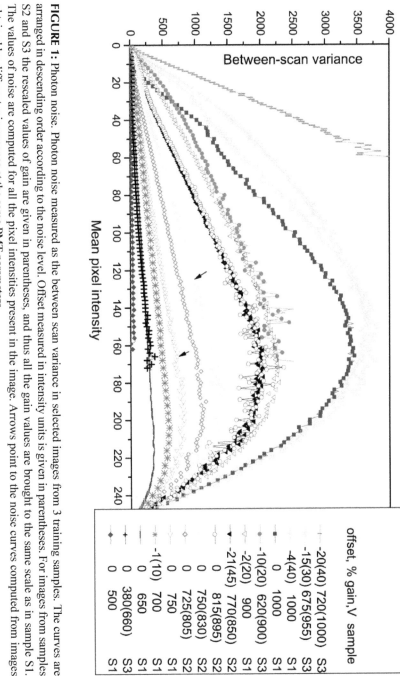

FIGURE 1: Photon noise. Photon noise measured as the between scan variance in selected images from 3 training samples. The curves are arranged in descending order according to the noise level. Offset measured in intensity units is given in parentheses. For images from samples S2 and S3 the rescaled values of gain are given in parentheses, and thus all the gain values are brought to the same scale as in sample S1. The values of noise are computed for all the pixel intensities present in the image. Arrows point to the noise curves computed from images obtained by different microscopes at the same PMT parameters.

The degree of the averaged image distortion due to clipping is charac-terized by the fraction of clipped pixels as a function of the pixel intensity. For each pixel the fraction is computed as a number of scans in which this specific pixel is clipped, divided by the total number of scans. Obviously for the pixels with the same intensity in two different images the frac-tion will be higher for the image with higher noise. The fall of between-scan variances at high intensities is explained by the fact that the fraction of clipped pixels approaches to 1 which results in saturation of the pixel value in the averaged image.

4.2.1.2 ESTIMATION OF ERRORS DUE TO IMAGE CLIPPING

Quantitative data are read off from the averaged image. The quantification procedure includes the detection of object (nucleus or cell) borders and the subsequent averaging of the values of all the pixels assigned to an object. As a result the data are represented by the mean intensity and coordinates of each object in the image.

The error due to image clipping arises in data extracted from confocal images in the event that these images are obtained by means of averaging the clipped single scans. In the presence of all the scans the error magni-tude can be estimated using the method based on the censoring technique [1]. Data errors due to clipping are defined as the absolute difference be-tween the true (unknown) value of the mean intensity and the mean in-tensity corrupted by clipping that is obtained from the observed averaged image.

To estimate data errors we first introduce a pixel error as a value of distortion of the pixel value in the averaged image caused by clipping. Due to clipping at the upper grayscale threshold, over-saturation, the pixel intensity is reduced by the value

$$\int_{c_a}^{\infty} (x - c_a) f_a(x) dx_t \qquad (1)$$

where c_a is the upper threshold, $f_a(x) = 1/(2\pi\sigma)\exp[(x-\mu)^2/\sigma^2]$ is the Gaussian distribution density. The parameters μ and σ are estimated for each pixel by the method of moments as described in [1]. After that the quantities (1) are averaged over all the pixels with equal intensities. Thus for any intensity k from the grayscale range $[0..c_a]$ the averaged error, U_k, is defined. We will name this type of error as upper error. The error in a data object is given by $(1/n)\Sigma U_k$, where the averaging is performed over all the pixels belonging to the object and N is the number of such pixels.

As a result of offset adjustment a certain portion of intensities is subtracted from an image and any pixel value smaller than the subtraction threshold is clipped and set to zero. This type of distortion of single scans, under-saturation, yields the overestimated values of pixel intensities in the averaged image. In this case the pixel error is given by

$$\int_0^{c_b} (c_b - x)f_b(x)dx_t \tag{2}$$

where c_b is threshold defined by the value of offset, $f_b(x) = \rho/\lambda(x/\lambda)^{\rho-1}e^{-(x/\lambda)^\rho}$ is the Weibull distribution density. The distribution parameters are estimated analogously to those of the Gaussian model (1). The error estimates are averaged over all the image pixels with equal intensities; the averaged pixel error, lower error, is denoted as L_k for any $k \in [0..c_a]$. The data error is also defined in this case as the averaged value of all the pixel errors computed for all the pixels assigned to an object, $(1/n)\Sigma L_k$.

Note, that the magnitude of pixel error is uniquely defined by the level of image noise for a given mean intensity.

Theoretically the method works at any degree of clipping but in practice its application is limited: for example the error estimation is infeasible if the true mean values of a pixel are clipped in all individual scans.

4.2.1.3 CONSTRUCTION OF THE REGRESSION MODEL

The method described in the previous section allows to precisely estimate and correct errors in images and data but its application requires the availability of all the confocal scans. In this section we construct a linear regression model for prediction of magnitude of error in the data extracted from the averaged image based solely on information about microscope parameters. The information about image acquisition is normally contained in scanning protocols saved by the microscope software. Among the microscope parameters the adjustment of PMT gain and offset exerts the greatest influence on the error magnitude and these parameters are incorporated into the regression system as independent variables. As a learning sample we use confocal images scanned at different combinations of gain and offset, for which all the scans are saved along with the averaged images.

The regression algorithm is implemented in several steps. First, for all the elements of the learning sample pixel errors, U_k and L_k, are estimated for all the intensities that are present in the images. Then the regression functions are constructed for each intensity value from the grayscale range. As signal and hence the degree of image distortions depends linearly on offset and exponentially on gain, independent parameters are chosen as the values of offset and exponent of gain. The regression function involves the total estimated distortion caused by under- and over-saturation, $E_k = U_k + L_k$, as the dependent variable. For each intensity level, $k \in [0..c_a]$, the linear regression function is defined as

$$E_k = \beta_{\text{offset},k}\text{offset} + \beta_{\text{gain},k}\ln(gain) + \beta_{0,k} \tag{3}$$

The regression coefficients $\beta_{\text{offset},k}$, $\beta_{\text{gain},k}$ and $\beta_{0,k}$ are estimated by the least squares method, minimizing

$$\sum_j \left[E_{kj} - \beta_{\text{offset},k}\text{offset}_j - \beta_{\text{gain},k}\ln(gain_j) - \beta_{0,k}\right]^2 \tag{4}$$

$k = 0, ..., c_a,$

where the summation is done over all the images, elements of the learning sample. Each term includes the values of offset and gain that were applied for the acquisition of the corresponding image.

Normally a pixel can be noticeably corrupted either by under- or over-saturation and hence only one kind of error, either U_k or L_k, can take considerable values (see Figure 2).

The results of regression estimation are applied to predict the size of errors in data extracted from an averaged image non-belonging to the learning sample, for which single scans are not saved. As a first step, the regression equation (3) is used for prediction of error size in each pixel in the averaged image. Next the predicted values are averaged over all the pixels within the area of each object detected in the image. The errors estimated in this way are used to correct the data distortions that arise due to clipping of single scans both at the highest and the lowest mean intensities.

4.2.2 IMAGE AND DATA ACQUISITION

Three learning samples were obtained from three confocal microscopes. All the images are two-dimensional of a size 1024 × 1024 pixels and have 8-bit grayscale resolution.

4.2.3 S1 SAMPLE

The images were obtained by Leica TCS SP5 confocal system (Institute of Cytology RAS, St.Petersburg, Russia). In the scanning experiments we used a specimen prepared from the lily of the valley (Convallaria) root, that is highly autofluorescent in wide spectrum. We also scanned three expression patterns of *Drosophila* embryo, that were also scanned on different microscope system when the S2 sample was constructed. The specimens were scanned 8 times using HCX PL APO 20.0x/0.70 IMM Lbd.BL objective and three lasers (Argon 488 nm, HeNe 543 nm, HeNe 633 nm) with different values of PMT gain and offset listed in Table 1. The power of Argon laser was normally set to 30% of it's maximal value. To check the effect of laser power variation on the noise, one specimen was scanned at

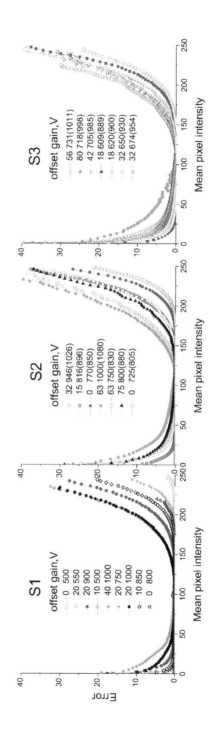

FIGURE 2: Estimated error values. Error values computed by the method (1-2) at different combinations of the PMT parameters for some of the images from 3 training samples. The offset values are measured in intensity units, the PMT voltage (gain) is given in the original and rescaled form (in parentheses). For explanations see text.

10, 20, 30, 40, 50, and 60 percent laser power at the same values of gain and offset.

TABLE 1: Learning sample S1

Offset\ gain,V	500	550	600	650	700	750	800	850	900	1000
0	I(2)	I	I(5)	I	I(2),II(2),III	II	I,II(5),III(3)	II	I,II(3),III	II(3)
-1%(10)	I(2)	I	I	I	I(2),II(2),III	II	I,II(2),III	II	I,II(2),III	-
-2%(20)	I(2)	I	I	I	I(2),II(2),III	II	I,II(3),III	II	I,II(3),III	II
-4%(40)	-	I	-	-	-	-	II	-	II	II

Combinations of PMT gain and offset used for the acquisition of 76 images. The gain values in parentheses are corrected to combine all the samples (see text). Offset values in parentheses are measured in intensity units subtracted from the image. Images are obtained in 3 microscope channels using different lasers (I - 488 nm, II - 543 nm; III - 633 nm). The table entries are the channel numbers in which images at the corresponding combination of PMT parameters are obtained; if more than one image is obtained the number of images is given in parentheses.

4.2.4 S2 SAMPLE

The experiments were performed at JUC "Chromas" of the St.Petersburg State University, Russia. Eight wild-type (OregonR) *Drosophila* melanogaster blastoderm embryos were immunostained for the expression of hb, gt and eve segmentation genes as described in [4-6]. We used fluorescent labels Alexa Fluor 488 (Invitrogen) for detection of Hb and Cad and Alexa Fluor 555 for detection of Eve and Gt proteins. The embryos were imaged with a HCX PL APO lambda blue 20.0x/0.70 IMM Lbd.BL objective of a Leica TCS SP5 confocal system using Argon 488 and HeNe 543 lasers. Each embryo was stained for the expression of 2 genes, each staining was scanned several times with different values of PMT gain and offset, and for each experiment a series of 8 individual scans was saved together with the averaged image.

In total 59 averaged images (see Table 2) were obtained. To test whether the properties of lasers change with time six stainings were stored and scanned anew with the same values of PMT parameters several months after all the other series of experiments were performed.

TABLE 2. Learning sample S2

offset\gain,V	725-750 (805-830)	770-800 (850-880)	815-850 (895-930)	900 (980)	950 (1030)	1000 (1080)
0	II(2)	II(2)	II(2)	-	-	-
-5%(11)	-	II(4)	II(2)	-	-	-
-10%(21)	II(3)	II(2)	I,II	I	I	I
-15%(38)	II(2)	II(3)	I,II	I(2)	I(2)	I
-20%(50)	II(3)	II(3)	I,II	I	I	I
-25%(63)	II(2)	II(2)	I	I(2),II	I	I
-30%(75)	II	II	-	-	-	-

The sample is composed of 76 images. For notations see caption to Table 1.

TABLE 3: Learning sample S3

Offset\gain,V	380 (660)	600-640 (980-920)	650-680 (930-960)	700-740 (980-1020)	780-800 (1060-1080)	815-850 (1095-1130)	860-920 (1140-1200)
0-8% (0-16)	I	I(2),II	I(2),II(2)	-	-	-	-
-10-12% (20-24)	-	-	II	II(4)	-	II	-
-25-28% (50-56)	-	-	-	II	II(4)	II(3)	-
-30-35% (60-70)	-	-	-	-	-	II	II(2)
-40-45% (80-90)	-	-	III(2)	III(2)	-	-	-

The sample is composed of 29 images. For notations see caption to Table 1.

4.2.5 S3 SAMPLE

12 embryos were immunostained for expression of one of four segmentation genes *gt, eve, hb* and *bcd* applying the same method as described
above for construction of S2. Each embryo was scanned several times with
different combinations of gain and offset settings (see Table 3). Fluorescent labels used were Alexa Fluor 488 (*bcd*), Alexa Fluor 555 (*eve, gt*),
and Alexa Fluor 647 (*hb*). Embryo images were taken with the 20X Plan

Apo dry objective (numerical aperture 0.7) of a Leica TCS SP2 confocal system at Stony Brook University, NY, USA.

The quantitative gene expression levels in nuclei are extracted from the images belonging to S2 and S3 with the use of a nuclear mask as described in [6,7]. The mask is a binary image in which all the pixels located within a nucleus are white and the rest pixels are black. The mask is superposed on the image and the values of pixels belonging to a nucleus are averaged. As a result each nucleus in the expression pattern is characterized by x and y coordinates and mean intensity level.

4.3 RESULTS

4.3.1 TRAINING OF THE SYSTEM

The regression system is trained on images acquired from three different confocal microscopes. The learning samples S1 and S2 contain images scanned by two different microscopes Leica TCS SP5, the third sample S3 is obtained with a microscope Leica TCS SP2. Details of image acquisition are given in the Methods section. The values of gain and offset used for acquisition of all the samples are presented in Tables 1, 2 and 3. To bring the offset values of different microscopes to the common scale we measure offset in intensity units subtracted from an image. In this way we calculate that 1% offset for microscopes used to acquire S1, S2 and S3 samples corresponds to 10, 2.5 and 2 intensity units, respectively.

First of all, we analyze the photon noise as a function of PMT parameters in all the samples. The noise is estimated as the between-scan variance computed for each element of all the learning samples. The typical behavior of the between-scan variance is shown in Figure 1 and discussed in detail in the Algorithm section. The measured variances are given in Figure 1 for selected values of pixel intensity and PMT parameters that makes it possible to compare the noise level in images obtained with different microscopes and at different conditions. As expected, due to different properties of electronic devices included into the microscope configuration, the noise is not equally defined by the PMT voltage in different microscopes. For example, images obtained at zero offset and equal gain, 750V, from samples S1 and S2 have different level of noise (labeled as errors in the figure). For all our experiments the noise level coincides in images obtained with the same microscope in different channels using different lasers. The power of the laser used for excitation of specimen is another factor that influences the image noise. Although this parameter is normally kept unchanged from experiment to experiment the output laser power may slowly decrease with time as the laser tube ages. We compared the noise in images scanned on different days, even separated by long intervals (up a year), and have established that the noise has not noticeably changed with time. The results of these tests (data not shown) allowed us to assemble all the images obtained by the same microscope into one learning sample. However, the system trained on a learning sample can be only used to predict the error magnitude in data acquired with the same microscope. To be able to predict errors in any data we need to standardize all our training data obtained with three different microscopes and combine them into one sample.

The regression system uses the values of PMT gain and offset as independent variables which means that these parameters uniquely define the predicted error magnitude. To bring the values of these parameters to common scale we represent offset as measured in intensity levels subtracted from an image (see Tables 1, 2 and 3), and further need to find the way how to standardize the values of gain for different PMTs used in different microscopes. As it was already mentioned above, the image noise is unequally defined by the PMT voltage in different microscopes, and even using different lasers in the same microscope, while the noise level completely defines the value of pixel error. Hence it is sufficient to associate the gain values with the level of between-scan noise in images from different learning samples. As noise is known to exponentially increase with increase of gain, to bring the gain values of two samples to correspondence we used additive correction for the gain value in one of the samples. The correction in sample S2 with respect to sample S1 is found to be 80V, such that, for example, the gain value 800V in sample S1 corresponds to 880V in sample S2, which means that these values of gain generate the same level of between-scan noise in images. The correction shift between the gain values in samples S1 and S3 is 280V as the PMT of microscope used to acquire S3 produces much higher noise. For example the level of between-scan noise almost coincide in images from S1 and S3 obtained at zero offset and gain 380V and 650V, respectively (see Figure 1).

Taking into account these corrections we create a common sample consisting of all the learning data obtained from all the microscopes.

TABLE 4: Results of the regression estimation

Intensity	#	R^2	β_0	β_{offset}	β_{gain}
0	132	0.941	-0.274	0.722	0.027
10	132	0.936	-2.067	0.117	0.032
20	132	0.901	-1.441	0.060	0.020
30	132	0.862	-0.980	0.037	0.013
50	132	0.791	-0.483	0.017	0.006
150	98	0.901	-0.829	0.007	0.013
170	98	0.948	-1.757	0.022	0.031
190	94	0.967	-2.915	0.048	0.061
200	89	0.972	-3.624	0.070	0.084
210	89	0.969	-3.892	0.098	0.110
220	85	0.950	-3.446	0.124	0.140
230	73	0.898	-1.835	0.179	0.168
245	37	0.847	-7.655	0.198	0.416
248	12	0.895	-2.569	-0.025	0.383

Results of the regression estimation for selected values of the mean pixel intensity. For each mean intensity the number of valid cases, determination coefficient R^2, and estimates of the regression coefficients betas are presented. The significant values of betas are given in bold.

4.3.2 REGRESSION ESTIMATION

The combined learning sample is used to fit the regression model (3) introduced in the Methods section. The regression estimation is separately performed for each intensity level $k \in [0..c_a]$. The value of dependent variables for each element of the learning sample is computed as the sum of upper and lower pixel errors U_k and L_k, if intensity level k is present in the image. All the images are 8-bit files, and hence the upper level c_a is equal to 255. However the highest intensities are present just in few images and the upper error values can be estimated for intensities not exceeding ~250. Besides, the highest pixel values are usually very strongly clipped which may lead to unreliable estimates. Examples of error estimates for different combinations of gain and offset are shown in Figure 2.

The regression coefficients $\beta_{offset,k}$, $\beta_{gain,k}$ and $\beta_{0,k}$ are estimated by the least square method (4). The results of regression estimation are summarized in Table 4, for the sake of space the estimated values of the coefficients are only given for selected values of intensities k. The regression results are visualized for low and high mean intensities in Figure 3. A close to 1 value of the determination coefficient R^2 is an evidence of adequacy of the regression model and its good prediction properties.

To cross-validate the accuracy of prediction we performed a so-called leave-one-out test. The test uses a single observation from the original sample as validation data and the remaining observations as training data. This procedure is repeated so that each observation in the sample is used once as validation data.

The test was slightly modified since some images were obtained at the same values of PMT parameters; all such images were together excluded from the training dataset to form the validation sample. At each step of the test procedure we apply the regression system to predict pixel errors for an image from the validation sample. For each pixel value the accuracy of prediction is characterized by the absolute difference between the computed and predicted error values. The cross-validation results are presented in Figure 4. The test confirms high accuracy of error estimation for samples S1 and S2, while for the images from S3 the deviation in error estimation attains 5 units in absolute value. The lower accuracy of error estimation for sample S3 is explained by much higher noise in the images from this sample.

The method was used to predict the sizes of errors in data on expression of 14 segmentation genes in *Drosophila* embryo generated in our previous work and stored in the FlyEx database (http://urchin.spbcas.ru/flyex/; http://flyex.uchicago.edu/flyex/). This dataset consists of about 5000 con-

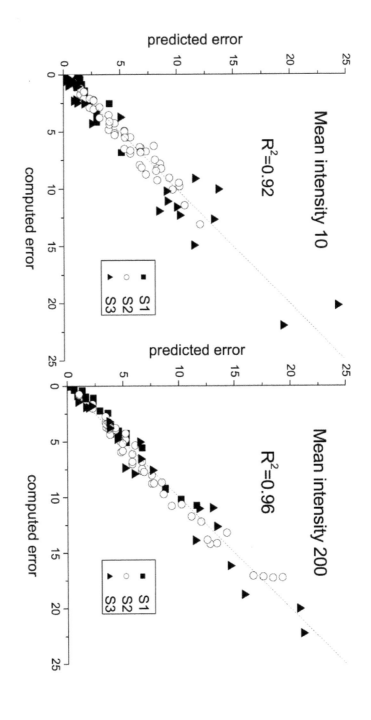

FIGURE 3: Results of the regression estimation. The predicted values of errors are plotted versus the computed ones for three training samples. The results for a low mean pixel intensity are presented in the left panel and for a high mean pixel intensity in the right panel.

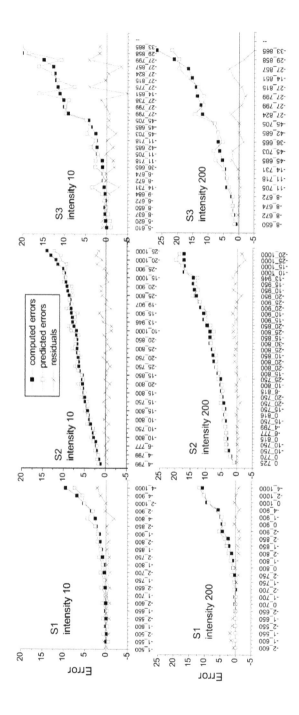

FIGURE 4: Results of the cross-validation test. The computed and predicted values of errors at low and high mean pixel intensities are presented for the images from 3 training samples. If there are several images obtained at the identical PMT parameters in the sample the data are only presented for one of such images. The x-axis is labeled by values of PMT offset and gain measured in percents and Volts, respectively.

focal images, of which 1263 were acquired by microscope and lasers used to generate S3 sample (see the Methods section), while the rest images were obtained by microscope and lasers applied to scan S2 sample [4-6]. The images were used to extract quantitative data on segmentation gene expression by the method presented in [6], however this data is corrupted by clipping because all the images were obtained under the standard image acquisition procedure which precludes saving of single scans along with the averaged image.

The data error is usually computed as an average of errors of all the pixels in a data object (in our case an embryo nucleus). However, at high mean intensities this approach is likely to produce unreliable estimates. Due to photon noise the intensities of some pixels in nucleus with high mean intensity may reach values exceeding 250, while estimates of pixel errors are inconsistent at such intensities. In this case it is rather recommended to estimate the data error as the error of a pixel with the intensity equal to the mean intensity of the data object. We have tested this simplified approach on the available data and have observed that the error estimates computed by both methods did not have noticeable differences.

In general, to apply the regression system for prediction of errors caused by clipping in data extracted from a series of 8-bit images scanned by any confocal microscope it is sufficient to bring the values of PMT parameters used for image acquisition to the common scale. For this purpose there is no need to create a full representative learning sample but just to run a specially designed experiment on the same microscope in the same channel and under the same conditions. To measure the value of offset subtracted from images the same staining is scanned twice using the same gain and two different values of offset. The mean difference between the images divided by the difference between the offset values will give the standard measure of offset. To standardize the gain all the confocal scans are saved for an image scanned at zero offset and any given value of gain. Then the between-scan noise is computed as described in and its values are put into correspondence with those computed for our combined sample and presented in Additional file 1, Table S1. The difference between the gain voltage that generates the same level of noise in an image from the combined sample and new experiment will give the correction shift for the gain.

Finally we come to the following scheme of the data error prediction algorithm:

1.   Bring the values of PMT parameters used for image acquisition to the standard scale.
2.   For any obtained image apply the regression system to to predict pixel errors using the standardized values of gain and offset as input parameters.
3.   Compute the sizes of errors caused by clipping by averaging the predicted values over all the pixels within the area of each object detected in the image, or just take the pixel error corresponding to the mean intensity in the object.

4.3.3 SOFTWARE TOOL

The algorithm for prediction of data errors in gene expression patterns is implemented as a software tool CorrectPattern freely available at http:// urchin.spbcas.ru/asp/2011/emm/. The main function of the program is to predict and correct errors due to pixel saturation in an input gene expression pattern. Input parameters of the program are the values of gain and offset used for image acquisition. The program provides a tool for automated parameter standardization. A user should provide a series of confocal scans obtained at zero offset for the gain standardization and two averaged images of the same specimen obtained with the same gain and different values of offset for offset standardization. The program computes and saves the corrections for gain and offset that are further used for standardization of the input parameters for any image obtained using the same microscope laser. Output data file is saved in the same format as the input file with the mean intensity values replaced by the corrected ones.

CorrectPattern is realized as a complex of programs in C, Java and JavaScript languages using the three-tier architecture. Program modules are installed on Linux server and use image processing libraries. The user interface is realized in the WEB browser in JavaScript language on the basis of AJAX technology. WEB server is used as an intermediary between the user interface and functional program modules.

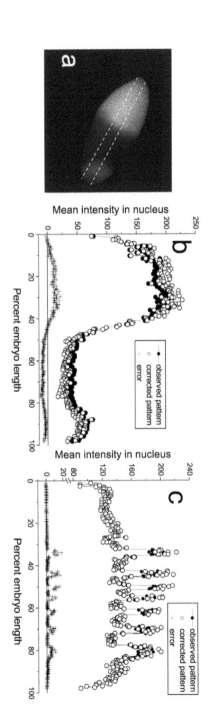

FIGURE 5: Correction of errors in quantitative data. a) Confocal image of *Drosophila* embryo stained for the expression of *hb* gene and scanned with offset -40% and gain 705V. The image belongs to S3. Quantitative data are extracted from the area outlined in the image. b) 1D expression pattern of *hb* gene extracted from the image (black circles) and corrected to eliminate the saturation effect (white circles). Differences between corrected and observed quantitative gene expression data are shown by crosses. c) Quantitative data on *ftz* gene expression extracted from the image taken by the same microscope as sample S2 with offset -2% and gain 1100V (the image is not shown).

4.3.4 EXAMPLES OF THE METHOD APPLICATION

4.3.4.1 CORRECTION OF GENE EXPRESSION PATTERNS

The method application is illustrated on two gene expression patterns. An example of error correction in the data extracted from an image belonging to S3 sample is shown in Figure 5b. Predicted error values reached 15-20 units at the highest mean intensities and 12 units at the minimal mean intensities present in the pattern.

An advantage of the proposed method is the opportunity to increase the dynamical range of low signal images and acquire accurate quantitative data from them. The second example illustrates the correction of an expression pattern of fushi tarazu (*ftz*) segmentation gene in *Drosophila* blastoderm obtained using a poor quality antibody. The specimen was scanned by the microscope and laser used to generate S2 sample.

According to the standard procedure for quantitative data acquisition [6] the gain and offset of the microscope photomultiplier should be adjusted so that the maximum level of gene expression corresponds to the maximum level of fluorescence intensity at an 8-bit scale. For immuno-chemical detection of the gene product we use rat antibody against Ftz [4] and the commercial secondary antibody anti-rat Alexa Fluor 488. Due to the long-term storage, the activity of the primary antibody decreased and even using very high antibody concentrations we had to raise the gain to almost maximum possible value to obtain images of intensity calibrated against the images of FlyEx embryos stained for the *ftz* expression. The noise level in such images is very high that gives rise to high errors in the data. The regression system was applied to correct these errors that reached considerable values of about 17 units at the intensity 200 as shown in Figure 5c.

Another source of errors in the data obtained with the use of the antibody at our disposal is a high non-specific background signal. We applied the method published in [8] to remove the background and normalize the data. Upon the application of this procedure the error magnitudes increased significantly up to 35 units at the intensity 150. This example shows that in cases when there is a need to use high values of PMT parameters the error correction method is very important to make the data suitable for analysis.

4.3.4.2 ESTIMATION OF ERROR SIZES IN THE FLYEX DATASET

The FlyEx dataset is a valuable source of information about mechanisms of pattern formation in early development. Besides confocal images of gene expression patterns it contains quantitative data extracted from these images, as well as a set of reference images and data representing the most typical expression pattern for a given developmental time. This data attracts attention of many scientific groups, which widely use FlyEx to study the mechanism of pattern formation, infer regulatory interactions in the segmentation genetic network and develop new mathematical models http://urchin.spbcas.ru/flyex/refs.jsp.

In our recent publication [1] we have shown that this data is corrupted by clipping. Notwithstanding the fact that the sizes of the data errors are small due to proper choice of microscope parameters, the errors should be removed as the quality of conclusions drawn critically depends on the data quality. The general-purpose method for correction of pixel saturation requires all the confocal scans to be saved without averaging. The regression system which we have developed allows us to circumvent this limitation, predict the error magnitude in the data extracted from the average image and apply the error correction procedure to the whole dataset.

The important corollary of extension of the method to predict the error sizes on the whole dataset is the opportunity to accurately estimate the sizes of these errors. Almost all the images from the FlyEx dataset were acquired with the gain values within the range from 450V to 680V and offset values under 45%. The predicted error values do not exceed 7-8% of the highest mean intensity present in expression pattern. The lowest value of mean intensity in nucleus is never zero due to inevitable presence of non-specific background staining [8] in the embryo. The background level is determined by the quality of antibodies used for staining, and in our data it varies between 2 and 100 units. The predicted errors take values not higher than 2-3 units in expression patterns with low background and are negligibly small in the case of high background. The detailed results of the error estimation are given in Additional file 2, Figure S1; Additional file 3, Figure S2 and Additional file 4, Figure S3. Errors at low intensities may slightly affect the estimation of background level but are small enough not to noticeably corrupt the pattern after background subtraction.

4.4 DISCUSSION

In our recent publication [1] we described a new method for estimation and correction of errors in the quantitative data extracted from clipped confocal images. The method was applied to the data on segmentation gene expression in *Drosophila*. A necessary requirement for the method application is availability of all the individual scans that usually are not saved but directly averaged by the microscope software to reduce the photon noise in images. Due to this requirement the method could not be used to correct errors in data obtained by the standard scanning procedure; the method only allows to determine the range of settings that provides acceptable level of errors in a specific microscope.

To extend the applicability of the method we have created a linear regression system and software to predict the magnitude of errors in the data obtained from a confocal image based on information about microscope parameters used for the image acquisition. The system was trained on three samples of images obtained from different microscopes with different combinations of the PMT gain and offset adjustments. As adjustment of PMT gain and offset to the same values in different microscopes produces different level of noise, the scales of these parameter were calibrated to achieve standardization. The standardized parameter values were used by the regression system as independent variables.

To estimate regression function each image in a training sample was saved together with all the individual scans. The computed errors were included as dependent variables into the regression model taking into account a known fact [3] that the error size depends linearly on the offset value and exponentially on the gain value. The system predicts the magnitude of errors in data extracted from an 8-bit image obtained by a confocal microscope using the values of standardized PMT parameters as input. The cross-validation tests demonstrated high accuracy of predictions.

It should be stressed that the standardization of the microscope parameters is very important as it puts into correspondence the properties of images obtained with different microscopes, lasers and under different experimental conditions. All what is needed is to perform a simple experiment to measure the photon noise via the estimation of between-scan variance and standardize the parameter values used for image acquisition against the scale utilized in the training sample. Upon that the regression system can be used on the data extracted from a series of images obtained at the same conditions.

We envisage one additional application of the regression system developed in this work. This system allows a user to extract more detailed quantitative information from the images, thereby increasing the accuracy of gene expression data. The confocal scanning experiment directed to the acquisition of quantitative gene expression data possesses certain specific features. For example, images used to acquire data on segmentation gene expression in *Drosophila* embryo are standardized against the image of an embryo exhibiting the pattern characteristic of maximal expression, that is normally observed at the late stages of development of a wild type embryo. The gain and offset values of the microscope photomultiplier are adjusted for this embryo and kept constant in all the series of scanning experiments. Because of this arrangement it may happen that images of embryos at early stages of development, and especially of mutants, are of very low contrast since the level of gene expression in these embryos is low. This is especially typical for images of expression patterns in embryos stained with the antibodies of poor quality, that give rise to a high nonspecific background. To be able to extract more detailed information from such images it is necessary to increase their intensity range by setting high values of PMT parameters; however this may lead to pixel saturation and errors in quantitative data extracted from the images. Our regression system provides means to estimate and correct errors in data obtained with an extended range of microscope parameters and hereby makes it possible to obtain more accurate quantitative information on gene expression.

4.5 CONCLUSIONS

- A regression system is created for error magnitude prediction in data obtained from an 8-bit confocal image. The prediction is based on information about microscope parameters used for image acquisition.
- The method demonstrates high prediction accuracy and was applied for correction of errors in the data on segmentation gene expression in *Drosophila* blastoderm stored in the FlyEx database (http://urchin.spbcas.ru/flyex/, http://flyex.uchicago.edu/flyex/).
- An important advantage of the developed prediction system is the possibility of error correction in data obtained from strongly clipped images, thereby permitting acquisition of higher dynamic range images, which would aid extraction of more detailed quantitative information.
- The system is realized as a software tool CorrectPattern freely available at http://urchin.spbcas.ru/asp/2011/emm/.

REFERENCES

1. Myasnikova E, Surkova S, Panok L, Samsonova M, Reinitz J: Estimation of errors introduced by confocal imaging into the data on segmentation gene expression in *Drosophila*. Bioinformatics 2009, 25:346-352.
2. Myasnikova E, Surkova S, Samsonova M, Reinitz J: Estimation of errors in gene expression data introduced by diffractive blurring of confocal images. Proceedings of 2009 13th Irish Machine Vision Conference (IMVIP 2009, IEEE) 2009, :53-58.
3. Pawley J: Fundamental limits in confocal microscopy. In Handbook of biological confocal microscopy. 3rd edition. Springer-Verlag New York Inc; 2006::20-41.
4. Kosman D, Reinitz J, Sharp D: Automated Assay of Gene Expression at Cellular Resolution. In Proceedings of the 1998 Pacific Symposium on Biocomputing. Edited by Altman R, Dunker K, Hunter L, Klein T. Singapore: World Scientific Press; 1997::6-17.
5. Kosman D, Small S, Reinitz J: Rapid Preparation of a Panel of Polyclonal Antibodies to *Drosophila* Segmentation Proteins. Development, Genes, and Evolution 1998, 208:290-294.
6. Janssens H, Kosman D, Vanario-Alonso C, Jaeger J, Samsonova M, Reinitz J: A high-throughput method for quantifying gene expression data from early *Drosophila* embryo. Development, Genes and Evolution 2005, 225(7):374-381.
7. Surkova S, Kosman D, Kozlov K, Myasnikova E, Samsonova A, Spirov A, Vanario-Alonso C, Samsonova M, Reinitz J: Characterization of the *Drosophila* segment determination morphome. Developmental Biology 2008, 313:844-862.
8. Myasnikova E, Samsonova M, Kosman D, Reinitz J: Removal of background signal from in situ data on the expression of segmentation genes in *Drosophila*. Dev Genes Evol 2005, 215(6):320-326.

There are several supplemental files that are not available in this version of the article. To view this additional information, please use the citation information cited on the first page of this chapter.

CHAPTER 5

SPACE: AN ALGORITHM TO PREDICT AND QUANTIFY ALTERNATIVELY SPLICED ISOFORMS USING MICROARRAYS

MIGUEL A. ANTON, DORLETA GOROSTIAGA,
ELIZABETH GURUCEAGA, VICTOR SEGURA,
PEDRO CARMONA-SAEZ, ALBERTO PASCUAL-MONTANO,
RUBEN PIO, LUIS M. MONTUENGA, AND ANGEL RUBIO

5.1 BACKGROUND

Alternative splicing (AS) is the process by which multiple mature mRNA sequences can be generated from the same precursor mRNA (pre-mRNA) upon the differential joining of exonic sequences limited by 5' and 3' splice sites. Through splicing mechanisms exons can be extended or shortened, skipped or included, and intronic sequences may even be retained in the mRNA sequences. AS is one of the most important sources of protein diversity in vertebrates, and at least half of human genes are alternatively spliced [1-3]. AS has been shown to be very relevant in a variety of human diseases, including cancer, and there is increasing interest in the use of AS in developing diagnostic tools and identifying new therapeutic targets [4-7].

This chapter was originally published under the Creative Commons Attribution License. Anton MA, Gorostiaga D, Guruceaga E, Segura V, Carmona-Saez P, Pascual-Montano A, Pio R, Montuenga LM, and Rubio A. SPACE: An Algorithm to Predict and Quantify Alternatively Spliced Isoforms Using Microarrays. Genome Biology *9,R46 (2008), doi:10.1186/gb-2008-9-2-r46.*

Two main strategies are pursued to identify and characterize AS events in expressed genes under both physiological and pathological conditions. On the one hand, expressed sequence tag (EST) alignment and mapping against known proteins or the whole genome may be used to identify different mRNA isoforms expressed in cell lines or tissues [8]. On the other hand, by performing analyses of splicing microarrays, the detection of new isoforms of a gene [9] and quantification of the relative concentrations for known isoforms may be obtained [10].

The most important manufacturers of commercial array platforms intended for the analysis of the expression of alternatively spliced isoforms are Affymetrix, Jivan Biotechnology (based on Agilent technology) and Exonhit (which can work both with Affymetrix and Agilent technologies). The strategy for Jivan and Exonhit includes two types of probes: exon probes or junction probes. Affymetrix uses only exon probes. Exon probes are complementary sequences to each of the known transcribed exons of a given gene, while junction probes include a segment of complementary nucleotides for each of the two sides of a known exon-exon junction in the mature mRNA of the gene. When designing expression arrays, exon probes are usually selected to meet a number of quality criteria. Since these probes may be located anywhere within the exon, probe specificity and affinity can usually be optimized to maximize their performance in the array hybridization step. In contrast, the selection for junction probes has very little room for optimization, as the nucleotide sequence at both sides of the junction is fixed and needs to be included in the junction probe. The only way to optimize junction probe quality is by changing the number of nucleotides at either side of the junction, and thus the total length of the probe. The overall performance of junction probes in any array is therefore remarkably lower than exon probes owing to poorer signal, frequent cross-hybridization, etc. This suboptimal probe quality leads to large variations in the signal levels for different probes of the same transcript, which may differ by as much as several orders of magnitude. Furthermore, the obvious phenomenon of cross-hybridization of a single probe with different transcripts that share half of the probe contributes to make the interpretation of junction-based arrays a real challenge [11]. A potential way to overcome this hurdle is to examine the probes mapped to a gene as a whole instead of analyzing individual probes one at a time.

Several tools and strategies have been proposed to deal with the complex bioinformatics analysis of splicing microarrays [9-16]. Cuperlovic-Culf et al.[17] provide a good and up-to-date comparative review of the traits and the performance of each available tool. Three of these strategies [9,13,15] can predict the existence of novel isoforms, but none of them is able to infer the intron/exon structure of the gene. In addition, only three of the previous works [10,12,14] provide a method to measure the relative concentrations of known transcripts. The aim of the present study is to develop a tool to measure the concentrations and structure of different transcripts from the output data of splicing microarrays (containing exon probes together with junction probes). The algorithm we propose here, which we have called SPACE (splicing prediction and concentration estimation), can (1) predict the number of different transcripts (some of them possibly unknown), (2) predict the structure of these transcripts and (3) measure their relative concentrations.

Our algorithm applies 'non-negative matrix factorization' (NMF) to the matrix of data. NMF is a factorization for non-negative multivariate data that allows us to find parts-based linear representations [18]. The main characteristic of NMF is its use of non-negative constraints. Given a matrix of non-negative data V, NMF finds an approximate factorization $V \approx W \cdot H$ into matrices with non-negative elements W and H [19]. In this work, we show that, when applied to splicing microarray data, NMF separates the data matrix for each gene into the product of two positive components corresponding to the structure of the gene transcripts and their individual concentrations, respectively. We have also developed an algorithm to determine the internal dimension of the factorization since previous attempts by other groups did not perform well in this particular application. We show that the internal dimension of the factorization is an estimate of the number of transcripts of each gene.

In summary, SPACE allows for the discovery of the structure of the expressed transcripts of a given gene, as well as for the determination of the relative concentration of each spliced isoform. It also makes the prediction/detection of new, previously unknown, alternatively spliced forms possible.

5.2 RESULTS AND DISCUSSION

We have applied the NMF algorithm described in the materials and methods section to both synthetic and real microarray datasets. For each gene, NMF of the expression matrix is performed $V \approx W \cdot H$.

Here H gives the relative concentration of each transcript while W gives the gene structure, that is, which probes hybridize to each transcript. The mathematical model used shows that the maximum value of each row of the W matrix is the affinity of the corresponding probe. Using this information it is possible to discern whether a probe hybridizes against a transcript (the corresponding entry of the W matrix is close to the row maximum) or not (the entry is close to zero).

5.2.1 SYNTHETIC DATASET

We have prepared a synthetic dataset to test the NMF algorithms. We generated this dataset as follows. Probe expression is proportional to the sum of concentrations of the transcripts that share the probe. The proportional constant, that is, the affinity, is a random number obtained from a distribution that mimics the distribution of real microarray data. Transcript concentrations are also random numbers. For junction probes, we simulated that these probes partially hybridize (20%) with each transcript that shares one of the sides of the junction. Hybridization is complete (100%) if the (exon or junction) probe matches perfectly with the transcript.

Among the possible parameters to consider while testing the algorithm, we have selected three: (i) noise level (five levels of additive noise and five levels of multiplicative noise); (ii) number of microarrays (5, 10, 25, 50, 100 and 200 arrays); (iii) position of the probes (only junctions J, only exons E, both of them J_E, junctions plus several probes per exon J_2E, J_3E). To make the simulation closer to a biological reality, we have borrowed the structure of eight different genes (*CASP2, HNRPA2B1, BCL2L1, BIRC5, TERT, VEGF, BAX* and *WT1*). These genes were selected just as examples and only by structural criteria among a larger list of genes whose AS is changed in cancer. Using the structure of these genes, simulated affinities, simulated concentrations, random noise and assuming a small amount of cross-hybridization for junction probes we have built the simulated expression data matrix. Table 1 shows the basic structural characteristics of these genes. The specific splicing structure is shown in Additional file 1 (Figures S1-S8).

TABLE 1: Description of the synthetic dataset genes used to test the SPACE algorithm

Gene name	Transcripts	Exons
CASP2	2	12
HNRPA2B1	2	12
BCL2L1	2	3
BIRC5	3	4
TERT	4	16
VEGF	4	8
BAX	4	6
WT1	4	10

The basic structural characteristics (number of transcripts and number of different exons) of the genes used to generate the synthetic dataset are shown. The SPACE algorithm has been tested with these genes.

Each simulation has been run 200 times (we tested 20 different combinations of probe affinity and transcript concentrations and each combination was run 10 times in the background of different random noises). The NMF iteration was run 3,000 times for each point to achieve convergence.

This study confronted us with a formidable problem: if each individual combination of parameters was checked, the number of conditions would be enormous (five additive noise levels × five multiplicative noise levels × six different numbers of arrays × five probe selections × eight genes = 6,000). These conditions should then each be simulated 200 times, and for each one we would need 3,000 iterations. In order to avoid this problem, we selected a 'central point' for each of the parameters and only one parameter was changed for each simulation experiment to analyze its effects. The 'central point' was selected with the following conditions: standard deviation of additive error was 5% of the median of the signal, standard deviation of multiplicative error was 7% of the signal, number of arrays was 40 and probes were located at exons and junctions (J_E). We performed

FIGURE 1: Influence of noise on the estimation of relative transcript concentrations of BIRC5 gene (synthetic data). BIRC5 gene structure is shown in Figure 6a and in Additional file 1 (Figure S4). (a) Additive noise effect on estimation of relative transcript concentrations. The y-axis shows the MAE between the relative concentration of transcripts without noise and that estimated by the algorithm under the effect of different degrees of additive noise (MAE %). Additive noise is in the form of $y + \varepsilon$ with $\varepsilon \sim N(0, \sigma_\varepsilon^2)$. The units of the x-axis are the variances σ_ε^2 of the additive error added to the simulated concentrations (10, 100, 1,000, 10,000, 100,000). These variances represent roughly 0.5%, 2%, 5%, 15% and 50% of the energy of the signal, respectively. (b) Multiplicative noise effect on estimation of relative transcript concentrations. The y-axis shows the MAE between simulated and estimated relative concentrations under the effect of different degrees of multiplicative noise (MAE %). Multiplicative noise is in the form of $y \cdot e^\eta$ with $\eta \sim N(0, \sigma_\eta^2)$. The units of the x-axis represent the variances σ_η^2 of the multiplicative error (5×10-5, 0.0005, 0.005, 0.05, 0.5). These variances represent roughly 0.7%, 2%, 7%, 25% and 100% of the energy of the signal, respectively. The different degrees of additive and multiplicative noise are tested while the other parameters are in the 'central point' condition (40 arrays and probes at exons and junctions). This means that there is always a component of additive and multiplicative noise in the form of $y \cdot e^\eta + \varepsilon$. Errors are represented by boxplots. A boxplot is a graphical representation of the variability of a random signal. They are composed by a box and a whisker. The box extends from the lower quartile to the upper quartile values and there is an additional horizontal line that shows the median. The whiskers are vertical lines extending from each end of the boxes to show the extent of the rest of the data. Outliers are data with values beyond the ends of the whiskers and are represented by crosses.

these simulation experiments on the gene *BIRC5* (results for other genes are similar, data not shown). We selected the mean absolute error (MAE) to measure the quality of the concentration estimation and the average Hamming distance to measure the quality of the predicted structure. We defined the Hamming distance as the proportion of probes in a gene that were mistakenly assigned (or unassigned) to a transcript. In Figures 1 and 2 we show an example of these simulations for gene *BIRC5*, a gene with five exons and three alternative transcripts. We now discuss the results of this prediction. We show the results only for the gene *BIRC5*, but the overall trend in the other genes is similar.

Additive and multiplicative error are well rejected by the algorithm (Figure 1). Only for very large variances does the quantification have large errors (MAE of the expression matrix for large variances is about 20%). The structure of the gene is correctly predicted in almost any case for low variances as shown in Figure 2. This figure also shows that the median of the Hamming distance is null for small variances, that is, each probe is perfectly assigned to each transcript for most of the simulations. The boxplots in panels B and C display the specificity and sensitivity of the SPACE algorithm for additive error. Both specificity and sensitivity have been calculated using a threshold of 0.5, that is, a probe is considered to belong to a transcript if its entry in the G matrix is larger than 0.5 and is not considered to belong to the transcript if the entry is smaller than 0.5. These figures show that both specificity and sensitivity are very good for small errors, and their performance worsens for larger errors. The same applies to multiplicative error in panels E and F.

Figure 3 shows the different results (for gene BIRC5) obtained when the number of arrays per experiment is changed. As expected, the structure is better estimated as the number of arrays per experiment increases. As a side effect, concentration estimation is also improved (Figure 3a).

Figure 4 shows the results for experiments changing the number and type of probes for BIRC5. Adding new exon probes does not significantly improve the performance of the concentration estimation but helps to estimate the structure of the transcripts. Junctions probes seem to be more informative than exon probes in this simulation. Nevertheless, this result should be taken cautiously since junction probes tend to be of inferior

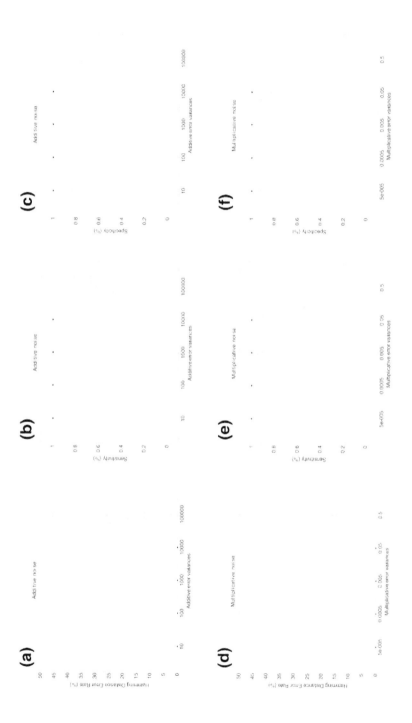

FIGURE 2: Influence of noise on splicing structure prediction for BIRC5 gene (synthetic data). (a) The effect of additive noise on splicing structure prediction. The y-axis shows the Hamming distance error rate between real and predicted pre-mRNA splicing structures. This measure represents the proportion of probes in a gene that were mistakenly assigned (or unassigned) to each transcript. The units of the x-axis are the variances σ_ε^2 of the additive error as explained in Figure 1a. (b) Sensitivity of the SPACE algorithm under additive noise. Sensitivity is defined as the proportion of probes that belong to each transcript that are correctly assigned in the predicted structure. (c) Specificity of the SPACE algorithm under additive noise. Specificity is defined as the proportion of probes that do not belong to a particular transcript that are correctly unassigned in the predicted structure. (d) Multiplicative noise effect on splicing structure prediction. The y-axis shows the Hamming distance error rate between real and predicted pre-mRNA splicing structures. The units of the x-axis are the variances σ_η^2 of the multiplicative error as explained in Figure 1b. (e) Sensitivity of SPACE under multiplicative noise. (f) Specificity of SPACE under multiplicative noise. The Hamming distance error rate is calculated in the form of $HD = (FP + FN)/N$, the sensitivity is calculated as $SN = TP/(TP + FN)$ and the specificity is calculated as $SP = TN/(TN + FP)$.

quality. This probe quality factor has not been taken into account in generating these synthetic data.

The simulation results when comparing different genes are particularly interesting (Figure 5). These simulations have been performed using the "central point." It can be seen that the accuracy of structure prediction decreases with the number of transcripts (see Table 1). For the same number of transcripts, structure predictive accuracy increases by increasing the number of probes with different hybridization patterns. The hybridization pattern is defined as the binding capability of a probe with each of the transcripts of a gene, that is, a logical vector that shows whether the probe belongs to each transcript or not.

The predicted and real structures and concentrations for genes that have an error close to the median of the simulations are shown in Figure 6. The comparison between real and predicted structure of BIRC5 (Figure 6b) clearly shows an almost perfect matching.

To further study the limits of the algorithm we provide a simple simulation: an experiment with gene CASP2 (that has two transcripts Casp2L and Casp2S) in which three arrays have been constructed under one condition and three arrays under a second condition. The number of arrays is low and quantification errors, as Figure 7d shows, are larger than in the central

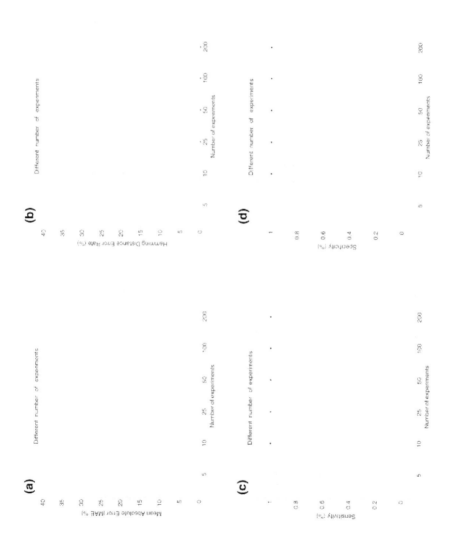

FIGURE 3: Influence of the number of arrays for BIRC5 gene (synthetic data). (a) Effect of the number of arrays on estimation of relative transcript concentrations. The y-axis shows the MAE (%) between simulated and estimated relative concentration of transcripts. The x-axis shows the different number of arrays used in the simulations. (b) Effect of the number of arrays on splicing structure prediction. The y-axis shows the Hamming Distance Error Rate between real and predicted pre-mRNA splicing structures. (c) Sensitivity of SPACE under different numbers of arrays. (d) Specificity of SPACE under different number of arrays.

point. Since this study is focused on AS, we considered that the overall concentration of the gene is constant (that is, the sum of the concentrations of the transcripts), and we have changed the relative abundance of each isoform. This sort of change cannot be detected by an expression array. We performed three different experiments in which the concentration ratio between each condition was 5:1 (Figure 7), 2:1 and 1.5:1 (see Additional file 1, Figure S10). These experiments show that splicing structure degrades as concentrations become closer to 1:1.

To improve the robustness of the performance figures, we have randomly selected 100 genes from the human genome with transcripts that range from 2 to 5. The number of exons of the selected genes ranges from 1 to 74.

We generated synthetic data for these genes assuming similar conditions to those selected for the central point: (1) junction probes and exon probes are included, (2) certain partial hybridization occur with junction probes (20%) and (3) additive and multiplicative noise with the same variance of the noise used in BIRC5 have been simulated.

A summary of the results obtained for these genes is shown in Table 2. In this table, the median value of each error measurement is shown in bold face between the lower and upper quartiles for equal number of transcripts. It can be noticed that the median values of MAE, Hamming distance, sensitivity and specificity corroborate the general trend found in the selected genes that performance decreases if the number of transcripts per gene increases. A more detailed description of the simulation results is available in Additional file 1 (Figures S11-S22).

FIGURE 4: Effect of the location of probes for BIRC5 gene (synthetic data). (a) Effect of the location of the probes on the estimation of relative transcript concentrations. The y-axis shows the MAE (%) between simulated and estimated relative concentration of transcripts. The x-axis shows the different location of probes along the transcripts of the gene. J: the gene is represented by all its junction probes; E: the gene is represented by exon probes located in all its exons; J_E: all junction and exon probes are present in the array; J_2E or J_3E: all junctions and two or three probes per exon, respectively, are present in the array. (b) Effect of the location of the probes on splicing structure prediction. The y-axis shows the Hamming distance error rate between real and predicted pre-mRNA splicing structures. (c) Sensitivity of SPACE with varying location of the probes. (d) Specificity of SPACE with varying location of the probes.

TABLE 2: Summary of simulation results for 100 random genes (synthetic data)

Number of transcripts	MAE	Hamming distance	Sensitivity	Specificity
2	3.1%-4.6%-7.0%	0%-0%-0%	1 - 1 - 1	1 - 1 - 1
3	3.7%-5.1%-6.8%	0%-0%-0.85%	0.99 - 1 - 1	1 - 1 - 1
4	4.1%-5.2%-7.0%	0%-1.4%-2.9%	0.96 - 0.99 - 1	1 - 1 - 1
5	4.0%-5.4%-8.3%	1.5%-3.4%-9.6%	0.88 - 0.95 - 0.98	0.96 - 0.99 - 1

The median value of MAE, Hamming distance, sensitivity and specificity for the simulation performed with 100 random genes is shown in bold for an equal number of transcripts. The variability of median values of each error measurement are indicated by the lower and upper quartiles at both sides of the median. It should be noted that the error increases as more transcripts are added to the simulation.

5.2.2 REAL DATASETS

To analyze the performance of the algorithm against biological data, we have tested SPACE on real datasets from large series of microarray hybridization experiments. We used two sets: one generated from an experiment using the Affymetrix platform (Wang dataset) [10] and another generated from an Agilent platform (Johnson dataset) [9].

The Wang dataset used Affymetrix technology to quantify the relative concentration of two transcripts of the CD44 gene. A large number of probes (184 perfect match (PM) and an equal number of mismatch (MM) probes) were available for this gene. On the other hand, in the Johnson

FIGURE 5: Influence of the gene structure and number of expressed transcripts in a comparative splicing analysis between different genes (synthetic data). (a) Estimation of the relative transcript concentrations for different genes. The y-axis shows the MAE (%) between simulated and estimated relative concentration of transcripts. The x-axis shows the different genes used in the simulation (CASP2, HNRPA2B1, BCL2L1, BIRC5, TERT, VEGF, BAX and WT1). The structure of the different transcripts of these genes and the location of probes is shown in Additional file 1 (Figures S1-S8). (b) Prediction of the splicing structure for different genes. The y-axis shows the Hamming distance error rate between real and predicted pre-mRNA splicing structures. (c) Sensitivity of SPACE for different genes. (d) Specificity of SPACE for different genes.

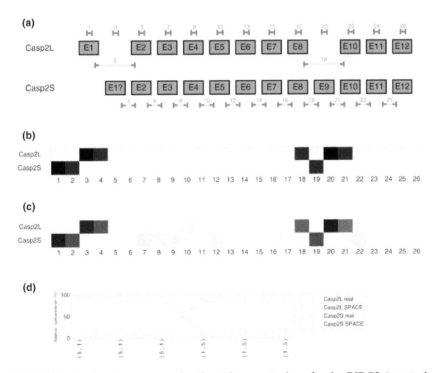

FIGURE 6: Predicted structure and estimated concentrations for the BIRC5 (apoptosis inhibitor survivin) gene in the 'central point' (synthetic data). (a) Structure of the different transcripts of the BIRC5 gene and location of probes used in the simulation. (b) Representation of the real and predicted splicing structures for the BIRC5 gene given by the probes used. In the graphic representing the real splicing structure the probes that match perfectly with the transcripts are represented by a white box (100% matching) and no hybridization is shown by a black box (0% matching). Gray levels show intermediate matching values. We have assumed that junction probes which include one side of the junction hybridize partially (20%). (c) Estimated relative concentrations of the three isoforms of BIRC5 gene. In each of the three graphics simulated and estimated relative concentration of each isoform is represented.

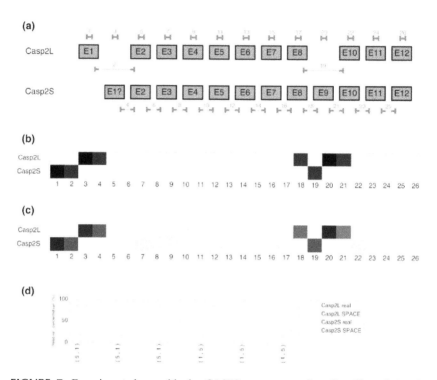

FIGURE 7: Experiment done with the CASP2 gene, transcripts Casp2L and Casp2S (synthetic data). Three arrays were performed with a concentration ratio between the two isoforms of CASP2 gene equal to 5:1 and another three with the opposite ratio 1:5. (a) Structure of the two transcripts of CASP2 gene and location of probes in the microarray. (b) Real structure of CASP2 gene indicated by probes. Probes that match perfectly are represented in white (100%), no hybridization in black (0%) and partial hybridization by different shades of gray (20%). (c) Predicted splicing structure for CASP2 gene with the alternating concentration ratio 5:1. If compared with the real structure of CASP2 transcripts (b), a strong similarity is noticed. (d) Real and estimated relative concentrations of the two isoforms of CASP2 gene in the experiment.

dataset Agilent technology was used to monitor the junctions of 10,000 multi-exon genes across 52 diverse samples. In this dataset, we applied SPACE to the analysis of those genes whose splicing prediction was validated by Johnson et al. using reverse transcription polymerase chain reaction (RT-PCR).

5.2.3 WANG DATASET

In this set, we used the 184 PM probes to measure two transcripts of CD44 spiked by Wang et al. in their experiment [10]. Even though we performed the simulations using two transcripts, we noticed that results assuming three transcripts were better. Results with three transcripts are shown in Figure 8. After performing the factorization, we noticed that the entries of the row of the W matrix corresponding to the third transcript had almost the same constant value. We concluded that our algorithm is mimicking the proposed method in [10] to remove background noise. Concentrations of the two spiked transcripts are well estimated (MAE = 4.47%). The structures of the transcripts are well predicted except for junctions 7-10, 11-12 and 16-17 and for exon 10 (which is incorrectly included in the first transcript by the algorithm).

We have tried the algorithm with other parameters (using two transcripts and the difference between PM and MM measurements, three transcripts using both PM and MM probes) and results are similar.

Surprisingly, when we considered only MM probes for the algorithm, it was also possible to predict the concentration of the transcripts with worse yet still reasonable accuracy (MAE = 14.06%). Even though our results are similar to Wang et al.'s (in fact, their results are so good that there is little room for improvement), we obtained them without any a priori knowledge of gene structure. Therefore, the relevant and major contribution of SPACE is that it can estimate concentration without knowing the structure of the gene. In addition, if part of the structure is known it can be readily included in our algorithm, as will be stated in the discussion. SPACE can be considered as a generalization of Wang et al.'s algorithm that is able to both predict structure and concentration of transcripts and is also able to take advantage of partial knowledge of the structure.

5.2.4 JOHNSON DATASET

The analysis performed by Johnson et al. to interpret their data can be considered as a variation of an analysis of variance (ANOVA) type II test using medians instead of means. For each log transformed expression, the median of probe expressions in different tissues (to remove the affinity effect) and the median of probe expression in each gene (to remove the gene level effect) were subtracted to obtain a residual. In a second step, they performed a discretization of the residues. Genes with large residues for several tissues are further analyzed since these patterns are probably due to AS.

As explained in the introduction, junction probes cannot meet quite so stringent quality criteria as exon probes. In addition, the number of probes per gene is much smaller than in the Wang dataset.

Furthermore, this dataset was obtained from real tissues instead of spiked transcripts. Therefore the results are not expected to be as accurate as in the previous dataset. The relative concentrations of transcripts were predicted using the SPACE algorithm. In the following we briefly describe the results for each gene and discuss the performance of our algorithm when compared with the RT-PCR analyses.

The first analyzed gene is OCRL (ENSG00000122126). Two splice variants were detected using RT-PCR in Johnson et al.'s paper. Therefore, we used an internal dimension of two transcripts to apply the algorithm.

Estimated relative concentrations (Figure 9c) match Johnson et al.'s expression results obtained by RT-PCR. Comparing the colormap shown in Figure 9b with the gene structure (for Ensembl release 40) shown in Figure 9a, it is possible to infer that the variant 1 predicted by the algorithm corresponds to ENST00000371113 and variant 2 corresponds to a group of three transcripts (the set of probes included in the array is not able to distinguish among the three of them). In these cases, the structure predicted by SPACE matched that of Ensembl known transcripts. The colormap structure suggests the existence of a cassette splicing event for exon 25 in ENST000037112.

Results for APP gene (ENSG00000142192) are shown in Figure 10. Estimated relative concentrations using three transcripts match Johnson et al.'s PCR results. The structure prediction is less accurate in this case. According to Johnson et al., this gene has three transcripts: a long form, one with a single exon cassette (exon 7) and one with a double exon cassette (exons 7 and 8). The algorithm is able to predict the structure of the long isoform and the double cassette isoform correctly but not the single cassette isoform. The third form shows what seems to be a splicing event

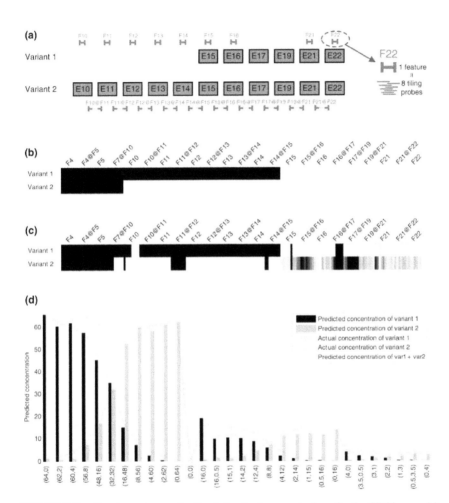

FIGURE 8: Predicted structure and estimated spiked concentrations for the CD44 gene. (a) Structure of the CD44 gene spiked transcripts and probe positions represented by features. A gene feature is either an exon or a junction. Exon features are represented by F followed by the number of the corresponding exon and junction features are represented by the two exon features to which they belong joined by @ symbol. Each feature is made up of eight probes following a tiling strategy. As 23 features have been measured this makes a total of 184 probes. Probes corresponding to exon features F4, F5 and junction features F4@F5, F7@F10 do not match any of the spiked transcripts and therefore are not shown in (a). (b) Expected hybridization pattern of all probes for each of the two variants of CD44 gene. (c) Splicing structure prediction for CD44 gene applying the SPACE algorithm. (d) Estimated concentrations of the two variants of CD44 gene compared to spiked concentrations. The y-axis shows the predicted and actual concentration of each variant. The x-axis indicates the experiments and actual concentrations of each variant pair.

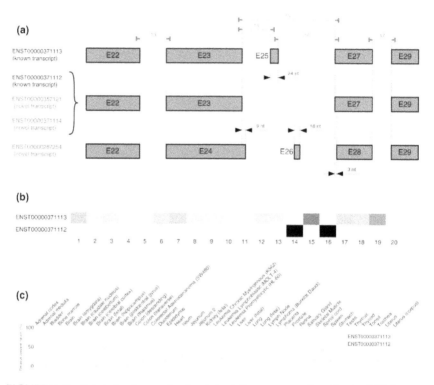

FIGURE 9: Predicted structure and estimated relative concentrations for the OCRL gene. (a) Structure of the different transcripts (known and novel) of OCRL gene according to Ensembl 40, as well as the real location of the probes in the microarray. As can be seen in the figure, given probes cannot distinguish between a group of three isoforms (one known and two novel). (b) Predicted splicing structure for OCRL gene given by probes. The SPACE algorithm only detect two isoforms that match with known transcripts of OCRL gene ENST00000371113 and ENST0000037112. (c) Estimated relative concentrations of the two isoforms detected of OCRL gene.

related to probes 16 and 17, the 3' region of the gene. VEGA (a curated genome database) shows that there is a short transcript (APP-012) that has experimental evidence and involves precisely these two probes. This result may be a coincidence but a careful study of these probes shows that they do not correlate well with the others, suggesting either a splicing event or an artifact of the array. The estimated concentrations of the long and double cassette isoforms match the results obtained by PCR in the original article.

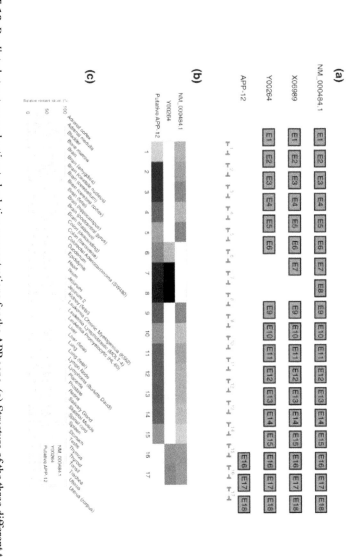

FIGURE 10: Predicted structure and estimated relative concentrations for the APP gene. (a) Structure of the three different transcripts of the APP gene proposed by Johnson et al. to be present in the samples as well as a short isoform APP-12 that match our results, the real locations of the probes in the microarray are also indicated. (b) Predicted splicing structure for the APP gene given by probes. SPACE detects three isoforms that match with transcripts NM_000484.1, Y00264 and APP-12 of the APP gene. (c) Estimated relative concentrations of the three isoforms detected for the APP gene.

Results for HMGCR gene (ENSG00000113161) using two transcripts are shown in Figure 11. In this case, the estimation of the concentrations is not as clear as in the other genes. Indeed, PCR concentrations in Johnson et al.'s paper show only significant differences in the concentrations for a tissue that was not hybridized in the set of arrays (peripheral leukocytes). Structure prediction shows the expected results (exons 12 and 13, where there is a cassette, have the smallest affinities along the probes for variant 2).

TABLE 3: Prediction of number of transcripts per gene (synthetic data)

Gene name	Number of transcripts					
	1	2	3	4	5	6
CASP2	0%	100%	0%	0%	0%	0%
HNRPA2B1	0%	95%	5%	0%	0%	0%
BCL2L1	0%	90%	10%	0%	0%	0%
BIRC5	0%	0%	95%	5%	0%	0%
TERT	0%	0%	10%	85%	5%	0%
VEGF	0%	0%	0%	90%	10%	0%
BAX	0%	0%	20%	80%	0%	0%
WT1	0%	0%	100%	0%	0%	0%

To apply NMF factorization, it is necessary to know, or estimate, the number of transcripts of a particular gene present in the samples under study. The estimations of the number of transcripts using SPACE for the genes used in the synthetic dataset are shown. Each entry of the table shows the proportion of predicted number of transcripts for different simulated data. The column corresponding to the real number of transcripts is shown in bold.

5.2.5 PREDICTION OF NUMBER OF TRANSCRIPTS

In the previous sections we have shown the potential of SPACE in determining the concentration and splicing structure of the genes, assuming that the number of transcripts is known. However, when the number of transcripts is unknown, this method can also be used to make a accurate prediction. In this section we describe this novel application using the synthetic and real datasets.

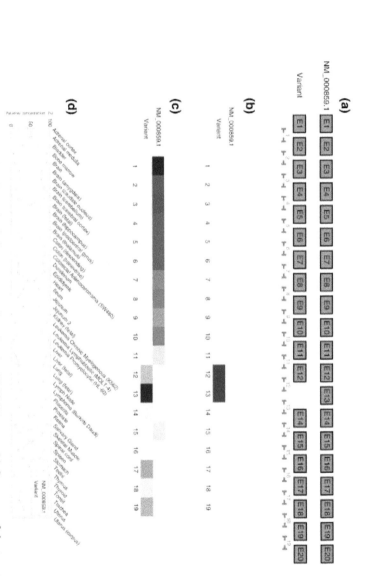

FIGURE 11: Predicted structure and estimated relative concentrations for the HMGCR gene. (a) Structure of the two transcripts of the HMGCR gene, NM_000859.1 and a variant with a cassette in exon 13, as well as the real locations of the probes in the microarray. (b) Predicted splicing structure for the HMGCR gene given by probes. If compared with the gene structure in (a), it can be seen that the cassette is detected but also more things that do not match with that model. (c) Real and estimated relative concentrations of the two isoforms of HMGCR gene.

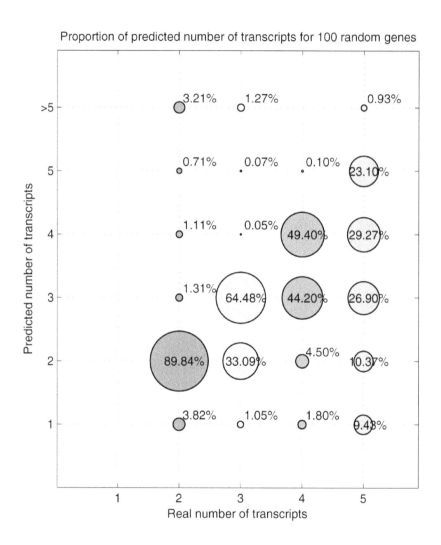

FIGURE 12: Proportion of predicted number of transcripts for the simulation performed with 100 genes (synthetic data). The 100 genes used have been randomly selected from the human genome with two to five transcripts. Each of the genes have been simulated 200 times for different noise, concentrations and affinities. The area of each circle represents the proportion of times the corresponding predicted number is chosen by the algorithm for a given number of transcripts. The algorithm tends to underestimate the number of transcripts as the real number of transcripts increases.

We have estimated the number of transcripts using the algorithm presented in the materials and methods section. In brief, this algorithm searches for the dimension that optimally splits the error figure for different numbers of transcripts into two groups: signal and noise.

The estimated number of transcripts for the initial list of eight genes used in the synthetic dataset is shown in Table 3. The estimated number is correct for all of the genes except WT1. In this case, the algorithm predicts three transcripts (instead of four) in 100% of the simulations. If WT1 structure is analyzed, it can be seen that two of their transcripts share all but one of the probes. At this level of noise, the algorithm finds it more likely to have three transcripts than four. The estimation of the number of transcripts improves dramatically for the other genes since they have a larger number of probes with different hybridization patterns.

In Figure 12, we further extend the analysis of predicting the number of transcripts for the simulation performed with 100 random genes of the human genome (synthetic data). This figure shows that the accuracy of the prediction decreases as the number of transcripts increases. It can also be noticed that the algorithm tends to underestimate the real number of transcripts when it does not make a correct guess. We describe two possible reasons that we have found for such a behavior. In some cases, there are very similar transcripts that can be discerned by only one probe (similar to the WT1 case). On the other hand, if some of the transcripts have very low concentrations they are considered to be noise.

The estimation performed in the Wang real dataset predicts three transcripts. As already explained, the SPACE algorithm behaves better using this result as the number of transcripts when compared with using the real number is two (the third transcript is measuring the background noise).

The results for the three genes in the Johnson dataset are as follows. In the OCRL gene (ENSG00000122126), SPACE estimated that two transcripts are present. In the case of the APP gene (ENSG00000142192) two transcripts are predicted (but the likelihood for three transcripts is similar).

Finally, SPACE predicted two transcripts for the HMGCR gene (ENSG00000113161).

5.3 DISCUSSION

We have described a method to predict the number, structure and concentration of gene transcripts using splicing microarray data.

Our simulations show that the SPACE algorithm method is able to predict unknown structures and to measure the relative concentration of alternatively spliced isoforms from synthetic data. The method is robust against multiplicative and additive noise. Structure prediction performance of SPACE increases with the number of arrays.

As expected, performance diminishes with an increasing number of transcripts per gene. The accuracy of the prediction is closely related to the number of probes that have different hybridization patterns along the transcripts. This fact is even more evident when analyzing the results of the simulation: additional exon probes do not improve the discriminative power of the array as much as including probes that provide new hybridization patterns. A good microarray design to detect splicing should have as many discriminative probes per gene as possible. An algorithm was previously proposed to select discriminative probes subject to constraints based on the Hamming distance [20]. This algorithm can be readily applied to splicing arrays.

Our simulation experiments also show that junction probes tend to be better at distinguishing between transcripts than exon probes. In contrast, partial hybridization with adjacent exons does not seem to be a problem. A major drawback associated with these types of splicing array is the poor thermodynamic quality of many junction probes. SPACE is useful in interpreting the output data from splicing arrays containing the usual proportion of suboptimal junction probes. In fact, we have also shown that the SPACE algorithm performs successfully with real splicing microarray datasets. Our analysis of the Wang dataset [10], based on an Affymetrix platform, estimated very similar transcript relative concentrations to those obtained in Wang et al.'s paper. Our data were also close to the real concentration of the spikes. Moreover, compared with previous studies, our algorithm provides a method to predict the structure of unknown isoforms. This feature is unique and novel since previous algorithms have only dealt with the probes that may be related to splicing events but do not predict any structure. In addition, our study provides a method to estimate relative concentrations of known and novel isoforms.

The SPACE algorithm deals with all of the probes as a whole and, as shown by the simulations, it is able to reject additive and multiplicative noise. The gene structure predicted by SPACE when using the Johnson dataset [9], based on an Agilent platform, performed better for the genes with two transcripts than for the genes with three transcripts, in which the structure of the third transcript was less accurately estimated. Concentration estimations were similar to those reported in Johnson et al.'s paper by means of RT-PCR.

The SPACE algorithm has certain limitations that must be taken into account when applying it. First, it assumes that the probe signal levels have been derived from a linear model. If the probe signals are not proportional to the concentration of the transcripts, both the structure and the concentrations can be predicted incorrectly. Second, we have tested that the factorization works better (that is, error diminishes) if there is variability in W and H. For the W matrix this means that the error figures improve if the array includes several probes that are able to distinguish between different transcripts. For the H matrix, the prediction power improves if we include several different experimental conditions. If only one experimental condition is performed (for example, in a design which includes several replicas of a single sample), this algorithm is not able to discover a mixture of transcripts.

The level of noise affects the ability of the algorithm to discover new transcripts. If the concentration of a particular transcript is very low, it may be masked by the noise background, and SPACE, in this case, would not be able to discern this transcript.

We have used the maximum value of a row in the W matrix to estimate the affinity of the probe and convert the W matrix into the product of matrices AG. The maximum value is a statistical operation that is strongly affected by outliers. However, NMF, using Kullback-Leibler (KL) distance, is robust against outliers and it is not likely to have an outlier in the final factor matrices. A side effect of this selection is that a probe does not hybridize with any of the transcripts of the gene, a row of zeroes in the G matrix, so this method will give a one in some of them. A way to overcome this problem is to identify whether any of the 'discovered' transcripts is, in fact, background noise and reject the probes in which the computed entries for G are closer to the value of the transcript that models the noise. This is the approach the we used for the Wang dataset.

As described in the materials and methods section, since only 15% of the exons in genes with AS are specific for a unique transcript, the algorithm fosters to fill the G matrix. Small transcripts that only have a few probes that hybridize on them tend to be predicted incorrectly. Many of the errors in the simulation of 100 genes correspond to genes with small transcripts.

This method can be further improved if some constraints are imposed on the NMF model. For example, if the structure is partially known, its corresponding W matrix can be used as an initial value for the optimization. The structure of W is maintained along the optimization by the algorithm since the described multiplicative update rule retains the null entries of both the W and H matrices.

On the other hand, the use of NMF multilayer [21] algorithms is a novel optimization technique that uses several matrices, instead of two, to perform the optimization. All of the matrices in these algorithms have positive entries. The original work of Wang et al. proposes the factorization affinities, features, property matrix and concentrations, that is, Y = AFGT + E. This factorization fits perfectly with a multilayer algorithm. In addition, some of these matrices are sparse (the affinity and the features matrix). Imposing sparsity constraints [22] and null entries may help the algorithm to improve the results.

5.4 CONCLUSION

In this paper we have presented SPACE, an algorithm based on NMF that is able to predict the number of transcripts, gene structure and concentration of known and unknown versions of splice forms. The validity of the results have been tested in both synthetic and biological data, and SPACE has shown its ability to reject additive and multiplicative noise.

It can be stressed that the algorithm not only predicts the concentration, but also the structure of the genes. This characteristic makes it completely novel.

5.5 MATERIALS AND METHODS

5.5.1 PROBE AND GENE STRUCTURE MODEL

In our algorithm, we assume that there is a linear relationship between the intensity of a probe y (an exon probe or a junction probe) and the concentration of the targets x measured by this probe (as proposed by Li and Wong [23]):

$$y = a \cdot x + e \tag{1}$$

where a is the affinity of the probe and e is an error term.

Let the matrix $Y_{i \times j}$ be the set of measurements of all the probes included in a gene. Its dimensions are i rows (probes included in the gene) and j columns (different arrays).

Taking into account that each probe may belong to a particular transcript and not to others, we can extend this model by using a 'property' matrix G. Combining the information for different transcripts, we can derive the following:

$$Y = A \cdot G \cdot T + E \tag{2}$$

where, $A = (a_{i, i})$ is an i × i diagonal matrix of unknown affinities. The matrix $T = (t_{k, j})$ represents the concentration of each k gene transcript (rows) in the j array (columns). The property matrix $G = (g_{i, k})$ relates the probes with the different transcripts depending on whether the probe belongs to the transcript or not. This model was proposed by Wang et al. [10]. In their paper, an additional matrix of features F is included, but it can be avoided without loss of generality. The proposed value for each entry in this case is

$$g_{i,k} = \begin{cases} 1 & \text{probe } i \text{ is included in transcript } k \\ 0 & \text{probe } i \text{ not included in transcript } k \\ \alpha & \text{probe } i \text{ partially hybridizes with transcript } k \text{ (useful for junction probes)} \end{cases}$$

In Wang et al.'s method, this matrix is binary (that is, if no perfect sequence identity is obtained, it is considered that there is no hybridization). We propose that, since partial hybridization does occur in junction probes, a proper selection of parameter a can take this fact into account. The value of a (partially hybridized probe) ranges from 0 to 0.6 depending on the length of the probes, their composition and manufacturing of the array [11]. Finally, E = $(e_{i,j})$ matrix represents the error term.

5.5.2 MODEL FITTING AND MINIMIZATION

In expression (2), both A (the affinity matrix) and T (the concentration matrix) are unknown. A natural way to find these unknowns is to minimize some function of the error term. Wang et al. proposed to minimize the sum of the squared differences between the measurements (Y matrix) and the estimation (A·G·T matrix) subject to the condition that the unknowns (affinities and concentrations) must be positive as follows:

$\min(\|Y - A \cdot G \cdot T\|_2)$

subject to $a_{i,i} \geq 0$ and $t_{k,j} \geq 0$

where $\| \ \|_2$ is the Fröbenius norm of the matrix (that is, the sum of its entries squared).

This optimization function may be ineffective if it is applied to splicing arrays. Since junction probes cannot be selected to have similar affinities (the position of the probe cannot be arbitrarily selected), their affinities vary by several orders of magnitude. This minimization function is proportional to the error squared and results are skewed to model probes with large affinities. Instead of this error function, we propose to use the Kompass family of divergence functions [24]:

$$D_{Ko}(Y, AGT) = \sum_{i=1}^{p} \sum_{j=1}^{n} \left(Y_{ij} \frac{Y_{ij}^{\beta-1} - (AGT)_{ij}^{\beta-1}}{\beta(\beta-1)} + (AGT)_{ij}^{\beta-1} \frac{(AGT)_{ij} - Y_{ij}}{\beta} \right)$$

(3)

This family has an additional parameter β that must be set. It can be easily shown that the dimension of the summation term is ε_{ij}^{β}. If $\beta = 2$, this function reverts into expression (2) and it is the selection of choice for additive Gaussian noise. If β tends to zero, it reverts into the Itakuro-Saito entropy (especially useful for multiplicative noise) and if β tends to one, it reverts to the KL entropy that can be used for noise that includes additive and multiplicative terms.

5.5.3 BLIND GENE STRUCTURE PREDICTION

The standard non-negative matrix factorization can be directly applied to the matrix Y, yielding two matrices W and H:

$$Y_{ij} \approx W_{ik} \cdot H_{kj} \approx (A_{ii} \cdot G_{ik}) \cdot T_{kj}$$

(4)

It is straightforward to identify W with AG and H with T (the relative concentrations of the k transcripts). This assignation is done because of their respective dimensions. There is an intuitive interpretation of NMF for the analysis of splicing: the first matrix represents the predicted structure of the gene (whether a particular probe belongs to a particular transcript or not) and the second is the relative abundance of the transcript.

Each of the matrices in the Wang et al. equation (2) has an interesting property: all of the entries are non-negative. In addition, most of the exons (in genes that are alternatively transcribed) are shared by several transcripts, that is, the G matrix usually has many ones and few zeros. If the NMF of the Y matrix is performed (Y = WH), non-negativity is ensured (first condition), and using certain algorithms, the W (the AG factor) matrix has few zero entries (second condition). Therefore, NMF provides a reasonable estimation of A, G and T and, as shown in the simulation, they are indeed very close to the real values of these matrices.

One simple way to obtain the A and G matrices from W is to consider that the affinity of a probe is the maximum value for the corresponding row in W. In this case, we obtain

$$a_{ii} = \max_{k}(W_{ik})$$

$$G = A^{-1}W \tag{5}$$

Here G will be a matrix whose entries lie between zero and one. The algorithm to perform the optimization (expressed in Matlab compact form) for the Kompass generalized divergence function [24] is

$$H \leftarrow H.* \left(W' * \left(V./(W*H)^{2-\beta} \right) \right)./(W' * (W*H)^{\beta-1})$$
$$W \leftarrow \left(W.* \left(V./(W*H)^{2-\beta} \right) * H' \right)./((W*H)^{\beta-1} * H'))^{1+\alpha w}$$
$$W \leftarrow W.* diag\{1./sum(W,1)\} \tag{6}$$

where .* represents element wise multiplication and αW is a small constant (around 0.005) that modifies (increases for positive values and decreases for negative values) the sparsity of matrix W. We have used for β the value of one (that is, we used KL divergence to perform the estimations).

5.5.4 IMPLEMENTATION ISSUES

We used Matlab 7.1 with the statistics toolbox on a Pentium IV 3.2 GHz PC. The Matlab code is available as Additional file 2, so that other researchers can validate our calculations. The time required to compute each gene depends on the number of probes, the number of transcripts and the number of arrays. For BIRC5 (11 probes, 3 transcripts and 40 arrays), it took 1.2 seconds to perform the factorization using 5,000 iterations. During the last 4,000 iterations the error function did not change, but we decided to use a large number of iterations to ensure convergence. Proper termination conditions may decrease the computing time by a factor of five (about 0.2 seconds per gene). Using these improvements, the computing time for 25,000 genes is less than 2 hours. The algorithm can be easily computed in parallel in a cluster if a shorter computing time is desired. On the other hand, the estimation of the number of transcripts is a computer-intensive task: the factorization has to be performed twice, for original data and randomized data, for up to 10 transcripts. On average, it takes about 20 times more time to estimate the number of transcripts than to perform a single factorization.

5.5.5 DEALING WITH NON-UNIQUENESS OF THE DECOMPOSITION

Non-negative factorization is not unique: in some cases there are several W-H pairs that reconstruct the same initial matrix. Other factorizations such as singular value decomposition (SVD) have additional orthogonality constraints that make them unique in most cases (if the singular values are different). When applying NMF to splicing analysis, there can be several structures and concentrations that are compatible with probe measurements for a particular gene. Let us illustrate this fact with an example. Let us assume that (1) all of the probes have the same affinity, (2) the real structure of the transcript is as shown in Figure 13a and (3) two measurements are performed (in the first the concentration of the transcripts are t_{11} = 1, t_{12} = 0 and in the second t_{21} = 0, t_{22} = 1).

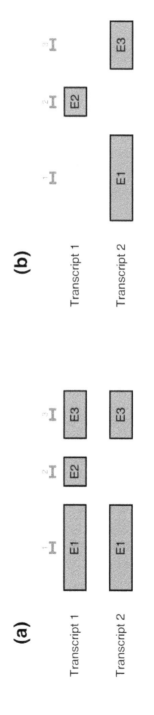

FIGURE 13: Example of the non-uniqueness of the splicing structure prediction. (a) Structure of a generic gene and proposed probe pattern. (b) Possible splicing structure prediction obtained by SPACE with the same probe pattern.

The matrix equation in this case is

$$\underbrace{\begin{bmatrix} 1 & 1 \\ 1 & 0 \\ 1 & 1 \end{bmatrix}}_{Y} = \underbrace{\begin{bmatrix} 1 & 0 & 0 \\ 0 & 1 & 0 \\ 0 & 0 & 1 \end{bmatrix}}_{A_1} \underbrace{\begin{bmatrix} 1 & 1 \\ 1 & 0 \\ 1 & 1 \end{bmatrix}}_{G_1} \underbrace{\begin{bmatrix} 1 & 0 \\ 0 & 1 \end{bmatrix}}_{T_1} = \underbrace{\begin{bmatrix} 1 & 1 \\ 1 & 0 \\ 1 & 1 \end{bmatrix}}_{G_1} \underbrace{\begin{bmatrix} 1 & 0 \\ 0 & 1 \end{bmatrix}}_{T_1}$$

(7)

This gene structure and concentrations give the matrix of measurements shown in the equation. Let us consider the alternative structure shown in Figure 13b.

In this case, one possible matrix equation for the same measurements could be

$$\underbrace{\begin{bmatrix} 1 & 1 \\ 1 & 0 \\ 1 & 1 \end{bmatrix}}_{Y} = \underbrace{\begin{bmatrix} 0 & 1 \\ 1 & 0 \\ 0 & 1 \end{bmatrix}}_{G_2} \underbrace{\begin{bmatrix} 1 & 0 \\ 1 & 1 \end{bmatrix}}_{T_2}$$

(8)

or, in a more general case,

$$\underbrace{\begin{bmatrix} 1 & 1 \\ 1 & 0 \\ 1 & 1 \end{bmatrix}}_{Y} = \underbrace{\begin{bmatrix} 1-\alpha & 1 \\ 1 & 0 \\ 1-\alpha & 1 \end{bmatrix}}_{G_2} \underbrace{\begin{bmatrix} 1 & 0 \\ \alpha & 1 \end{bmatrix}}_{T_2}$$

(9)

It can be seen that the same probe expression matrix Y is compatible with two different combinations of properties and concentrations (G_1 and T_1 versus G_2 and T_2). Probe expression does not provide sufficient information to decide between the two versions. However, although it is necessary to determine which of the two versions is real, probe expressions do not provide the necessary information for this.

However, if the data belong to the same gene, it is more likely that an exon is shared by several transcripts. In Ensembl release 40, only 15% of the exons of alternatively spliced genes are specific for a unique transcript. We have computed this number by performing a query against the Ensembl 40 MySQL database. Therefore, if two structures are valid, the structure that shares more exons among transcripts is preferable. From the point of view of G matrices, it is preferable to have a full matrix (probes are shared by many transcripts) rather than a sparse matrix (probes are specific to a few transcripts).

NMF factors tend to be sparse (in this example, the NMF factorization will probably give the structure shown in Figure 13b instead of the first). In terms of the factorization, it is necessary to ease the filling of the matrix W. This can be done by adding a penalty term related to the sparseness of matrix W or, as proposed in [24], selecting the αW parameter in (6) to be a small negative value. We have selected -0.005 as proposed in [24].

5.5.6 GETTING THE NUMBER OF TRANSCRIPTS

There is still a free parameter to be set in the algorithm: the number of transcripts k for a particular gene. In this case, this number is the internal dimension k of the factorization. This problem is indeed a 'dimensionality reduction problem', that is, how many dimensions explain the behavior of a set of data. In this work we have tried previous algorithms to select this number (cophenetic correlation coefficients [25,26] and SVD-based selection [27]). Cophenetic coefficients, although promising at first, did not work as expected and only gave good results when k is very small (one or two transcripts).

We have used a variation of the method proposed by Zhu and Ghodsi [28] (also used for NMF by Fogel et al. [29]). Intuitively, the idea of the algorithm is to compare how the error decreases compared with random data. If the improvement is not larger than what is obtained with random data, then there is no need to add a new transcript.

Zhu and Ghodsi [28] propose to select the dimension that maximizes a likelihood function of the singular values. The intuitive idea of their algorithm is to find the gap in the scree plot of the singular values. In this algorithm, the singular values are assumed to belong to two groups with different means and the same variance and a maximum likelihood criteria (MLC) finds the most likely partition of these groups.

In NMF there are no singular values associated with the rows and columns of matrices W and H, but the increment in the error function can be interpreted as the variation explained by the additional dimension. The input vector to the MLC is calculated as follows.

Let Obj_l be the objective function of the factorization selecting l transcripts

$$Obj_l = D_{Ko}(Y, W_{il}, H_{lk}) \tag{10}$$

where D_{Ko} is the Kompass divergence among Y, W and H.

Let $\overline{Obj_l}$ be the objective function of the factorization selecting l transcripts

$$\overline{Obj_l} = D_{Ko}(\hat{Y}, \hat{W}_{il}, \hat{H}_{lk}) \tag{11}$$

where $\hat{Y}$ is a shuffled version of the Y data matrix (data for each row is randomized) and W and H are the results of the optimization.

Let FC_l be the natural logarithm of the fold change between the objective function for l transcripts and the objective function for l - 1 transcripts:

$$FC_1 = \log\left(\frac{obj_1}{obj_{l-1}}\right) \tag{12}$$

and, for convenience, it is assumed that

$$Obj_\oslash = D_{Ko}(Y, R) \tag{13}$$

where R is a positive random matrix with unit variance. On the other hand

$$\overline{FC}_1 = \log\left(\frac{\overline{obj}_l}{\overline{obj}_{l-1}}\right) \tag{14}$$

We consider

$$\Delta l = FC_1 - \overline{FC}_1 \tag{15}$$

that is, how much the error function diminishes compared with random data. To improve the normality of the data (needed for MLC), a Box-Cox [30] transformation is performed on this vector to have standard kurtosis.

Intuitively Δl will be large if the dimension is smaller than the real number of transcripts. In this case, adding a new transcript diminishes the error more than what would be expected for random data.

Finally, the MLC algorithm will find the gap in the Δl that splits the dimensions in two groups: true transcripts and noise.

REFERENCES

1. Maniatis T, Tasic B: Alternative pre-mRNA splicing and proteome expansion in metazoans. Nature 2002, 418:236-243.
2. International Human Genome Sequencing Consortium: Initial sequencing and analysis of the human genome. Nature 2001, 409:860-921.
3. Brett D, Pospisil H, Valcarcel J, Reich J, Bork P: Alternative splicing and genome complexity. Nat Genet 2002, 30:29-30.
4. Caceres JF, Kornblihtt AR: Alternative splicing: multiple control mechanisms and involvement in human disease. Trends Genet 2002, 18:186-193.
5. Faustino NA, Cooper TA: Pre-mRNA splicing and human disease. Genes Dev 2003, 17:419-437.
6. Garcia-Blanco MA, Baraniak AP, Lasda EL: Alternative splicing in disease and therapy. Nat Biotechnol 2004, 22:535-546.
7. Pajares MJ, Ezponda T, Catena R, Calvo A, Pio R, Montuenga LM: Alternative splicing: an emerging topic in molecular and clinical oncology. Lancet Oncol 2007, 8:349-357.

8. Hui L, Zhang X, Wu X, Lin Z, Wang Q, Li Y, Hu G: Identification of alternatively spliced mRNA variants related to cancers by genome-wide ESTs alignment. Oncogene 2004, 23:3013-3023.

9. Johnson JM, Castle J, Garrett-Engele P, Kan Z, Loerch PM, Armour CD, Santos R, Schadt EE, Stoughton R, Shoemaker DD: Genome-wide survey of human alternative pre-mRNA splicing with exon junction microarrays. Science 2003, 302:2141-2144.

10. Wang H, Hubbell E, Hu JS, Mei G, Cline M, Lu G, Clark T, Siani-Rose MA, Ares M, Kulp DC, Haussler D: Gene structure-based splice variant deconvolution using a microarray platform. Bioinformatics 2003, 19(Suppl 1):i315-i322.

11. Castle J, Garrett-Engele P, Armour CD, Duenwald SJ, Loerch PM, Meyer MR, Schadt EE, Stoughton R, Parrish ML, Shoemaker DD, Johnson JM: Optimization of oligonucleotide arrays and RNA amplification protocols for analysis of transcript structure and alternative splicing. Genome Biol 2003, 4:R66.

12. Shai O, Morris QD, Blencowe BJ, Frey BJ: Inferring global levels of alternative splicing isoforms using a generative model of microarray data. Bioinformatics 2006, 22:606-613.

13. Srinivasan K, Shiue L, Hayes JD, Centers R, Fitzwater S, Loewen R, Edmondson LR, Bryant J, Smith M, Rommelfanger C, Welch V, Clark TA, Sugnet CW, Howe KJ, Mandel-Gutfreund Y, Ares MJ: Detection and measurement of alternative splicing using splicing-sensitive microarrays. Methods 2005, 37:345-359.

14. Fehlbaum P, Guihal C, Bracco L, Cochet O: A microarray configuration to quantify expression levels and relative abundance of splice variants. Nucleic Acids Res 2005, 33:e47.

15. Le K, Mitsouras K, Roy M, Wang Q, Xu Q, Nelson SF, Lee C: Detecting tissue-specific regulation of alternative splicing as a qualitative change in microarray data. Nucleic Acids Res 2004, 32:e180.

16. Bingham J, Sudarsanam S, Srinivasan S: Profiling human phosphodiesterase genes and splice isoforms. Biochem Biophys Res Commun 2006, 350:25-32.

17. Cuperlovic-Culf M, Belacel N, Culf AS, Ouellette RJ: Data analysis of alternative splicing microarrays. Drug Discov Today 2006, 11:983-990.

18. Lee DD, Seung HS: Learning the parts of objects by non-negative matrix factorization. Nature 1999, 401:788-791.

19. Lee DD, Seung HS: Algorithms for non-negative matrix factorization. Adv Neural Inf Process Syst 2001, 13:556-562.

20. Klau GW, Rahmann S, Schliep A, Vingron M, Reinert K: Integer linear programming approaches for non-unique probe selection. Discrete Appl Math 2006, 155:840-856.

21. Cichocki A, Amari S, Zdunek R, Kompass R, Hori G, He Z: Extended SMART algorithms for non-negative matrix factorization. In Proceedings of the Internationl Conference on Artificial Intelligence and Soft Computing (ICAISC 2006): 25-29 June 2006; Zakopane, Poland. Berlin: Springer; 2006::548-562. [Lecture Notes in Artificial Intelligence, vol. 4029.]

22. Pascual-Montano A, Carazo JM, Kochi K, Lehmann D, Pascual-Marqui RD: Non-smooth nonnegative matrix factorization (nsNMF). IEEE Trans Pattern Anal Mach Intell 2006, 28:403-415.

23. Li C, Wong WH: Model-based analysis of oligonucleotide arrays: expression index computation and outlier detection. Proc Natl Acad Sci USA 2001, 98:31-36.

24. Cichocki A, Zdunek R, Amari S: Csiszár's divergences for non-negative matrix factorization: family of new algorithms. In Proceedings of Independent Component Analysis and Blind Signal Separation (ICA 2006): 5-8 March 2006; Charleston, SC. New York: Springer; 2006::32-39. [Lecture Notes in Computer Science, vol. 3889.]

25. Brunet JP, Tamayo P, Golub TR, Mesirov JP: Metagenes and molecular pattern discovery using matrix factorization. Proc Natl Acad Sci USA 2004, 101:4164-4169.

26. Pascual-Montano A, Carmona-Saez P, Chagoyen M, Tirado F, Carazo J, Pascual-Marqui R: bioNMF: a versatile tool for non-negative matrix factorization in biology. BMC Bioinformatics 2006, 7:366.

27. Kim PM, Tidor B: Subsystem identification through dimensionality reduction of large-scale gene expression data. Genome Res 2003, 13:1706-1718.

28. Zhu M, Ghodsi A: Automatic dimensionality selection from the scree plot via the use of profile likelihood. Comput Stat Data Anal 2006, 51:918-930.

29. Fogel P, Young S, Hawkins D, Ledirac N: Inferential, robust non-negative matrix factorization analysis of microarray data. Bioinformatics 2007, 23:44-49.

30. Box GEP, Cox DR: An analysis of transformations. J Roy Statist Soc B (Statist Meth) 1964, 26:211-252.

There are several supplemental files that are not available in this version of the article. To view this additional information, please use the citation information cited on the first page of this chapter.

CHAPTER 6

LINK-BASED QUANTITATIVE METHODS TO IDENTIFY DIFFERENTIALLY COEXPRESSED GENES AND GENE PAIRS

HUI YU, BAO-HONG LIU, ZHI-QIANG YE, CHUN LI, YI-XUE LI, AND YUAN-YUAN LI

6.1 BACKGROUND

Identification of differentially expressed genes (DEGs) is a key step in comprehending the molecular basis of specific biological processes and screening for disease markers. This methodology looks at absolute changes in gene expression levels, and treats each gene individually. However, genes and their protein products do not perform their functions in isolation, but in coordination [1], and the dynamic switch of a gene from one community to another always implies altered gene function [2,3]. Therefore, gene coexpression analysis was developed to explore gene interconnection at the expression level from a systems perspective [4-10], and 'differential coexpression analysis (DCEA)', as a complementary technique to the traditional 'differential expression analysis' (DEA) [11,12], was designed to investigate molecular mechanisms of phenotypic changes through identifying subtle changes in gene expression coordination [11-14].

This chapter was originally published under the Creative Commons Attribution License. Yu H, Liu B-H, Ye Z-Q, Li C, Li Y-X, and Li Y-Y. Link-Based Quantitative Methods to Identify Differentially Coexpressed Genes and Gene Pairs. BMC Bioinformatics 12,315 (2011), doi:10.1186/1471-2105-12-315.

B

$$dC_a(DCe) = \sum_{x=2}^{3} C_3^x \left(\frac{2}{5}\right)^x \left(1 - \frac{2}{5}\right)^{3-x} = 0.35$$

(Suppose two out of the five links are DCLs, and they are a-b and a-c.)

$$dC_a(DCp) = \sqrt{(x_{ab} - y_{ab})^2 + (x_{ac} - y_{ac})^2 + (x_{ad} - y_{ad})^2} / \sqrt{n_a}$$

$$= \sqrt{(-0.7 - 0.8)^2 + (0.8 - (-0.9))^2 + (0.7 - 0.8)^2} / \sqrt{3}$$

$$= 1.31$$

$$dC_a (WGCNA)$$

$$= \frac{s_x(a)}{\max\{ s_x(a), s_x(b), s_x(c), s_x(d)\}} - \frac{s_y(a)}{\max\{ s_y(a), s_y(b), s_y(c), s_y(d)\}}$$

$$= \frac{(-0.7 + 0.8 + 0.7)}{1.8} - \frac{(0.8 - 0.9 - 0.8)}{1.2}$$

$$= -0.14$$

(for simplicity, the original correlation values are used here. s() is a sum of all associated correlations. For instance, $s_x(a) = x_{ab} + x_{ac} + x_{bd} = -0.7 + 0.8 + 0.7$

$$dC_a (LRC) = abs [log_{10}(c_x(a)) / (c_y(a))]]$$

$$= abs (log_{10}(1/3))$$

$$= 0.48$$

($c_x(a)$ and $c_y(a)$ are the number of solid links (PCCs larger than 0.8) connecting gene a in the respective network.)

$$dC_a (ASC) = (c_x(a) + c_y(a)) / 2$$

$$= (0 + 2) / 2 = 1$$

$c'_x(a)$ and $c'_y(a)$ are the number of solid links connecting gene a in only one of the two networks.)

A

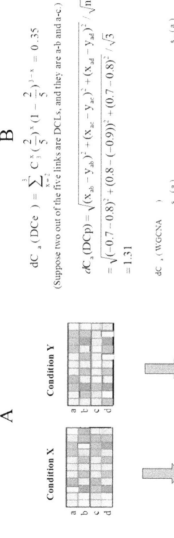

Condition X   Condition Y

The correlation threshold is 0.8; link **c-d** does not meet the half-thresholding criterion: its two correlation values x_{cd}=0.2 and y_{cd}=0.2, both smaller than 0.8. Therefore, link c-d is not shown in the coexpression networks.

FIGURE 1: A simplified illustration of the framework of five DCEA methods. A), two expression data matrices from two contrastive experimental conditions (X and Y) involving genes a, b, c, and d, visualized using shades from red (high expression level) to blue (low expression level), are transformed to a pair of coexpression networks. In the coexpression networks, gene pairs with absolute expression correlation values larger than 0.8 are connected with solid lines, while the rest with dashed lines. The line thickness is proportional to the absolute coexpression value. Red color highlights a negative coexpression value, and the grey-shaded node, gene a, is the one whose differential coexpression (dC) calculations are to be illustrated. B), different DCEA methods calculate the dC measure of gene a in different ways (see Results and Discussion for details).

In a typical DCEA workflow, a pair of gene expression datasets under two conditions, such as disease and normal, are transformed to a pair of coexpression networks in which links represent transcriptionally correlated gene pairs (Figure 1A), and then the differential coexpression is calculated for each gene (Figure 1B). After surveying three previously proposed DCEA methods (Figure 1B): 'Log Ratio of Connections' (LRC) [15], 'Average Specific Connection' (ASC) [12], and 'weighted gene coexpression network analysis' (WGCNA) [16-19], we realize that although DCEA methods have been used more and more frequently in transcriptome studies [11,12,15,17,20,21], they have not been well developed, and the most crucial issue in DCEA - the choice of differential coexpression measure, is far from settled.

In LRC [15], the differential coexpression of a gene is defined as the absolute logarithm of the ratio of its two connectivities - the numbers of links connecting the gene in two coexpression networks. This method does not distinguish the coexpression neighbors of a gene, and hence may fail if the connectivities of a gene in two networks are close while the gene neighbors are rather different. This defect is overcame in the average specific connection (ASC) method [12], which compares the 'specific connections' that exist in only one network. In simply dealing with the numbers of neighboring genes, however, both LRC and ASC fail to achieve a more precise characterization of differential coexpression that would be attainable if the quantitative expression correlation values were not discarded. The third method, WGCNA [16-19], goes beyond ASC and LRC as it compares the sums of expression correlation values associated with a gene between two conditions, which is essentially the comparison of weighted connectivities of a gene. We therefore classify all these three methods into a gene connectivity-based type. Because these connectivity-based methods do not quantify coexpression changes link by link, they cannot precisely estimate the differential coexpression of a gene. As a result, they fail to distinguish dramatically changed links from those relatively trivial ones, and they also cannot detect a special type of coexpression change - correlation reversal between positive and negative, which is never rare [22,23] and probably has important biological significance [24,25].

Since coexpression is in essence a property of gene pairs (links), it should be more reasonable to design link-based DCEA methods that concentrate directly on the coexpresssion change of each gene pair. In this work, we develop two link-based DCEA algorithms for identifying differentially coexpressed genes (DCGs) and differentially coexpressed gene pairs or links (DCLs). Based on the exact coexpression changes of gene pairs, these methods take into account both the gene neighbor identity information and the quantitative coexpression change information. It was demonstrated on simulated datasets that both novel methods had an improved performance over the existing methods to retrieve predefined differentially regulated genes and gene pairs. We furthermore applied the methods to a publicly available expression dataset on type 2 diabetes (T2D) and provided additional information to characterize T2D-related genes. The novel methods for DCEA analysis have been implemented in an academically available R package DCGL [26].

6.2 RESULTS AND DISCUSSION

6.2.1 NOVEL "HALF-THRESHOLDING" STRATEGY IN CONSTRUCTING GENE COEXPRESSION NETWORKS

There are currently two accepted strategies, namely hard-thresholding [11,12] and soft-thresholding [16-19], for inferring gene coexpression network from expression correlation values. The hard-thresholding one,

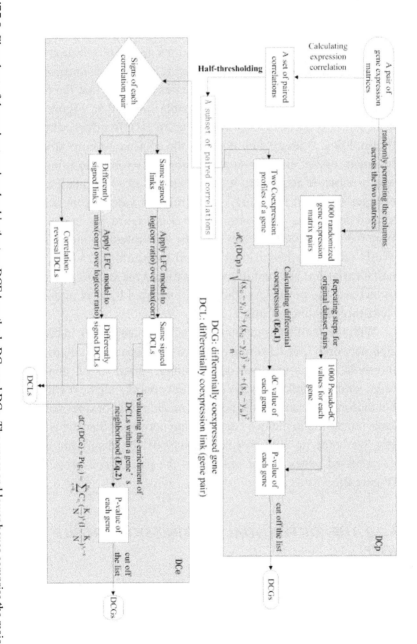

FIGURE 2: Flowchart of the main steps involved in the two DCEA methods DCp and DCe. The upper and lower boxes comprise the major steps for DCp and DCe respectively, while outside the boxes are a few steps of the shared pre-processing process.

adopted by LRC and ASC, keeps a link in the coexpression network as long as the coexpression value exceeds a predefined threshold (solid lines in Figure 1A). The soft-thresholding strategy, adopted by WGCNA, keeps all possible links and raises the original coexpression values to a power 'beta' so that the high correlations are emphasized at the expense of low correlations (its formula in Figure 1B uses the untransformed correlation values for illustration convenience). Note that the coexpression value pair associated with the invisible link c-d in Figure 1A are utilized in the WGCNA dC formula (Figure 1B). In effect, while the 'hard-thresholding' strategy dichotomizes the continuous correlation values to be coexpression and non-coexpression, it is robust to minor variations and meanwhile its sensitivity is impaired, as some small coexpression changes (link a-d in Figure 1A, correlation values from 0.7 to 0.8) are treated equally as large ones (link b-d in Figure 1A, correlation values from 0.8 to 0.2). On the other side, the 'soft-thresholding' strategy can be overly sensitive when using a low soft-threshold (i.e. a low power) since noisy variations are kept in its calculation. One way to get around this is to increase the power. Another way, proposed here, is to devise a novel "half-thresholding" strategy.

With the "half-thresholding" strategy, we keep a link in both coexpression networks if at least one of the two coexpression values exceeds the threshold. In this way, we ignore minor variations of 'non-informative links' whose correlation values in both networks are below the threshold, but thoroughly examine the possibly meaningful coexpression changes of links remaining in the two coexpression networks. Starting with this strategy, we come up with two novel methods for identifying differentially coexpressed genes and/or links from the pair of coexpression networks (Figure 2).

6.2.2 THE "DIFFERENTIAL COEXPRESSION PROFILE" METHOD (DCP)

We consider two gene expression datasets under two different conditions. For each dataset, we calculate the Pearson correlation coefficients (PCCs) between the expression profiles of all gene pairs. For gene i and gene j, let x_{ij} and y_{ij} denote their PCCs under the two conditions. Then the two datasets are encoded into a set of paired correlations CP = $\{(x_{ij}, y_{ij})\}$ over all gene pairs. We then filter out non-informative correlation pairs using the half-thresholding strategy. Specifically, a pair is kept if any of the two PCCs has a q-value lower than a cutoff, say 25%, where the q-value is a false discovery rate estimated from the p-value of PCC using the Benjamini-Hochberg method [27]. This results in a subset of correlation pairs, which are equivalent to two coexpression networks with identical structure but different link weights (PCCs).

For gene i, the PCCs between it and its n neighbors in the filtered set form two vectors, $X = (x_{i1}, x_{i2}, ..., x_{in})$ and $Y = (y_{i1}, y_{i2}, ..., y_{in})$ for the two conditions, which are referred to as 'coexpression profiles'. We define the differential coexpression (dC) of gene i with Eq. 1.

$$dC_i(DCp) = \sqrt{\frac{(x_{i1} - y_{i1})^2 + (x_{i2} - y_{i2})^2 + \cdots + (x_{in} - y_{in})^2}{n}} \tag{1}$$

This measure captures the average coexpression change between a gene and its neighbors. As this method is based on the differential coexpression profiles, it is denoted as DCp. An example calculation of DCp dC is shown in Figure 1B.

The dC value can be used to rank genes. To evaluate the statistical significance of dC, we perform a permutation test, in which we randomly permute the disease and normal conditions of the samples, calculate new PCCs, filter gene pairs based on the new PCCs, and calculate the dC statistics. The sample permutation is repeated 1000 times, and a large number of permutation dC statistics form an empirical null distribution. The p-value for each gene can then be estimated.

The major steps of the DCp algorithm is outlined in the upper box of Figure 2.

6.2.3 THE "DIFFERENTIAL COEXPRESSION ENRICHMENT" METHOD (DCE)

While DCp takes advantage of the coexpression changes of individual gene links, its final goal is to identify differentially coexpression genes (DCGs). To extend the findings from DCGs to differentially coexpressed links (DCLs), we devise another method, 'Differential Coexpression Enrichment', which first identifies DCLs, and then identifies DCGs. As the method is based on enrichment of DCLs, it is named DCe.

The filtered correlation set (determined with a cutoff ρ of expression correlation values or qth of the q values, as described in the DCp method details) represents the beginning links to be screened for DCLs. For a link or a pair of correlation values, we first determine the maximum (absolute) correlation and the log (absolute) correlation ratio. If the two correlation values of a link are same signed, we intuitively propose that the log correlation ratio may serve as a basic measure for the link's differential coexpression; in contrast, if the link has two differently signed correlation values, its differential coexpression is more likely to be reflected by the maximum correlation. We then separately deal with the same signed links and the differently signed links using the limit fold change (LFC) model [28]. LFC is a robust statistical method originally proposed for selecting differentially expressed genes (DEGs), by modeling the relationship between maximum expression and log expression ratio of genes. In coexpression analysis, we instead model the relationship between maximum coexpression and log coexpression ratio of links.

For the same signed links, as is illustrated in Figure 3A, we categorize them into bins according to their maximum coexpression values, and within each bin, select a fraction δ of links with highest log fold changes, and fit a curve $y = a + (b/x)$ over the boundary links. Links lying above the fitted curve are considered as DCLs. In most experiments of this work, we set $\delta = 0.1$, but the effect of tuning δ was tested in the following simulation study.

Among the differently signed links, those with both PCCs surpassing the cutoff ρ of correlation values or qth of the q-values are directly taken out as DCLs, specifically, correlation-reversal DCLs. In parallel to the same signed case, LFC model is applied to the remaining differently

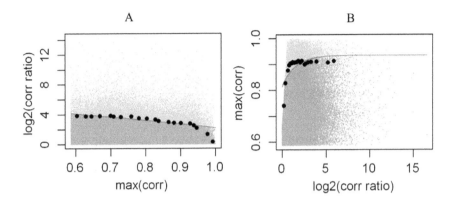

FIGURE 3: Limit fold change model applied to identify differentially coexpressed links (DCLs) from a simulated dataset pair in group C (dataset pair III). Each point represents a gene pair or a link characterized by its log correlation ratio and maximum absolute correlation value. A curve (gray line) y = a + (b/x) is used to fit the boundary outliers (black dots) determined by fraction δ, and points (light gray dots) lying above the fitted curves are considered DCLs. A, same signed links; B, differently signed links.

signed links with the roles of maximum coexpression and log coexpression ratio flipped due to our foresaid consideration (Figure 3B). Again, links above the fitted curve are considered as DCLs.

Suppose there are N links in each filtered coexpression network, from which we have determined K DCLs using the procedures above. For gene i with ni links of which k_i are DCLs, the p-value is calculated based on a binomial probability model (Eq. 2). The obtained p-value can be regarded as the dC measure of a gene, with a smaller value indicating a higher degree of differential coexpression. The enrichment step of DCe method is also illustrated in Figure 1B.

$$dC_i(DCe) = P(g_i) = \sum_{x=k_i}^{n_i} C_{n_i}^x \left(\frac{K}{N}\right)^x \left(1 - \frac{K}{N}\right)^{n_i-x}$$

(2)

The major steps of the DCe algorithm is outlined in the lower box of Figure 2.

6.2.4 COMPARING DIFFERENT DCEA METHODS IN SIMULATION EXPERIMENTS

In a simulation experiment, we first define two gene regulation networks, which are overall similar but have differences in a small portion of regulation relationships (gene links), then simulate two gene expression datasets based on the two networks, respectively. The predefined discrepant regulations are termed differentially regulated links (DRLs) and the associated genes are differentially regulated genes (DRGs). We evaluated DCp and DCe in terms of their capability to retrieve the predetermined DRGs and DRLs from the simulated data. Also included in the comparative evaluation were three representative DCEA methods that we reviewed in the Background: LRC [15], ASC [12], and WGCNA [16,19]. Note that the WGCNA has evolved into two slightly different versions, the 'signed' and the 'unsigned', and here we adopted the signed version and set its parameter beta at the default 12.

We first analyzed a pair of simulated datasets (dataset pair Z) from a published study [29], which were generated based on two yeast signaling networks using SynTReN [30]. A total of seven genes, *PHO2, FLO1, MBP1_SWI6, FLO10, TRP4, CLB5* and *CLB6*, were involved in the altered interconnection [29], therefore taken as DRGs. As Table 1 shows, the DCp dC score ranked all seven DRGs exclusively at the top, while the DCe p-value ranked six at the top and the other one at the 8th position; both methods had better performances than the other three methods. It was noticeable that *SWI4*, a gene falsely detected in the original study [29], puzzled WGCNA and ASC (which both ranked it at the 5th position), but not DCp and DCe (which ranked it at the 9th or 15th position).

Additionally, we used SynTReN to simulate three groups of dataset pairs (denoted data groups A, B, C) based on a predefined *E.coli* gene regulatory network of a total of 1300 genes [30]. Specifically, we selected a sub-network of 1000 genes as the original network, and exerted artificial perturbation on 10% of its links as if it was from a different condition.

The three groups had different perturbation types. For group A, we used regulation-elimination (removing a link between a pair of genes). For group B, we used regulation-switch (switching the regulation effect between activation and repression). For group C, we applied half regulation-elimination and half regulation-switch. For each group, we generated five dataset pairs, one simulated from the original network and the other from the perturbed network.

TABLE 1: The twenty yeast proteins involved in simulated dataset pair Z and the ranking of them by DCEA methods DCp, DCe, signed WGCNA, ASC, and LRC separately.

protein	DCEA methods				
	DCp	Dce	signed-WGCNA	ASC	LRC
PHO2	1	1	3	1	5
MBP1_SWI6	2	3	8	2	8
FLO1	3	2	1	10	7
FLO10	4	4	2	6	4
TRP4	5	5	4	7	9
CLB5	6	6	14	3	18
CLB6	7	8	18	4	19
ACE2	8	14	16	15	1
SWI4	9	15	5	5	16
CDC11	10	7	9	12	17
CDC10	11	11	10	13	10
SWI4_SWI6	12	16	6	8	12
HTB1	13	13	7	11	15
ACT1	14	12	13	14	6
CAF4	15	9	19	19	3
LEU2	16	17	11	9	13
SPT16	17	18	15	18	11
HO	18	10	17	16	2
CTS1	19	19	12	17	14
SNF6	20	20	20	20	20

The proteins are sorted by the DCp ranks. Bold refers to the truly differentially regulated genes (DRGs) in the simulation.

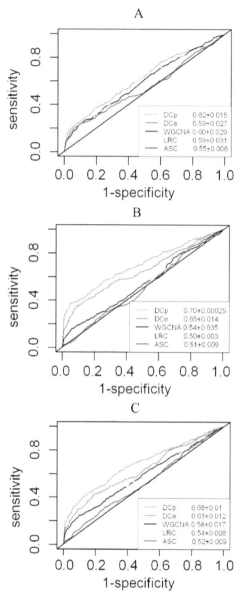

FIGURE 4: Receiver-operating-characteristic (ROC) curves showing the capabilities of five DCEA methods in retrieving predefined DRGs. To simulate change of regulatory relationships, 10% links were removed (A), 10% switched (B), 5% removed and 5% switched (C) in a 1000-node network. Each curve is averaged over five simulations. The numbers in legend are areas under the ROC curves (mean and std).

We applied every DCEA method on every dataset pair and plotted the Receiver-Operating-Characteristic (ROC) curves to show the balance of the five methods between sensitivity and specificity in identifying DRGs (Figure 4). Dataset group A, simulating regulation-elimination, seemed a tough problem for all methods, as none of the ROC curves was obviously far away from the diagonal line representing random assortment (Figure 4A). Nevertheless DCp performed better than the others. The advantage of DCp and DCe over the other methods was increased on group B which simulated regulation-switch, while the performances of ASC and LRC were not significantly different from a random guess (Figure 4B). On dataset group C with both regulation-elimination and regulation-switch included, DCp and DCe still outperformed other methods (Figure 4C). In all, DCp and DCe did better in retrieving DRGs, especially on data involving switched regulation relationships. The WGCNA method, which utilizes the continuous expression correlation values as DCp and DCe do, ranked immediately after DCp and DCe, ahead of LRC and ASC.

Since the signed WGCNA performed better than the other existing methods ASC and LRC and it actually gains more acknowledgement from users, we additionally performed a more comprehensive comparison of WGCNA against the novel methods, with different settings of the key parameter beta of WGCNA taken into account. It turned out that in general the signed WGCNA was more powerful than the unsigned WGCNA, but yet both were incomparable to DCp and DCe regardless of the choices of the parameter beta (Additional File 1: more parameter testing for gene-level evaluation). We also found that the WGCNA methods' performances deteriorate with the beta value, especially when beta exceeded eight, and that the WGCNA methods were relatively more competent for the regulation-elimination scenario (Additional File 1: more parameter testing for gene-level evaluation).

As all these DCEA methods except WGCNA involve a gene link filtering step, or a correlation value thresholding process, we repeated the performance comparison on various qth values (0.25, 0.2 and 0.1); additionally, as the perturbation rate of 10% was set arbitrarily, we also tried another two levels, 20% and 30%. It proved that DCp and DCe consistently outperformed the other three with DCp doing better than DCe in most situations (Additional File 1: more parameter testing for gene-level evaluation). We also found that for perturbation rates 10% and 20%, algorithm performances generally increased with more stringent qth values, and they dropped a little when the perturbation rate reached 30%. Finally, we tested the sensitivity of algorithms to sample sizes of datasets. At sample size five, six, seven, eight, nine, ten, fifteen and twenty, it was shown that the performances of all algorithms were basically stable, and that DCp and DCe were better than the others (Additional File 1: more parameter testing for gene-level evaluation).

TABLE 2: Fractions of DRLs or 'extended DRLs' in DCL sets.

Golden Standard	Dataset	Background	DCe-DCLs	ASC-DCLs
True	I	8.8e-4	1.5e-3	8.5e-4
DRLs	II	1.0e-3	3.7e-3	7.1e-4
	III	9.8e-4	3.4e-3	7.9e-4
Extended	I	0.44	0.49	0.45
true	II	0.43	0.67	0.43
DRLs	III	0.43	0.63	0.44

I, II, and III denote three dataset pairs from the data groups A, B, and C, respectively.

Finally, we compared the only two methods, DCe and ASC, which have the potential to retrieve DCLs, with respect to their capability to retrieve DRLs. It was found that, in three simulated dataset pairs (I, II, and III), DRLs always accounted for a tiny fraction of identified DCLs, but DCe outperformed ASC in enriching DRLs in DCLs (Table 2). As gene coexpression changes may spread from the perturbed links to adjacent ones, we took DRLs and their one-step adjacent links as 'extended DRLs'. Likewise, DCe enriched the extended DRLs (Table 2), which was statistically significant according to tests against empirical distribution of randomly sampled links (Additional File 2: significance in link-level evaluation). In an actual practice of DCe, in order to narrow down the identified DCLs for a follow-up examination, one can raise coexpression value cutoffs (ρ) or lower outlier fractions (δ). We found that raising ρ refined correlation-reversed DCLs efficiently, while lowering δ not only cut down the number of DCLs of the other two types (same sign and different sign) but improved the accuracy of identified DCGs (Additional File 3: reducing DCL scales). Besides, the identified DCLs could also be sifted according to their relevance with a selected gene list, for example, DCGs.

In summary, the results from simulation studies indicate that reasonably designed DCEA methods can retrieve pre-set differentially regulated genes and links from expression datasets. That is, based on the results from a series of rigorously designed simulation experiments, we provide a preliminary support to the anticipation that DCEA methods are capable of deciphering differential regulation or differential networking underpinning diseases [13].

6.2.5 UNIQUENESS OF DCP AND DCE COMPARED TO EXISTING DCEA METHODS

We attributed the improved performance of DCp and DCe mainly to the exploitation of the linkwise quantitative coexpression changes, which starts with our 'half-thresholding' strategy in coexpression network construction, and continues with the dC measures reflecting the linkwise coexpression changes (Eq. 1 and Eq. 2). Capturing the linkwise coexpression changes is much more reasonable than merely extracting the connectivity and/or neighbor identity, or getting the summed correlation values. That is why DCp and DCe outperformed existing methods LRC, ASC, and WGCNA in simulation studies.

We tried designing our methods based on coexpression changes of all possible links, i.e., discarding the half-thresholding, but found the performance was not comparable to the current version of DCp and DCe (data not shown). This suggests the necessity of the link prescreening step. However, it is not easy to determine the optimal coexpression threshold for each specific study, and further investigation on optimizing our half-thresholding procedure is necessary.

It is noticeable that DCp and DCe are especially good at identifying a special type of coexpression change, the coexpression reversal between positive and negative, which is why they have the greatest advantage in the simulated datasets involving regulation-switches. In previous studies,

TABLE 3: DCGs with existing evidence of T2D-relevance.

gene	DCP.dC	Expression fold change	Reported Relevance
Rapgef4	1.29	0.81	T2D-related
Nr5a1	1.28	0.56	T2D-related
Cd28	1.21	1.21	KEGG rno04940
Ucp2	1.18	1.31	T2D-related
Pparg	1.17	1.05	T2D-related; T2D-associated
RT1-Bb	1.16	0.74	KEGGrno04940
Cox6a2	1.15	0.57	KEGG rno00190
Bdnf	1.13	2.13	T2D-related
Gad1	1.12	0.23	KEGG rno04940
Prkaa1	1.12	1.03	KEGG rno04910
Prkab1	1.11	0.52	KEGG rno04910; T2D-related
RT1-Da	1.09	1.99	KEGG rno04940
Nos3	1.09	0.76	T2D-related
Cox6c1	1.09	0.62	KEGG rno00190
Inpp5d	1.08	2.36	KEGG rno04910; T2D-related
Arnt	1.07	0.71	T2D-related
Sstr5	1.06	1.77	T2D-related
Lipe	1.06	2.04	KEGG rno04910; T2D-related
Cacna1a	1.06	0.55	KEGG rno04930
RT1-Ba	1.05	1.71	KEGG rno04940
Tnfrsf1a	1.05	1.07	T2D-related
Il6	1.05	1.77	T2D-related
Gip	1.05	0.63	T2D-related
Cacna1c	1.05	0.74	KEGG rno04930
Mapk10	1.03	0.79	KEGG rno04930; KEGG rno04910; T2D-related
Nras	1.02	0.71	KEGG rno04910
Serpine1	1.02	2.07	T2D-related
Kcnj5	1.01	0.74	T2D-related
Hla-dma	1.01	2.84	KEGG rno04940

Included genes appeared in at least one of the following sources: KEGG T2D-related pathways (Rno04930: type II diabetes mellitus; rno04940: type I diabetes mellitus; rno04910: insulin signaling pathway; rno00190: oxidative phosphorylation); a self-compiled set of 425 T2D-related genes in rat; a list of 52 T2D-associated genes. The lists of "T2D-related" genes and "T2D-associated" genes are provided in "Additional File 6: 52 T2D-associated genes & 425 rat T2D-related genes".

negative correlation values were often flipped to positive values [12], set to zero [31], or crushed to a very narrow region on the right of zero[16,19], and these operations hindered coexpression reversals from emerging. In reality, coexpression reversal probably has biological significance. Taking the coexpression of p53 and Klf4 as an example, it was recently reported that the positive or negative correlation between these two genes determines the outcome of DNA damage—DNA repair or apoptosis [24]. We believe our attention to this special coexpression change will help to explore subtle mechanisms involved in tuning of molecular balances between opposite factors.

6.2.6 RE-ANALYZING A T2D DATASET FROM THE PERSPECTIVE OF DIFFERENTIAL COEXPRESSION

For an application of the novel DCEA methods, we downloaded dataset GSE3068 from the Gene Expression Omnibus (GEO) database. GSE3068 was designed to study type 2 diabetes (T2D) in rats. It involves 6,955 probesets interrogating 4,765 genes, and the twenty samples herein were divided equally into a T2D group and a normal group. Details on preprocessing this dataset are in "Additional File 4: Preprocessing GSE3068".

We applied DCp to GSE3068 to identify differentially expressed genes (DCGs) and obtained 337 (p-values cutoff 0.05, FDR < 65%) DCGs out of 4765 genes. We listed the 337 DCGs in "Additional File 5: 337 DCGs identified by DCp" regarding their dC scores, log fold changes, and potential relevance with T2D (T2D-associated or T2D-related genes are provided in Additional File 6: 52 T2D-associated genes & 425 rat T2D-related genes). The DCGs with T2D relevance deserving more attention were selected and shown in Table 3.

We then identified DCLs using DCe methods, and narrowed them down to 4046 DCLs that were connected with the 337 DCGs (Additional File 7: DCLs identified by DCe). As we believed that correlation-reversal was a noteworthy but neglected type of differential correlation, we took a close-up look at those correlation-reversed DCLs. Out of a total of 110 reversed DCLs (Additional File 8: network modules organized by solely correlation reversals), 73 were connected with the 337 DCGs. Figure 5 shows three subnetwork modules organized solely by reversed links. Subnetwork A (Figure 5A) and B (Figure 5B) included quite a number of previously reported T2D-related genes: Tcf4 and Dcc [32]; Cd3d [33], Uts2r [34] and Map2k1 [35]. Subnetwork C (Figure 5C) was the largest reversed DCL-organized module and it contained an interesting four-gene-circuit (including Arpc5l, Tra1, Mcm3ap, and Hspe1) of consistent negative-to-positive correlation reversal. Although not being previously reported to be related with T2D, the genes and reversed links included in Figure 5C, as well as other novel cases reported in the supplementary tables of DCGs and DCLs (Additional File 5: "337 DCGs identified by DCp.xls" and Additional File 7: "DCLs identified by DCe.xls"), should receive adequate attention for their distinct traits from the perspective of differential coexpression. Further studies on the transcriptional mechanisms and functional consequences involved in these DCGs and DCLs would be helpful for elucidating how the changed coordination contributes to the pathogenesis of T2D.

Since GSE3068 had been thoroughly analyzed from the differential expression perspective [36], we investigated the relationship between the two gene expression properties, differential expression (DE) and differential coexpression (DC) in this dataset. We first examined the consistency between the DCp-identified 337 DCGs and the 119 previously reported DEGs, of which 36 could be corrected by oral administration of vanadyl sulfate (VS). It was found that the overlapping of DCGs with the 119 DEGs was not significant (hypergeometric test p = 0.22), but that with the 36 VS-corrected DEGs was significant (hypergeometric test p = 0.01). This indicates that differential expression and differential coexpression are somewhat related to each other at least in the T2D context.

We then looked at the disagreement between DCGs and DEGs. A previous differential coexpression analysis on human cancer using the ASC method reported a quite low level of overlapping between DCEA and DEA results (3%) [12]. In our case, DCGs and DEGs had only 3% (DCGs) or 8.4% (DEGs) in common, with the rest majority genes in disagreement. For instance, both Pparg and Tspan8 had been found to play key roles in T2D pathogenesis [37-39], but they were identified by DCEA and DEA respectively. Pparg had an expression fold change of 1.05 in the GSE3068 dataset (Table 3), so its relevance with T2D was not discerned by DEA.

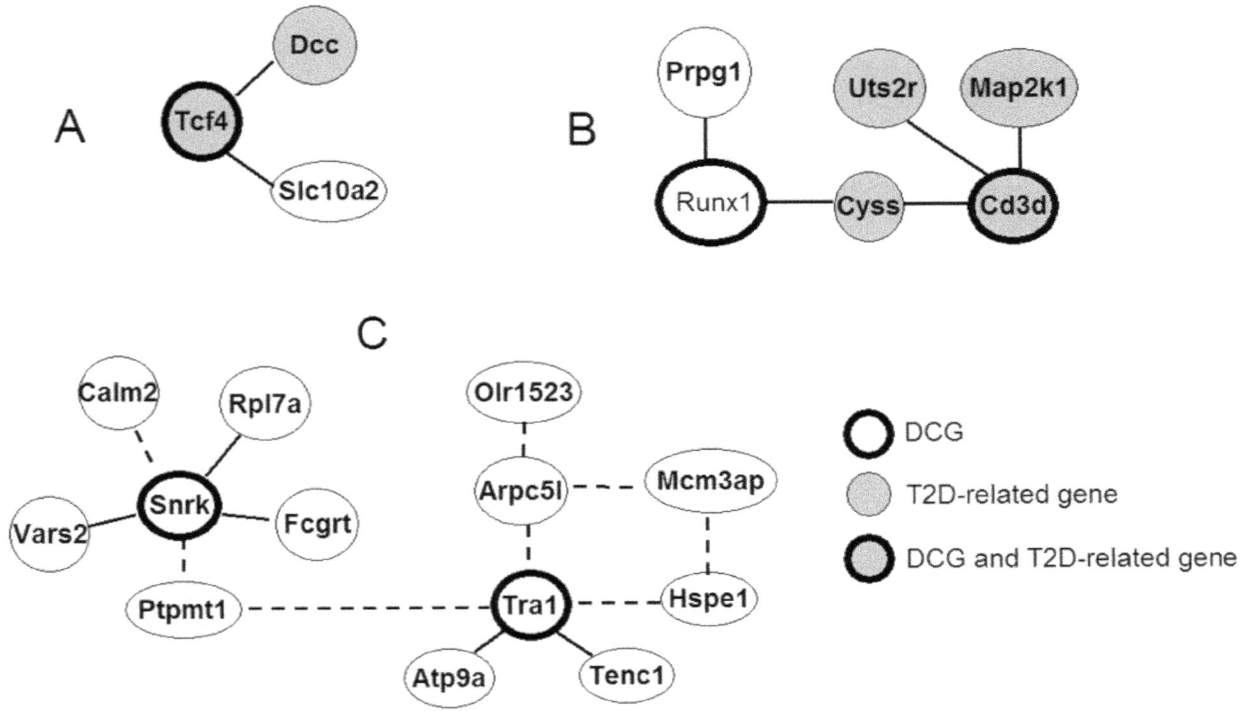

FIGURE 5: Genes were organized into network modules via correlation-reversal relationships. A solid link connects a pair of genes with positive correlation in normal state but negative in disease state, while a dashed link connects genes with negative correlation in normal state but positive in disease state.

From the perspective of DC, however, it stood out since its dC value (1.17) was ranked 28th of all 4765 genes. On the contrary, Tspan8, with a large expression change (2.3) but a minor coexpression change (dC = 0.57), was identified as a DEG but not a DCG.

According to our brief comparison at the gene level, DEA and DCEA are both powerful techniques to find out useful information from expression data. They are significantly related and mutually complementary. Similar conclusion was made at the pathway level based on the observed interplay of gene differential expression and gene differential coexpression in mouse mammary gland tumor [40].

6.3 CONCLUSIONS

In this work, we pointed out the critical weakness of current popular differential coexpression analysis methods, and developed two novel link-based algorithms, DCp and DCe. DCp and DCe differ from previous methods primarily in that they are designed to make use of link-specific correlation change values directly while previous methods mainly focuses on the gene connectivity. A novel strategy to filter links in coexpression networks, the half-thresholding strategy, is also proposed as a necessary pre-processing step of the two novel methods.

Based on the results from a series of rigorously designed simulation experiments, we proved that reasonably designed DCEA methods were able to discriminate pre-set differentially regulated genes and links; in another word, we provided a preliminary support to the anticipation that DCEA methods are capable of deciphering differential regulation or differential networking underpinning diseases [13]. Of the five DCEA methods we surveyed, we proved the overall performances of our DCp and DCe against three existing algorithms, and identified WGCNA as the best of the existing three. It is noticeable that while the existing methods were somewhat comparable to link-based methods in case of pure regulation-elimination perturbations, they were significantly outperformed when regulation-switch perturbations were introduced. Regulation-switch is believed to be an relevant phenomenon in fine-tuning of signal transduction [24].

Applying DCp and DCe to a real expression dataset designed for T2D study, we identified 337 DCGs and their associated 4046 DCLs, which may serve as a useful resource for identification and characterization of T2D relevant genes. We also analyzed the relationship between DEA and DCEA in this example, and pointed out that DEA and DCEA are significantly related and mutually complementary techniques to make discoveries from expression data.

Recently, differential coexpression analysis is being appreciated as a significant step towards the differential networking analysis of complex diseases [41], and the area of DCEA is undergoing rapid development as various solutions to set-wise differential coexpression problems are being proposed [20,21,42]. We believe that our methodological improvement will benefit the development of DCEA and help extend it to a broader spectrum of biomedical studies.

REFERENCES

1. Rachlin J, Cohen DD, Cantor C, Kasif S: Biological context networks: a mosaic view of the interactome. Mol Syst Biol 2006, 2:66.
2. Han JD, Bertin N, Hao T, Goldberg DS, Berriz GF, Zhang LV, Dupuy D, Walhout AJ, Cusick ME, Roth FP, et al.: Evidence for dynamically organized modularity in the yeast protein-protein interaction network. Nature 2004, 430(6995):88-93.
3. Huang Y, Li H, Hu H, Yan X, Waterman MS, Huang H, Zhou XJ: Systematic discovery of functional modules and context-specific functional annotation of human genome. Bioinformatics 2007, 23(13):i222-229.
4. Oldham MC, Horvath S, Geschwind DH: Conservation and evolution of gene coexpression networks in human and chimpanzee brains. Proc Natl Acad Sci USA 2006, 103(47):17973-17978.
5. Butte AJ, Tamayo P, Slonim D, Golub TR, Kohanel sS: Discovering functional relationships between RNA expression and chemotherapeutic susceptibility using relevance networks. PNAS 2000, 97(22):12182-12186.
6. Lee HK, Hsu AK, Sajdak J, Qin J, Pavlidis P: Coexpression analysis of human genes across many microarray data sets. Genome Res 2004, 14(6):1085-1094.
7. Stuart JM, Segal E, Koller D, Kim SK: A gene-coexpression network for global discovery of conserved genetic modules. Science 2003, 302(5643):249-255.
8. Zhou XJ, Kao MC, Huang H, Wong A, Nunez-Iglesias J, Primig M, Aparicio OM, Finch CE, Morgan TE, Wong WH: Functional annotation and network reconstruction through cross-platform integration of microarray data. Nat Biotechnol 2005, 23(2):238-243.

9. Obayashi T, Hayashi S, Shibaoka M, Saeki M, Ohta H, Kinoshita K: COXPRESdb: a database of coexpressed gene networks in mammals. Nucleic Acids Res 2008, (36 Database):D77-82.

10. D'Haeseleer P, Liang S, Somogyi R: Genetic network inference: from co-expression clustering to reverse engineering. Bioinformatics 2000, 16(8):707-726.

11. Carter SL, Brechbuhler CM, Griffin M, Bond AT: Gene co-expression network topology provides a framework for molecular characterization of cellular state. Bioinformatics 2004, 20(14):2242-2250.

12. Choi JK, Yu U, Yoo OJ, Kim S: Differential coexpression analysis using microarray data and its application to human cancer. Bioinformatics 2005, 21(24):4348-4355.

13. de la Fuente A: From 'differential expression' to 'differential networking' - identification of dysfunctional regulatory networks in diseases. Trends Genet 2010, 26(7):326-333.

14. Hudson NJ, Reverter A, Dalrymple BP: A differential wiring analysis of expression data correctly identifies the gene containing the causal mutation. PLoS Comput Biol 2009, 5(5):e1000382.

15. Reverter A, Ingham A, Lehnert SA, Tan SH, Wang Y, Ratnakumar A, Dalrymple BP: Simultaneous identification of differential gene expression and connectivity in inflammation, adipogenesis and cancer. Bioinformatics 2006, 22(19):2396-2404.

16. Mason MJ, Fan G, Plath K, Zhou Q, Horvath S: Signed weighted gene co-expression network analysis of transcriptional regulation in murine embryonic stem cells. BMC Genomics 2009, 10:327.

17. Fuller TF, Ghazalpour A, Aten JE, Drake TA, Lusis AJ, Horvath S: Weighted gene coexpression network analysis strategies applied to mouse weight. Mamm Genome 2007, 18(6-7):463-472.

18. van Nas A, Guhathakurta D, Wang SS, Yehya N, Horvath S, Zhang B, Ingram-Drake L, Chaudhuri G, Schadt EE, Drake TA, et al.: Elucidating the role of gonadal hormones in sexually dimorphic gene coexpression networks. Endocrinology 2009, 150(3):1235-1249.

19. Zhang B, Horvath S: A general framework for weighted gene co-expression network analysis. Stat Appl Genet Mol Biol 2005., 4 Article17

20. Watson M: CoXpress: differential co-expression in gene expression data. BMC Bioinformatics 2006, 7:509.

21. Cho SB, Kim J, Kim JH: Identifying set-wise differential co-expression in gene expression microarray data. BMC Bioinformatics 2009, 10:109.

22. Yu H, Yu F-D, Zhang G-Q, Shen X, Chen Y-Q, Li Y-Y, Li Y-X: DBH2H: vertebrate head-to-head gene pairs annotated at genomic and post-genomic levels. Database 2009, 2009:bap006.

23. Li YY, Yu H, Guo ZM, Guo TQ, Tu K, Li YX: Systematic analysis of head-to-head gene organization: evolutionary conservation and potential biological relevance. PLoS Comput Biol 2006, 2(7):e74.

24. Zhou Q, Hong Y, Zhan Q, Shen Y, Liu Z: Role for Kruppel-like factor 4 in determining the outcome of p53 response to DNA damage. Cancer Res 2009, 69(21):8284-8292.

25. Zhou L, Lopes JE, Chong MM, Ivanov II, Min R, Victora GD, Shen Y, Du J, Rubtsov YP, Rudensky AY, et al.: TGF-beta-induced Foxp3 inhibits T(H)17

cell differentiation by antagonizing RORgammat function. Nature 2008, 453(7192):236-240.

26. Liu BH, Yu H, Tu K, Li C, Li YX, Li YY: DCGL: an R package for identifying differentially coexpressed genes and links from gene expression microarray data. Bioinformatics 2010, 26(20):2637-2638.

27. Benjamini Y, Hochberg Y: Controlling the false discovery rate: a practical and powerful approach to multiple testing. Journal of the Royal Statistical Society Series B 1995, 57:289-300.

28. Mutch DM, Berger A, Mansourian R, Rytz A, Roberts MA: The limit fold change model: a practical approach for selecting differentially expressed genes from microarray data. BMC Bioinformatics 2002, 3:17.

29. Zhang B, Li H, Riggins RB, Zhan M, Xuan J, Zhang Z, Hoffman EP, Clarke R, Wang Y: Differential dependency network analysis to identify condition-specific topological changes in biological networks. Bioinformatics 2009, 25(4):526-532.

30. Van den Bulcke T, Van Leemput K, Naudts B, van Remortel P, Ma H, Verschoren A, De Moor B, Marchal K: SynTReN: a generator of synthetic gene expression data for design and analysis of structure learning algorithms. BMC Bioinformatics 2006, 7:43.

31. Southworth LK, Owen AB, Kim SK: Aging mice show a decreasing correlation of gene expression within genetic modules. PLoS Genet 2009, 5(12):e1000776.

32. Hollis-Moffatt JE, Hook SM, Merriman TR: Colocalization of mouse autoimmune diabetes loci Idd21.1 and Idd21.2 with IDDM6 (human) and Iddm3 (rat). Diabetes 2005, 54(9):2820-2825.

33. Ghabanbasani MZ, Buyse I, Legius E, Decorte R, Marynen P, Bouillon R, Cassiman JJ: Possible association of CD3 and CD4 polymorphisms with insulin-dependent diabetes mellitus (IDDM). Clin Exp Immunol 1994, 97(3):517-521.

34. Sidharta PN, Wagner FD, Bohnemeier H, Jungnik A, Halabi A, Krahenbuhl S, Chadha-Boreham H, Dingemanse J: Pharmacodynamics and pharmacokinetics of the urotensin II receptor antagonist palosuran in macroalbuminuric, diabetic patients. Clin Pharmacol Ther 2006, 80(3):246-256.

35. Hancock AM, Witonsky DB, Gordon AS, Eshel G, Pritchard JK, Coop G, Di Rienzo A: Adaptations to climate in candidate genes for common metabolic disorders. PLoS Genet 2008, 4(2):e32.

36. Willsky GR, Chi LH, Liang Y, Gaile DP, Hu Z, Crans DC: Diabetes-altered gene expression in rat skeletal muscle corrected by oral administration of vanadyl sulfate. Physiol Genomics 2006, 26(3):192-201.

37. Prokopenko I, McCarthy MI, Lindgren CM: Type 2 diabetes: new genes, new understanding. Trends Genet 2008, 24(12):613-621.

38. Altshuler D, Hirschhorn JN, Klannemark M, Lindgren CM, Vohl MC, Nemesh J, Lane CR, Schaffner SF, Bolk S, Brewer C, et al.: The common PPARgamma Pro-12Ala polymorphism is associated with decreased risk of type 2 diabetes. Nat Genet 2000, 26(1):76-80.

39. Grarup N, Andersen G, Krarup NT, Albrechtsen A, Schmitz O, Jorgensen T, Borch-Johnsen K, Hansen T, Pedersen O: Association testing of novel type 2 diabetes risk alleles in the JAZF1, CDC123/CAMK1D, TSPAN8, THADA, ADAMTS9, and NOTCH2 loci with insulin release, insulin sensitivity, and obesity in a popu-

lation-based sample of 4,516 glucose-tolerant middle-aged Danes. Diabetes 2008, 57(9):2534-2540.

40. Mentzen WI, Floris M, de la Fuente A: Dissecting the dynamics of dysregulation of cellular processes in mouse mammary gland tumor. BMC Genomics 2009, 10:601.

41. Fuente Adl: From 'differential expression' to 'differential networking' - identification of dysfunctional regulatory networks in diseases. Trends Genet 2010, 26(7):326-333.

42. Choi Y, Kendziorski C: Statistical methods for gene set co-expression analysis. Bioinformatics 2009, 25(21):2780-2786.

There are several supplemental files that are not available in this version of the article. To view this additional information, please use the citation information cited on the first page of this chapter.

CHAPTER 7

DIMENSION REDUCTION WITH GENE EXPRESSION DATA USING TARGETED VARIABLE IMPORTANCE MEASUREMENT

HUI WANG AND MARK J. VAN DER LAAN

7.1 BACKGROUND

Gene expression microarray data are typically characterized by large quantities of variables with unknown correlation structures [1,2]. This high dimensionality has presented us challenges in analyzing the data, especially when correlations among variables are complex. Including many variables in standard statistical analyses can easily cause problems such as singularity and overfitting, and sometimes is not even doable. To manage this problem, the dimensionality of the data will often be reduced in the first step. There are multiple ways to achieve this goal. One is to select a subset of genes based on certain criteria such that this subset of genes is believed to best predict the outcome. This gene selection strategy is typically based on some univariate measurement related to the outcome, such as t-test and rank test [3,4]. Another strategy is to use a weighted combination of genes of lower dimension to represent the total variation of the data. Representa-

This chapter was originally published under the Creative Commons Attribution License. Wang H and van der Laan MJ. Dimension Reduction with Gene Expression Data Using Targeted Variable Importance Measurement. BMC Bioinformatics *12,312 (2011), doi:10.1186/1471-2105-12-312.*

tive approaches are principle component analysis (PCA) [5] and partial least squares (SLR) [6-9]. Machine learning algorithms such as LASSO [10,11] and Random Forest [12] have embedded capacity to select variables while simultaneously making predictions, and can be used to accommodate high dimensional microarray data.

As always, there is no one-size-fits-all solution to this problem, and one often needs to resort to a mix-and-match strategy. The univariate-measurement based gene selection is a very popular approach in the field. It is fast and scales easily to the dimension of the data. The output is usually stable and easy to understand, and fulfills the objectives of the biologists to directly pursue interesting findings. However, it often relies on over-simplified models. For instance, the univariate analysis evaluates every gene in isolation of others, with the unrealistic assumption of independence among genes. As a result, it carries a lot of noise and the selected genes are often highly correlated, which themselves create problems in subsequent analysis. Also, due to the practical limit of the size of the gene subset, real informative genes with weaker signals will be left out. In contrast, PCA/PLS constructs a few gene components as linear combinations of all genes in a dataset. This "Super Gene" approach assumes that the majority of the variation in the dataset can be explained by a small number of underlying variables. One then uses these gene components to predict the outcome. These approaches can better handle the dependent structure of genes and their performances are quite acceptable [13]. But it is harder to interpret gene components biologically, and to assess the effect of individual genes one needs to look at the weight coefficients of the linear combination. Machine learning algorithms are very attractive variable selection tools to deal with large quantities of genes. They are prediction algorithms with embedded abilities to select gene subsets. However, whether or not a gene is chosen by a learning algorithm may not be the best measurement of its importance. Machine learning algorithms are constructed to achieve an optimal prediction accuracy, which often overlooks the importance of each variable. Consequently, small changes in data or tuning parameters may result in big changes in variable rankings and the the selected gene subsets are instable. For example, Random Forest, a tree-based non-parametric method, has a variable importance measurement that greatly contributes to its popularity. This measurement is sensitive to the parameter choices of trees in the presence of high correlations among variables, because different sets of variables can produce nearly unchanged prediction accuracy [14,15]. Another example is LASSO—one of the most popular regularization algorithms. Assuming a sparse signal, LASSO handles the high dimensionality problem by shrinking the coefficients of most variables towards zero [16]. A recent implementation of LASSO is in the GLM-NET R package [17]. The package uses a coordinate descent algorithm and can finish an analysis of 20,000 variables within a few seconds. To us, its result is somewhat sensitive to the choice of the penalizing parameter λ. Different λs may result in gene subsets with little overlapping. In the mean time, variable importance measurements are not readily available in LASSO. One can simply rank genes by their coefficients, but this can be quite subtle. Although permutation tests may be used to derive p-values, how to perform the permutation is a tricky matter due to selection of tuning parameters. For small p-values, it is still computationally infeasible. In this paper, inspired by concepts of counterfactual effects from the causal inference literature, we propose a targeted variable importance measurement [18,19] to rank genes and reduce the dimensionality of the dataset. Counterfactuals are usually defined in the context of treatment to disease. It is the outcome a patient would have had a treatment been assigned differently, with everything else held the same. Hence counterfactuals are "counter"-fact and apparently impossible to be observed. But it can be estimated statistically. Suppose that we have an outcome Y, a binary treatment A, and the confounding variables W of A, and we have worked out correctly an estimate $\hat{E}$ of the conditional expectation of Y given A and W. A common way to estimate the counterfactual effect of A is to compute the difference between the $\hat{E}(Y_i|A_i = 1, W_i)$ and the $\hat{E}(Y_i|A_i = 0, W_i)$ for every observation and then average over all observations, referred to as the G-computation method [20]. Although counterfactuals may not be completely relevant to gene microarray data, thinking about the data in this way is very helpful for us to assess the importance of a gene. Our VIM definition uses the concepts of counterfactuals and the estimation framework is built on the methodology of targeted maximum likelihood estimation (TMLE) [21]. By tailoring this recently developed technique specifically to gene expression data, we hope to introduce to the community an alternative strategy to carry out gene selection in addition to current methods. Our approach takes the advantage of prediction power of learning algorithms while targeting at the individual importance of each variable. Its mathematical property has been studied in [22], and we will focus on its application. In brief, our approach consists of two-stages. In the first stage, we predict the outcome given all genes. In the second stage, we improve the first stage by modeling the mechanism between an individual gene and its confounding variables. Both stages can be very flexible ranging from using univariate analysis to refined learning algorithms. When machine learning algorithms are used, we have the flexibility to determine how to make predictions without restricting ourselves to explicit models and distributions. In the meanwhile, as in the case of the univariate analysis, we return to a simple and well interpretable measure of the importance of each gene. This importance measurement is derived in the presence of the confounding variables of a gene, and hence can help to exploit the redundant information among correlated genes. It is generally also more stable than the variable importance produced by machine learning algorithms. In addition, our approach provides a simple way for statistical inference based on asymptotic theories, and is well suited for the exploratory analysis of microarray data.

7.2 METHODS

Suppose the observed data are i.i.d. $O_i = (Y_i, A_i, W_i)$, where Y is a continuous outcome, A is the gene of interest, W is a set of confounding variables of A, and i = 1,..., n indexes the observation. Let $\Psi(a)$ represents the variable importance measurement (VIM) of A. One can define the VIM of A as the marginal effect of A on the outcome Y at value A = a relative to A = 0 adjusted for W, and then averaged over the distribution of W [18]:

$$\Psi(a) = E_W[E(Y|A = a, W) - E(Y|A = 0, W)]$$

Consider the semiparametric regression model:

$$E(Y|A, W) = \beta A + F(W)$$

where f(W) is a function of W. With this parameterization, we have $\Psi(a) = \beta a$. We can then view β as an index of the VIM of A. In the above model, the only assumption we make is the linearity of A. The definition of the VIM of A is closely related to the definition of the counterfactual effect in causal inference [23]. Although β can not be directly interpreted as an causal effect without proper assumptions [19], it serves well as a surrogate of the magnitude of the causal relationship between the outcome and a gene. The motivation of this parameterization is that by selecting more causally related genes, the resulting prediction function will be better generalized to new experiments with the same causal relation between the outcome Y and A, but a different joint distribution of W. If in a next experiment, the technology or the sampling population is somewhat different, but the causal mechanism is still the same, then a prediction function that uses the correlates of the true causal variables will perform poorly while a prediction function using the true causal variables will still perform nicely. This idea will be illustrated in our simulations.

Our goal is to estimate β. In [22], this estimation problem was addressed in the framework of targeted maximum likelihood estimation (TMLE). TMLE is an estimating equation and efficient estimation theory based methodology [24], and is particularly useful when it comes to semiparametric models. Estimators from the traditional method such as MLE perform well for parametric models, however, they are generally biased relative to their variances especially when the model space is large. This is because the MLE focuses on doing a good job on the estimation of the whole density rather than on the parameter itself. TMLE is designed to achieve an optimal trade-off between the bias and the variance of the estimator. It uses an MLE framework, but instead of estimating the overall density, TMLE targets on the parameter of interest and produces estimators minimally affected by changes of the nuisance parameters in a model. In Additional File 1 we provide a brief overview of this methodology with a demo simulation example. The formal mathematical formulation of TMLE can be found in the original paper by van der Laan and Rubin [21]. The implementation of TMLE to estimate β is fairly simple and consists of two stages. First, we estimate $E(Y | A, W)$ without any parametric restriction. We then regress the residual of Y and $E(Y | A = 0, W)$ onto βA to conform with our semiparametric regression model. This will yield an initial estimator of β and fitted values of $E(Y | A, W)$, denoted by $\beta_n^{(0)}$ and the $Q_n^{(0)}$. In the second stage, we update these initial estimates in a direction targeted at β. This involves regressing the residuals of Y and the fitted $Q_n^{(0)}$ on the clever covariate $A - E(A|W)$. The $E(A|W)$ evaluates the confounding of A with W, and we name it the "gene confounding mechanism". It needs to be estimated if unknown. Let us denote the coefficient before the clever covariate as ε. The updated TMLE estimate of β is $\beta_n^{(0)} + \varepsilon_n$, where ε_n is the estimated value of ε. The variance estimate of β can be computed from its efficient influence curve. Below is a step-by-step implementation of our algorithm, and we refer to it as the TMLE-VIM procedure.

1. Obtain the initial estimator $Q_n^{(0)}$ and $\beta_n^{(0)}$. Use your favorite algorithm here, for example, linear regression, LASSO, Random Forest, etc.

2. Obtain the $g_n(W)$ estimate for the gene confounding mechanism $E(A|W)$. As in the case of $Q_n^{(0)}$, a broad spectrum of algorithms can be used. In this paper, we use LASSO (in the GLMNET R package) for its optimal speed.

3. Compute the "clever covariate":

$$r(A,W) = A - g_n(W)$$

4. Fit regression $Y' = Y - Q_n^{(0)} \sim \varepsilon r(A,W)$.
5. Update the initial estimate $\beta_n^{(0)}$ with

$$\beta_n^{(1)} = \beta_n^{(0)} + \varepsilon_n$$

and update the initial fitted values $Q_n^{(0)}$ with

$$Q_n^{(1)} = Q_n^{(0)} + \varepsilon_n r(A, W)$$

6. Compute the variance estimate σ_n^2 for $\beta_n^{(1)}$ according to its efficient influence curve:

$$\sigma_n^2 = \frac{\sum_i r(A_i, W_i)^2 (Y_i - Q_n^{(1)})^2}{(\sum_i r(A_i, W_i) A_i)^2}$$

where i indexes the i-th observation.
7. Construct the test statistic:

$$T(A) = \frac{\beta_n^{(1)}}{\sigma_n}$$

$T(A)$ follows the standard Gaussian distribution under the null hypothesis $\beta = 0$ when the sample size n goes to infinity.

The TMLE estimator $\beta_n^{(1)}$ is a consistent estimator of β when either the $Q_n^{(0)}$ or the $g_n(W)$ is consistent. When the $Q_n^{(0)}$ is consistent, it is also asymptotically efficient. The derivations of the clever covariate, the efficient influence curve, the TMLE estimate and its mathematical properties can be found in [22] and [18]. Upon the construction of the test statistic, a p-value can be calculated for the adjusted marginal effect of A and used as an index of the variable importance.

In the application to dimension reduction, for each variable in the dataset, we compute a TMLE-VIM p-value. We then reduce our variable space based on these p-values. There are two notions. First, in principle, a separate initial estimator $Q_n^{(0)}$ should be fitted for every gene A by forcing A as a term in the algorithm used. This can become quite time consuming. To solve the problem, instead of estimating $E(Y|A, W)$ for each A, we obtain a grand estimate $G_n(V)$ for $E(Y|V)$. Here V represents all variables in the dataset. Then for every A in V, we carry out the regression $Y \sim \beta A$ with the offset $G_n(A = 0)$ to get $\beta_n(0)$ and $Q_n^{(0)}$. Second, when obtaining the $g_n(W)$, we want to be attentive to how closely W is correlated with A. The independence between W and A results in zero adjustment to the initial estimator, while a complete association causes β to be unidentifiable. A simple option is to use all variables less than a pre-defined correlation with A as W. In [22], they authors suggest 0.7 as a conservative threshold. Instead of applying a universal cutoff, we can also set individualized correlation threshold for each A. Below we provide a data adaptive procedure to do it. One first defines a sequence of correlation cutoffs δ. For each choice of δ, one computes the corresponding TMLE p-value for A. One then sets a p-value threshold λ, and chooses the maximum δ that has produced a p-value less than λ. The degree of the protection is determined with the value of λ. In general, the smaller the λ is, the more the protection. The value of λ can be either fixed a priori or chosen by cross validation. We refer to it as the TMLE-VIM(λ) procedure. It allows us to adjust for the confounding in the dataset adaptively and flexibly, and protect the algorithm against the harm from high correlations among variables. It works best when many variables are closely correlated in a complex way. However, it does require more time to run, especially when λ needs to be chosen by cross validation. In many cases, a universal cutoff of 0.7 will work fine. In Additional File 1 we provide the mathematical formulation of the TMLE-VIM(λ) procedure. Once we have all the variables ranked by their p-values. the candidate list can be truncated by either applying a p-value threshold or taking the top k ranked variables. Both of them are sound practices. In our simulations and data analysis, we usually truncate the list at a p-value threshold 0.05.

7.3 RESULTS AND DISCUSSION

7.3.1 SIMULATION STUDIES

We performed two sets of simulations. The first set of simulations investigates how TMLE-VIM responds to changes in the number of confounding variables, the correlation level among variables, and the noise levels.

The second set studies the TMLE-VIM with more complex correlation structures and model misspecification. The performance of the dimension reduction procedure was primarily evaluated by the achieved prediction accuracy using a prediction algorithm on the reduced sets of variables, illustrated in the following analysis flow:

Compute VIM $\rightarrow$ Reduced variables $\rightarrow$ Prediction Algorithm

Two prediction algorithms, LASSO and D/S/A (Deletion/Substitution/ Addition) [25], were used. D/S/A searches through the variable space and selects the best subset of covariates by minimizing the cross validated residual sum of squares. In our simulations, LASSO and D/S/A predictions are often similar. We used D/S/A in simulation I as it provides convenience to count what variables are included in the prediction model. LASSO was used in simulation II for its faster speed. We also used multivariate linear regression (MVR) as a comparison to machine learning algorithms when applicable.

7.3.2 PART I

In simulation I, we varied the number of non-causal variables (W), the correlation coefficient ρ among variables, and the noise level σ_e^2 to see how TMLE-VIM responds to them. For each simulated observation $O_i =$ (Y_i, A_i, W_i), where i indexes the i-th sample, the outcome Y_i was generated from a main effect model of 25 As:

$$Y_i = 2 \sum_{j=1}^{25} a_{ji} + e_j$$

where j indexes the j-th A, and e_i is a normal error with mean 0 and variance σ_e^2. Each A_j was correlated with m Ws, and hence the total number of Ws is $m_w = 25$ m. A_j and its associated W s were jointly sampled from a multivariate Gaussian distribution with mean 0 and variance-covariance matrix S, where S is a correlation matrix with an exchangeable correlation coefficient ρ. This simulation scheme resulted in 25 independent clusters of covariates. Within each cluster, the covariates are correlated at level ρ.

Simulations were run for combinations of:

- $m = (10, 20)$ corresponding to $m_w = (250, 500)$;
- $\sigma_e = (1, 5, 10)$;
- and $\rho = (0.1, 0.3, 0.5, 0.7, 0.9)$.

For each combination, we simulated a training set of 500 data points and a testing set of 5000 data points. The training set was used to obtain the prediction model while the testing set was used to calculate the L2 risk. We also calculated a cross-examined L2 risk using a testing set with a ρ other than that of the training set. This is to demonstrate that by identifying more causally related variables, TMLE-VIM is robust to the change of the joint distribution among the covariates As and Ws. In specific, for each prediction model obtained from a training set, we calculated the L2 risk on the testing set generated with $\rho = 0.1$ regardless of what ρ was used to generate the training set. As a benchmark, we also used univariate regression in parallel with TMLE-VIM to reduce the dimensionality of the dataset, denoted with UR-VIM. Once the variable importance was calculated, we cut short the variable list using a p-value threshold 0.05. Each combination was replicated 10 times and results took the average.

TMLE-VIM used LASSO to obtain both the initial estimator $Q_n^{(0)}$ and the gene confounding mechanism estimator $g_n(W)$. In the $g_n(W)$, W was all the variables excluding A. TMLE-VIM has demonstrated a consistent advantage over UR-VIM with respect to the final prediction error over a range of simulation settings. This is particularly the case when the joint distribution of the covariates changes and when predictions were made by MVR that lacks internal capacity of model selection. Smaller σ_e, larger m_w, and larger ρ tend to magnify this advantage. Also, TMLE-VIM risks have smaller standard errors than the UR-VIM risks. In Table 1, we present our simulation results for five different ρ values and two different mw values, with σ_e^2 fixed at 5. The following summary quantities are reported:

TABLE 1: The simulation I results

ρ		$m_w = 250$		$m_w = 500$	
		MVR	DSA	MVR	DSA
0.1	R_r	**0.2341** ; 0.2251	**0.2436** ; 0.2351	**0.4035** ; 0.3784	**0.4230** ; 0.3943
	R_A	1.0870	-	1.2136	-
	R_w	0.6522	-	0.8846	-
	RR_{DSA}	na	1.0130	na	1.0680
0.3	R_r	**0.2202** ; 0.2297	**0.2231** ; 0.2247	**0.2341** ; 0.2307	**0.2027** ; 0.1975
	R_A	1.0776	-	1.0684	-
	R_w	0.1528	-	0.0958	-
	RR_{DSA}	na	1.0345	na	1.0299
0.5	R_r	**0.2425** ; 0.3115	**0.1169** ; 0.1285	**0.4883** ; 0.5959	**0.1268** ; 0.1217
	R_A	1.0373	-	1.0331	-
	R_w	0.0355	-	0.0149	-
	RR_{DSA}	na	1.0251	na	1.0335
0.7	R_r	**0.3599** ; 0.5872	**0.1307** ; 0.2545	**0.8001** ; 0.9093	**0.1740** ; 0.2976
	R_A	1.0081	-	1.0000	-
	R_w	0.0275	-	0.0162	-
	RR_{DSA}	na	1.0693	na	1.1055
0.9	R_r	**0.2262** ; 0.7248	**-0.1364** ; 0.2805	**0.9390** ; 0.9885	**-0.5498** ; 0.2657
	R_A	0.8415	-	0.5502	-
	R_w	0.0364	-	0.0204	-
	RR_{DSA}	na	1.2630	na	1.6103

Bold fonts: testing set (a). Italic fonts: testing set (b).
na: not available. -: the same value as the previous entry.

- R_r = (UR-VIM risk - TMLE-VIM risk)/UR risk: the proportion of the risk reduction of TMLE-VIM relative to the UR-VIM risk. It measures by how much TMLE-VIM outperforms UR-VIM. The bigger the number, the more the advantage.
- R_A = TMLE-VIM N_A/UR-VIM N_A: the ratio of the number of As (N_A) in the TMLE-VIM list to the number of As in the UR-VIM list.
- R_w = TMLE-VIM N_w/UR-VIM N_w : the ratio of the number of Ws (N_w) in the TMLE-VIM list to the number of Ws in the UR-VIM list.

- RR_{DSA} = TMLE-VIM P_A/UR-VIM P_A: the ratio of the proportion of As (P_A) in the final D/S/A prediction model resulted from the TMLE-VIM procedure to that from the UR-VIM. It measures the relative chance of arriving at a truly associated variable in the final model through the path of TMLE-VIM, referenced to the UR-VIM.

The R_r was calculated on two different testing sets. One is the testing set generated with the same ρ as the corresponding training set, and we refer it to "testing set (a)"; the other is the testing set generated with ρ = 0.1, and we refer it to "testing set (b)". Testing set (a) shares the same correlation structure as the training set, while in testing set (b) all the variables are essentially independent of each other. Testing set (b) is a simple representation of the scenario that when a new experiment is conducted the overall joint distribution of the covariates changes while the causal mechanism remains the same. In Table 1, the bold R_r was calculated on testing set (a), and the Italic R_r was on testing set (b). We make a few points here about Table 1:

- The proportion of the risk reduction (R_r) of the TMLE-VIM relative to the UR-VIM is typically more than 20% for the MVR prediction and 10% for the D/S/A prediction. In some cases, the risk reduction of the MVR can be very significant. For example, when m_w = 500 and ρ = 0.7, the TMLE-VIM risk is close to only half of the UR-VIM risk. TMLE-VIM tends to deliver more advantages when m_w = 500 than when m_w = 250. When the correlation coefficient ρ increases, the TMLE-VIM performs increasingly better than the UR-VIM for the MVR prediction. For the D/S/A prediction, small or large ρs seem to benefit most from the TMLE-VIM. For intermediate ρ, the benefit is still there but reduced. We believe that how much the risk can be reduced by the TMLE-VIM is a combination of factors such as the number of As and Ws in the reduced candidate list, the correlation structures among covariates and the internal optimization procedures of D/S/A. The advantage of the TMLE-VIM over the UR-VIM does seem to be more significant on the testing set (b) than the testing set (a), in support of our hypothesis that by identifying more causally related variables the TMLE-VIM results generalize better to new experiments.
- Most R_A values are slightly higher than 1 while the R_W values are much smaller. This indicates that on average, in the TMLE-VIM list, the number of correctly identified As is slightly higher than that in the UR-VIM list, while the number of falsely associated W_s is much less. It is especially the case when the correlations are high among variables. The low counts of false positives is a major contributing factor that the prediction made on the TMLE-VIM candidate list is better than that on the UR-VIM.

- As to the number of As that are finally made into the D/S/A prediction model, the TMLE-VIM in most cases displays a slight advantage over the UR-VIM. A closer look reveals that the variables included in the D/S/A model only differs by one or two between the TMLE-VIM and the UR-VIM. But the prediction risk has a measurable difference. This probably implies that every single variable counts in making good predictions in these simulations.
- When $\rho = 0.9$, the situation seems to be losing its track. The TMLE-VIM did worse than the UR-VIM in terms of correctly identified variables as well as the prediction risk of the testing set (a). Considering the high correlations among variables, this could possibly be attributed to the overfitting in the $g_n(W)$. Indeed, in [22], the authors showed that TMLE deteriorates when adjusting for variables with correlation coefficients beyond 0.7. However, the RR_{DSA} indicates that the chance of including a correct variable in the final D/S/A model based on the TMLE-VIM list is higher than that on the UR-VIM. Further looking into the data, we found out that the number of As that made into the D/S/A model from the TMLE-VIM list is actually greater than that from the UR-VIM, while the number of Ws is much less. Henceforth, the D/S/A model built on the TMLE-VIM list is closer to truth, but somehow its prediction is worse than the model built on the UR-VIM list. This seems to suggest that when provided with the UR-VIM list, the D/S/A has offset its model for the missed As from highly correlated Ws, while for the TMLE-VIM, this can not be done since there are not many Ws in the list. It is the same reason that the UR-VIM underperforms the TMLE-VIM on the testing set (b) when those surrogates of As were lost. For the MVR, although the TMLE-VIM shows a dominant advantage over the UR-VIM with respect to the prediction accuracy, the TMLE-VIM only identified 77% (when $m_w = 250$) and 57% (when $m_w = 500$) of the As identified by the UR-VIM. The better prediction is merely due to the fact that the MVR breaks down when too many variables entered the model. This is particularly the case when $m_w = 500$.

Figure 1 presents a graphical representation of a typical example in simulation I with ($\sigma_e = 5$, $m_w = 250$, $\rho = 0.7$). Besides the advantage displayed by the TMLE-VIM relative to the UR-VIM, we also see much smaller differences between the TMLE-VIM risks of the testing set (a) and (b) compared to the UR-VIM because TMLE-VIM was able to detect more As. In summary, when confounder are properly adjusted in $g_n(W)$, TMLE-VIM improves not only the performance of relatively simple algorithms such as the MVR, but also the more complex learning algorithms with built-in capacities of variable selection. Interested readers can find all the original prediction risks and counts of As and Ws and their standard errors in Additional File 2 for this simulation.

FIGURE 1: A typical example in simulation I. This graph presents the average L2 risk of the final prediction model on the candidate list from the UR-VIM and the TMLE-VIM, for simulation I data with setup ($\sigma_e = 5$, $m_w = 10$, $\rho = 0.7$). In the left panel, the MVR risk is plotted against a series of p-value thresholds used to truncate the candidate list; the right panel plots the D/S/A risk. The testing set (a) predictions are grouped in solid blue lines, and the testing set (b) predictions are grouped in broken orange lines. Dots represent the UR-VIM, and triangles represent the TMLE-VIM.

7.3.3 PART II

Simulation II examines the TMLE-VIM on larger-scale datasets with much more complex correlation structures. The simulation consists of 500 samples and 1000 variables. We used a correlation matrix derived from the top 800 genes in a real dataset published in [26]. For these genes, the median absolute correlation coefficient was centered at 0.26, the 1st/3rd quartile being 0.16/0.37, and the maximum as high as 0.9977. Hence, simulation II tried to mimic the correlation structure in this real data set. The outcome Y was generated from two different models using 20 As. One is a linear model, and the other is polynomial.

Details of this simulation is provided in the Additional File 1. A test dataset of 5000 points were simulated to assess the L2 prediction risk. We repeated the simulation for 10 times and results took the average. In TMLE-VIM, we tried two different initial estimators. One is the univariate regression as simple as $Y \sim A$, and the other is the LASSO estimator. LASSO was also used to get the $g_n(W)$ and to make the final predictions. we adjusted universally in the $g_n(W)$ for the variables that are correlated with A with an arbitrary correlation coefficient less than 0.7. All the $Q_n^{(0)}$ and $g_n(W)$ models were main-term linear. Hence, with the polynomial outcome, we could examine how TMLE-VIM performs when mis-specified models were provided. To summarize the result, we computed a R^2 quantity, representing the proportion of explained variance relative to an intercept model. It is defined as 1 - mean risk/MST, where MST = $(1/n)$ $\Sigma_i(Y_i - \bar{Y})^2$ and $\bar{Y}$ is the mean of Y. Table 2 lists the R^2 and the number of true positives (T.P.) and false positives (F.P.) in the reduced list of candidate variables, for the UR-VIM, the TMLE - VIM($Q_n^{(0)}$ = UR), and the TMLE - VIM($Q_n^{(0)}$ = LASSO). Compared to UR-VIM, TMLE-VIM improved the prediction risk by providing LASSO a candidate list with more truly and less falsely associated variables for both the linear and polynomial simulations. It is worth noting that even with an initial estimator as simple as the univariate regression ($Q_n^{(0)}$ = UR), TMLE-VIM still achieves a significant increase in R^2 by modeling the $g_n(W)$.

TABLE 2: The simulation II results (p-value)

	Simulation					
	Linear			Polynomial		
	R^2	T.P	F.P.	R^2	T.P.	F.P.
UR-VIM	0.2887	13.8	605.3	0.1851	13.4	555.9
TMLE - VIM(Q_n^0 = UR)	0.4849	16.6	280.5	0.3245	14.7	255.5
TMLE - VIM(Q_n^0 = LASSO)	0.6289	19.7	29.1	0.4203	17.9	24
TMLE-VIM(λ)	0.6479	20	41.6	0.4498	19.2	105.9

The candidate variable list contains all variables with p-values less than 0.05.

The numbers in Table 2 were based on candidate lists that were cut short with a p-value threshold of 0.05. In Table 3, we provide the results based on the top 100 ranked genes. The numbers of UR-VIM and the TMLE - VIM ($Q_n^{(0)}$ = UR) are less satisfying than those in Table 2, while the TMLE - VIM($Q_n^{(0)}$ = LASSO) achieved comparable results. This suggests that the TMLE - VIM($Q_n^{(0)}$ = LASSO) p-values of As are among the smallest ones, and shortening the length of the list does not affect the final result. Regardless of the weakened results, The TMLE - VIM($Q_n^{(0)}$ = UR) still displays a non-ignorable advantage over the UR-VIM with respect to the prediction accuracy, while the number of correctly identified As is slightly smaller than that of the UR-VIM. We then looked at the correlation matrix among the top 100 selected genes, and it occurs that the correlation among them is the least for the TMLE - VIM($Q_n^{(0)}$ = LASSO), the most for the UR-VIM, and the TMLE - VIM($Q_n^{(0)}$ = UR) lies in between. This could explain why the TMLE - VIM($Q_n^{(0)}$ = UR) does a better job in prediction regardless of less As.

We also carried out the TMLE-VIM(λ) procedure with LASSO as the initial estimator, allowing the data select the correlation cutoff for variables to be adjusted in the $g_n(W)$. Results are also reported in Table 2 and Table 3. TMLE-VIM(λ) identified more As but also more Ws, and the prediction accuracy is only slightly improved. On the other hand, the correlations among the selected top 100 variables are quite small. It seems by data adaptively adjusting for the correlation levels in the $g_n(W)$, TMLE-VIM(λ) returns a more independent set of genes. The actual risks and standard errors are contained in Additional File 2.

TABLE 3: The simulation II results (top 100)

	Simulation					
	Linear			Polynomial		
	R^2	T.P	cor.	R^2	T.P.	cor.
UR-VIM	0.1444	9.0	0.2956	0.0862	8.2	0.3642
TMLE - VIM(Q_n^0 = UR)	0.1907	8.8	0.2534	0.1605	7.2	0.2590
TMLE - VIM(Q_n^0 = LASSO	0.6059	19.9	0.2289	0.4132	19.2	0.2234
TMLE-VIM(λ)	0.5916	20	0.1242	0.3859	17.7	0.0867

The candidate variable list contains the top 100 variables ranked by their p-values.

7.4 DATA ANALYSIS

Breast cancer patients are often put on chemotherapy after the surgical removal of the tumor. However not all patients will respond to chemotherapy, and proper guidance for selecting the optimal regimen is needed. Gene expression data have the potential for such predictions, as studied in [26]. The dataset from [26] contains the gene expression profiling on 22283 genes for 133 breast cancer patients. The outcome is the pathological complete response (pCR). This is a binary response associated with long-term cancer free survival. There are also 13 clinical variables collected in the dataset including the ER (estrogen receptor) status, which is a very significant clinical indicator for chemotherapy response.

The goal of the study is to select a set of genes that best predict the clinical response pCR. The first step is to reduce the number of genes worth of consideration, and we applied both UR-VIM and TMLE-VIM (with $Q^{(0)}$ = UR and $Q^{(0)}$ = LASSO) for this purpose. For the TMLE-VIM($Q^{(0)}$

= LASSO), the $Q_n^{(0)}$ was estimated by LASSO using the top 5000 ranked genes. We then took all the genes with the FDR-adjusted p-values less than 0.005 [27], as suggested in the original paper, and upon them we built a predictor using the Random Forest (tuning parameters mtry = number of variables/3, ntree = 3000 and nodesize = 1). The clinical covariates were treated in the same way as genes. To prevent the algorithm from breaking down, we only adjusted for the confounder with correlation coefficients less than 0.7 with A in the $g_n(W)$. We carried out a 10-fold honest cross validation. We divided the dataset into 10 subsets. Each subset was regarded as a validation set and the rest as the training set. We reperformed the entire analysis, i.e. VIM calculation → dimension reduction → Random Forest classifier, on all 10 training sets and predicted the outcome of the validation set using the classifier built on the training samples. We can then use these cross validated predictions to assess the true classification accuracy of our algorithm.

Analysis results are tabulated in Table 4. The UR-VIM produced a candidate list of 326 genes and one clinical variable the "ER status", while the list of the TMLE-VIM($Q^{(0)}$ = UR) consists of 660 genes and TMLE-VIM($Q^{(0)}$ = LASSO) 818 genes. The TMLE-VIM identified many more genes than the UR-VIM. Among all the identified genes, 429 overlap between the TMLE - VIM($Q_n^{(0)}$ = LASSO) and TMLE - VIM($Q_n^{(0)}$ = UR), 15 overlap between the UR-VIM and TMLE-VIM($Q^{(0)}$ = UR), 10 overlap between the UR-VIM and TMLE-VIM($Q^{(0)}$ = LASSO), and only 4 genes are shared among all three (please see Figure 2). The TMLE-VIM appeared to have selected almost a different set of genes than the UR-VIM.

TABLE 4: The analysis result of the breast cancer dataset

	Num. of genes in the candidate list	C.V. classification accuracy	Corr. level among the top 100 genes
UR-VIM	327	0.7669	0.43
TMLE-VIM($Q^{(0)}$ = UR)	660	0.7744	0.18
TMLE-VIM($Q^{(0)}$ = LASSO)	818	0.7744	0.21

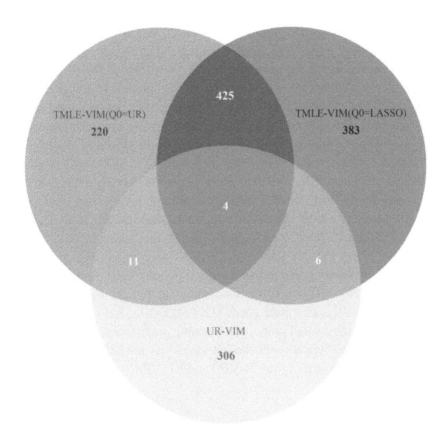

FIGURE 2: The venn diagram of the breast cancer data. This venn diagram shows the overlaps of identified candidate genes from the breast cancer dataset using the UR-VIM, the TMLE-VIM($Q^{(0)}$ = UR), and the TMLE-VIM($Q^{(0)}$ = LASSO).

The TMLE-VIM($Q^{(0)}$ = UR) and the TMLE-VIM($Q^{(0)}$ = LASSO) results are quite similar to each other regardless of the adequate difference between the initial estimators. It seems the modeling of the $g_n(W)$ had played a significant role and steered away the initial univariate estimates. Further investigation found out that genes in the UR-VIM list are highly correlated with the clinical indicator ER status, while the TMLE-VIM genes are not. Consequently, the TMLE-VIM genes are less correlated to each other than the UR-VIM genes. Looking at the first 100 ranked genes, the absolute median of the correlation coefficients for the UR-VIM is 0.43, while for the TMLE-VIM, it is about half of that number. Although the input variables to the Random Forest are different, The cross validated (CV) classification accuracy are quite similar among these three methods. We also passed all 22,000 genes to Random Forest and looked at its variable importance measurement. The Random Forest VIM (RF-VIM) is more similar to the UR-VIM: about 50% of them overlap but only a few overlap with the TMLE-VIM. The RF-VIM genes are also highly correlated with the ER status, albeit the less severity than the UR-VIM. Its OOB classification accuracy (0.7669) is comparable with all three other methods.

In summary, the UR-VIM and RF-VIM seemed to have identified genes that are strong predictors of the clinical variable ER status. The ER status is a strong indicator of the outcome pCR. Hence, the final prediction accuracy still seems quite good. The TMLE-VIM has identified a list of genes of which a small proportion is strong predictors of ER status and others are not associated with the ER status. Its prediction accuracy is slightly better than that of the UR-VIM and RF-VIM.

7.5 CONCLUSIONS

We have shown in this paper with extensive simulations that the TMLE based variable importance measurement can be incorporated into a dimension reduction procedure to improve the quality of the list of the candidate variables. It requires an initial estimator $Q_n^{(0)}$ and a gene confounding mechanism estimate gn(W). A consistent $Q_n^{(0)}$ ensures the consistency and the efficiency of the TMLE estimate. When $Q_n^{(0)}$ is not consistent, a correct specification of gn(W) can still produce consistent estimates while that estimate will not be efficient any more. We generally recommend to do as good a job as we can on obtaining the $Q_n^{(0)}$, as a better $Q_n^{(0)}$ means both a smaller bias and a smaller variance. Nevertheless, algorithms as simple as univariate regression are also valid choices, and in this case, we will rely solely on the goodness of $g_n(W)$. The computation of $g_n(W)$ directly affects the speed of the TMLE-VIM, as it has to be redone for every variable. Hence, one may want to choose an approach that is reasonably fast. In our study, we chose the GLMNET R Package as our primary tool to get $g_n(W)$, and it worked very well. In practice, one needs to balance the resources used for the initial estimator and the gene confounding mechanism. With a proper design of the two estimating stages, TMLE-VIM is a fairly fast procedure. It is also worth mentioning that the TMLE-VIM can sometimes be sensitive to the overfitting in the $Q_n^{(0)}$, and hence, caution needs to be exercised when choosing an aggressive algorithm.

A popular dimension reduction approach is the principle component analysis (PCA). The PCA computation does not involve the outcome, and so it could be less powerful when prediction is the primary goal. Its output is a linear combination of all the genes. Though not a gene selection approach, we still carried it out on our simulation I data as an interesting comparison to our approach. PCA demonstrates an intermediate performance with respect to the UR-VIM and the TMLE-VIM on small p-value cutoffs. This means a few top components carry all the prediction power. When the p-value cutoff is increased, and more components enter the candidate list, its results became quite unsatisfying. When the correlation structure changes among the genes, PCA has done a poor predicting job. The PCA results are contained in Additional File 3.

Usually, the reduced set of variables will serve as the input of a prediction algorithm to build a model. Such algorithms used in this article include MVR, LASSO, and D/S/A. We have noticed that in most of our simulations, the MVR prediction often achieves a similar risk as LASSO and D/S/A on the TMLE-VIM reduced set of variables. It suggests that further variable selection may not be necessary for the TMLE-VIM candidate list, and we can use simpler algorithms to get a good prediction. In fact, the TMLE-VIM can go beyond the scope of dimension reduction. It can be iteratively applied to the data until it converges to a list of several variables that are most likely to be causal to the outcome. In this case, one may want to use the Super Learner [28] as the prediction algorithm, which works more effectively with the TMLE-VIM. The Super Learner is an ensemble learner that combines predictions from multiple candidate learners with optimal weights. It has been shown in [29] that the Super Learner performs asymptotically equal to or better than any of its candidate learners. The Super Learner allows the data to objectively blend results from different algorithms rather than relying on a single algorithm chosen subjectively by an analyst. Hence it enjoys a greater flexibility to explore the model space and usually produces reasonable predictions consistently across a wide variety of datasets, and serves as a very good prediction algorithm for the TMLE-VIM. On the other hand, it is also more computationally demanding.

TMLE-VIM is a quite general approach. Besides gene expression data, TMLE-VIM can also be applied to genetic mapping problems. The genome-wide association studies (GWAS) can involve more than a million of genetic markers. In this case, only the univariate analysis seems to be feasible of ranking every marker. With the TMLE-VIM procedure, we can run more complex algorithms on a subset of top ranked markers, taking it as the initial estimator, and then evaluate every single marker. The variable importance of each marker is thus obtained through a multi-marker approach and being adjusted for its confounder. However, the GWAS in human beings is usually case-control data, and the current TMLE-VIM needs to be extended to accommodate such outcomes.

REFERENCES

1. West M, Blanchette C, Dressman H, Huang E, Ishida S, Spang R, Zuzan H, Olson JA, Marks J, J N: Predicting the Clinical Status of Human Breast Cancer using Gene Expression Profiles. Proc Natl Acad Sci USA 2001, 98:11462-11467.
2. Dudoit S, Fridlyand J, Speed T: Comparison of discrimination methods for the classification of tumors using gene expression data. JASA 2002, 97:77-87.
3. Hedenfalk I, Duggan D, Chen Y, Radmacher M, Bittner M, Simon R, Meltzer P, Gusterson B, Esteller M, Rafield M, Yakhini Z, A BD, Dougherty E, Kononen J, Bubendorf L, Fehrle W, Pittaluga S, Gruvberger D, Loman N, Johannsson O, Olsson H, Wilfond B, Sauter G, Kallioniemi O, Borg A, Trent J: Gene expression profiles in hereditary breast cancer. New England Journal of Medicine 2001, 244:539-548.
4. Dettling M, Buhlmann P: Boosting for tumor classification with gene expression data. Bioinformatics 2003, 19:1061-1069.

5. Ghosh D: Singular value decomposition regression modeling for classification of tumors from microarray experiments. Proceedings of the Pacific Symposium on Biocomputing 2002, :18-29.

6. Nguyen DV, Rocke DM: Tumor classification by partial least squares using microarray gene expression data. Bioinformatics 2002, 18:39-50.

7. Nguyen DV, Rocke DM: Multi-class cancer classification via partial least squares with gene expression profiles. Bioinformatics 2002, 18:1216-1226.

8. Huang X, W P: Linear regression and two-class classification with gene expression data. Bioinformatics 2003, 19:2072-2978.

9. Boulesteix A: PLS Dimension reduction for classification with microarray data. Statistical applications in genetics and molecular biology 2004, 3:1-33.

10. R T: Regression shrinkage and selection via the lasso. J Royal Statist Soc B 1996, 58:267-288.

11. Efron B, Hastie T, Johnstone I, R T: Least angle regression. Annals of Statistics 2004, 32:407-499.

12. L B: Random forests. Machine Learning 2001, 45:5-32.

13. Dai JJ, Lieu L, Rocke D: Dimension Reduction for Classification with Gene Expression Microarray Data. Statistical Applications in Genetics and Molecular Biology 2006, 5:Article 6.

14. Strobl C, Boulesteix AL, Zeileis A, Hothorn : Bias in random forest variable importance measures: illustrations, sources and a solution. BMC Bioinformatics 2007, 8:25.

15. Strobl C, Boulesteix AL, Kneib T, Augustin T, Zeileis A: Conditional variable importance for random forests. BMC Bioinformatics 2008, 9:307.

16. Rosset S, Zhu J, Hastie T: Boosting as a regularized path to a maximum margin classifier. Journal of Machine Learning Research 2004, 5:941-973.

17. Friedman J, Hastie T, Tibshirani R: Regularization paths for generalized linear models via coordinate descent. Journal of Statistical Software 2010., 33

18. Yu Z, van der Laan MJ: Measuring treatment effects using semiparametric models. [http://www.bepress.com/ucbbiostat/paper136] webcite U.C. Berkeley Division of Biostatistics Working Paper Series 2003.

19. van der Laan MJ: Statistical inference for variable importance. Int J Biostat 2006, 2:Article 2.

20. Robins JM: A new approach to causal inference in mortality studies with sustained exposure periods - application to control of the healthy worker survivor effect. Mathematical Modelling 1986, 7:1393-1512.

21. van der Laan MJ, Rubin DB: Targeted maximum likelihood learning. Int J Biostat 2006, 2:Article 11.

22. Tuglus C, van der Laan MJ: Targeted methods for biomarker discovery, the search for a standard. [http://www.bepress.com/ucbbiostat/paper233] webcite U.C. Berkeley Division of Biostatistics Working Paper Series 2008.

23. Rubin DB: Estimating Causal Effects of Treatments in Randomized and Nonrandomized Studies. Journal of Educational Psychology 1974, 66:688-701.

24. Bickel PJ, Klaassen CAJ, Ritove Y, Wellner JA: Efficient and adaptive estimation for semiparametric models. Baltimore: The Johns Hopkins University Press; 1993.

25. Sinisi SE, van der Laan MJ: Deletion/Substitution/Addition algorithm in learning with applications in genomics. Statistical Applications in Genetics and Molecular Biology 2004, 3:Article 18.

26. Hess KR, Anderson K, Symmans WF, Valero V, Ibrahim N, Mejia JA, Booser D, Theriault RL, Buzdar AU, Dempsey PJ, Rouzier R, Sneige N, Ross JS, Vidaurre T, Gomez HL, Hortobagyi GN, Pusztai L: predictor of sensitivity to preoperative chemotherapy with paclitaxel and fluorouracil, doxorubicin, and cyclophosphamide in breast cancer. J Clin Oncol 2006, 24:4236-4244.

27. Benjamini Y, Hochberg Y: Controlling the false discovery rate: a practical and powerful approach to multiple testing. J Royal Statist Soc B 1995, 57:289-300.

28. van der Laan MJ, Polley EC, Hubbard AE: Super Learner. Statistical Applications in Genetics and Molecular Biology 2007, 6:Article 25.

29. van der Laan MJ, Dudoit S, van der Vaart AW: The cross-validated adaptive epsilon-net estimator. Statistics and Decisions 2006, 24:373-395.

There are several supplemental files that are not available in this version of the article. To view this additional information, please use the citation information cited on the first page of this chapter.

PART III

GWAS

GENOME-WIDE ASSOCIATION STUDY OF STEVENS-JOHNSON SYNDROME AND TOXIC EPIDERMAL NECROLYSIS IN EUROPE

EMMANUELLE GÉNIN, MARTIN SCHUMACHER, JEAN-CLAUDE ROUJEAU, LUIGI NALDI, YVONNE LISS, RÉMI KAZMA, PEGGY SEKULA, ALAIN HOVNANIAN, AND MAJA MOCKENHAUPT

8.1 BACKGROUND

Adverse drug reactions are a major public health issue as they represent an important cause of morbidity and mortality [1]. Skin lesions are frequent expressions of adverse drug reactions with, for some drugs, up to 10% of cutaneous reactions observed [2,3]. SJS and TEN are severe cutaneous adverse reactions characterized by the development of acute exanthema which progresses towards limited (in SJS) or widespread (in TEN) blistering and erosion of the skin and mucous membranes [4]. SJS and TEN are thus considered to be two different forms of the same disease with TEN representing the most severe form [5]. The incidence of SJS/TEN is estimated to be of approximately 1-2 patients per million individuals per year

This chapter was originally published under the Creative Commons Attribution License. Génin E, Schumacher M, Roujeau JC, Naldi L, Liss Y, Kazma R, Sekula P, Hovnanian A, and Mockenhaupt M. Genome-Wide Association Study of Stevens-Johnson Syndrome and Toxic Epidermal Necrolysis in Europe. Orphanet Journal of Rare Diseases 6,52 (2011), doi:10.1186/1750-1172-6-52.

[6]. It is thus a very rare disease but with a high morbidity and mortality (reaching up to 45% in TEN) that requires intensive treatment.

SJS/TEN is not associated with a single drug or a single group of drugs but several different drugs have been involved. A limited number of drugs however are more often associated with the disease: antibacterial sulfonamides (especially sulfamethoxazole), allopurinol, which is the most frequent drug involved [7], carbamazepine, lamotrigine, phenobarbital, phenytoine, non-steroidal anti-inflammatory drugs (NSAIDs) of the oxicam type and neviparine [4,8]. Only a small number of individuals exposed to these "highly suspected" drugs develop the disease and a genetic susceptibility has been suggested [9-12].

An association with HLA was reported more than 20 years ago [12,13]. More recently, studies in the Han Chinese population have involved the HLA-B locus with very strong drug-specific associations: the HLA-B*1502 allele was found in all carbamazepine-induced SJS/TEN patients [14] and the HLA-B*5801 in all allopurinol-induced SJS/TEN patients [15]. An investigation of HLA-B associations in European samples did not detect the association between the HLA-B*1502 allele and carbamazepine-induced SJS/TEN but did report a strong but not complete association between HLA-B*5801 and the allopurinol-induced disease [16,17]. Furthermore, it has also been suggested that HLA genetic predisposition may not be the same for SJS and TEN and that it might thus be important to take into account the disease severity in association tests [18].

Apart from HLA, several other candidate genes have been tested for association with SJS/TEN, mostly genes involved in the immune response, in the inflammation process or in drug metabolism [18-21]. However, no association with these genes has been consistently found and it has been suggested that rather than focusing on candidate genes, a genome-wide association study might provide more insights into the genetic susceptibility of adverse drug reactions [21].

In the context of the RegiSCAR project (European Registry of Severe Cutaneous Adverse Reactions to Drugs and Collection of Biological Samples) DNA of 563 cases of SJS/TEN was collected. This is the largest available sample of SJS/TEN patients in the world with accurate medical information regarding the severity of the disease and the history of drug intake. These patients were genotyped at the Centre National de Genotypage (CNG) using Illumina 317 K chips and a Genome Wide Association Study (GWAS) was conducted against controls selected from the CNG European Reference Control Panel [22].

8.2 METHODS

8.2.1 PATIENTS

A total of 563 cases (226 males and 337 females, sex-ratio = 0.67) were included in this study. They were collected as part of the "European Registry of Severe Cutaneous Adverse Reactions" (RegiSCAR) (see http://regiscar. uni-freiburg.de/) in six countries (Austria (1 case), France (184 cases), Germany (331 cases), Israel (14), Italy (26) and The Netherlands (7)). All of them had a diagnosis of SJS (268 cases, 48%), SJS/TEN overlap (181 cases, 32%) or TEN (114 cases, 20%) validated as probable or definitive by the expert committee of RegiSCAR blindly from information on drug exposure.

For each patient, written informed consent was obtained and a blood sample was taken for genomic DNA extraction that was carried out at the CEPH-Fondation Jean Dausset (France) (http://www.cephb.fr) until 2005 and at the biobank of the CIC-Henri Mondor / Créteil after 2005.

The 563 patients were genotyped on Illumina 317 K chips in France at the CNG (http://www.cng.fr) for a total of 318,127 SNPs (among which 309,091 were located on autosomes). Stringent quality-control (QC) was performed using Plink v1.06 [23] that led to the exclusion of 68 out of the 563 cases for low genotyping (MIND > 0.05) and 15,088 SNPs (14,343 autosomal SNPs) for missing data (GENO > 0.05). After QC, a total of 495 affected individuals were considered in the analysis for whom genotypes on 303,039 SNPs (294,748 located on autosomes) were available. The total genotyping rate in these 495 remaining individuals was 0.95.

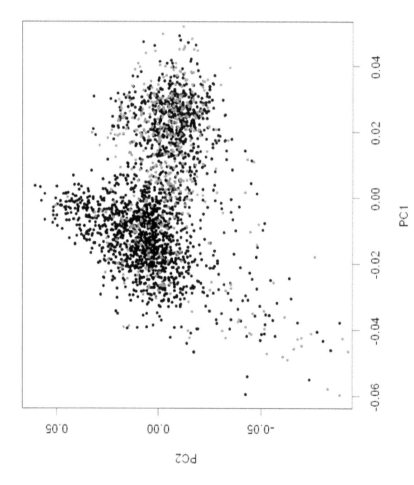

FIGURE 1: Plots of top two principal components for cases (in gray) and controls (in black).

8.2.2 CONTROLS

Controls were selected from the CNG European Reference Panel collected at the CNG [22]. This panel includes 5,847 unrelated individuals from 13 European countries who are all genotyped on Illumina 317 K chips. Since patients collected as part of the RegiSCAR project were mostly from France and Germany, we decided to consider only the 1,881 individuals from the Reference Control panel originating from these two countries (653 from Germany and 1,228 from France).

8.2.3 QUALITY CONTROL AND PRINCIPAL COMPONENT ANALYSIS OF GENOTYPE DATA

A second quality control of the data was performed where markers that were missing in more than 5% of the controls or had significant different missing rates between cases and controls were removed. A total of 25,834 autosomal markers were excluded.

To assess the level of population stratification in the sample, a principal component analysis (PCA) was performed using the genotypes of the 495 cases and 1,881 controls for a sub-panel of 35,232 markers obtained after pruning based on linkage disequilibrium using Plink (option indeppairwise 50 5 0.1). The program smartPCA from the Eigenstrat package [24,25] was used with the default options. A total of 71 patients were excluded from the study as they were found to be outliers. Those were mostly individuals with a suspected African or Asian ancestry. Plots of the first two principal components (PCs) are provided in Figure 1.

As shown in the flowchart in Figure 2, a total of 268,914 autosomal SNPs were kept for the association testing in a sample of 424 SJS/TEN patients and 1,881 controls.

8.2.4 DRUG EXPOSURE

Information on drug exposure was collected by interviewing patients about their drug intakes and by consulting medical records. Reactions were considered to be potentially caused by allopurinol, carbamazepine or phenytoin if the drug has been taken between 4 and 10 days before the onset of the disease and if the drug has not been started more than 42 days before onset. This resulted in a group of 57 patients (13.4%) with potentially allopurinol-induced reactions, a group of 25 patients (5.9%) with potentially carbamazepine-induced reactions and a group of 19 patients (4.5%) with potentially phenytoin-induced reactions. Because of the low number in the latter two groups we decided to only compare the allopurinol group and the group with other drugs to the controls in this study. Note that because of the stringent time criterion used to determine allopurinol as culprit drug there might be some patients with an allopurinol-induced reaction left in the group of patients with a culprit drug other than allopurinol. For this reason, this analysis was further enforced by an additional analysis where any allopurinol exposed patient (irrelevant of time of usage) was assigned to the allopurinol group resulting in a higher sensitivity with respect to allopurinol-induced reaction while the specificity is decreased.

8.2.5 ASSOCIATION TESTING

Association was tested using the PC-corrected Armitage-trend test implemented in Eigenstrat [24]. A sensitivity analysis was performed to determine the number of PCs that should be accounted for and we decided to adjust only on the first two PCs (the genomic control coefficient lambda [26] was 1.023 when adjusting on the top 2 PCs, 1.019 when adjusting on the top 10 PCs and 1.016 when adjusting on the top 20).

Odds-ratios (OR) and confidence intervals of the different genotypes were computed using the logistic regression model implemented in Plink v1.06 [23].

To study the influence of the drug involved in the disease, a second study was performed where the multinomial logistic model implemented in Stata v10 [27] was used as in [28]. The disease status D was coded in 3 classes: 0 unaffected (controls), 1 affected with a suspected allopurinol-induced disease, 2 affected with another suspected drug induced disease. At each SNP, an additive model was considered and two ORs were

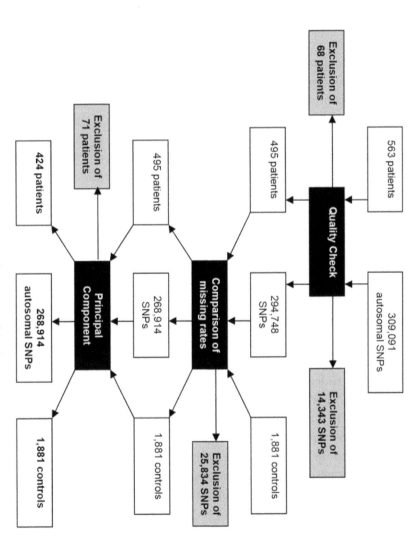

FIGURE 2: Flowchart summarizing the different steps of the quality control of the data.

computed to model the effect of the SNP among allopurinol-induced affected individuals ($OR_{allopurinol}$) and among other affected individuals ($OR_{otherdrugs}$). The null hypothesis tested was one of no effect of the SNP in any of the strata, i.e. $OR_{allopurinol} = OR_{otherdrugs} = 1$. It was tested using a 2 degrees-of-freedom chi-square test. The significance of the difference between $OR_{allopurinol}$ and $OR_{otherdrugs}$ was also tested in this model using the corresponding one degree-of-freedom chi-square test.

In the same way, a multinomial logistic model was used to determine whether genetic differences exist depending on the disease severity. A four class disease status was used: 0 unaffected (controls), 1 SJS, 2 SJS/TEN overlap and 3 TEN.

Furthermore, haplotype association was tested using Plink v1.06 [23].

For all the logistic regression modelling and haplotype association test-ing, an adjustment on the first two PCs obtained from the PCA of the genotype data was used.

8.3 RESULTS

The clinical details of the 424 patients remaining after quality-control are summarized in Table 1: 61.3% are female, 47.2% have a SJS phenotype, 34.4% a SJS/TEN overlap phenotype and 18.4% a TEN phenotype. A ma-jority of the patients was sampled from France (110 patients) and Ger-many (277 patients). History of drug intake was carefully monitored and a subgroup of 57 patients (13.4%) with potentially allopurinol-induced reactions was identified. There is no difference in phenotype distribution between males and females but there are some differences depending on the drug and the country where samples were collected. There is a lack of TEN among allopurinol-induced patients (test of homogeneity of the three disease forms in allopurinol induced versus non-allopurinol induced partients: chi-square = 7.89, p-value = 0.019) and a lack of TEN among patients from Germany as compared to patients from France (chi-square = 23.68, p-value = 7.2 10-6). The two variables however are not independent since allopurinol is more used in Germany than in some other countries and in particular France.

TABLE 1: Description of the sub-sample of 424 patients selected for the analysis

Number of patients	SJS	SJS/TEN	TEN	Total
Total	200 (47.2%)	146 (34.4%)	78 (18.4%)	424
Females	124	89	47	260 (61.3%)
Allopurinol-induced	33	21	3	57 (13.4%)
From France	41	32	37	110 (25.9%)
From Germany	145	97	35	277 (65.3%)
From Other countries	14	17	6	37 (8.7%)

The number of patients (proportion of the total of 424 patients in percents) with the different phenotypes is presented in the total sample, among the female patients and as a function of the country where patients were identified.

For association testing, the 424 patients were compared to the 1,881 controls from the reference panel originating from France (1,228 individuals) and Germany (653 individuals) who were also the closest controls to the cases on the PCA plot (Figure 1). We then tested each SNP for association controlling on the first two PCs of the PCA using Eigenstrat [24]. Results are presented in Figure 3. One genome-wide significant signal was detected in the MHC region on chromosome 6 where 6 SNPs showed False Discovery Rate adjusted q-values below 5% (see Figure 4 and Table 2). The most significant SNP, rs9469003, is located at position 31,515,807, ~85 kb upstream of the HLA-B locus and has an OR of 1.73 (95%CI = [1.44;2.08]). The five other SNPs are all located in a more telomeric region, 250 kb apart from rs9469003, in between positions 31,114,834 and 31,250,224 (see Figure 5). There is very limited Linkage Disequilibrium (LD) between these five SNPs and the top one and the six SNPs defined 13 haplotypes with a frequency of 1% or above. The haplotype CACGAC formed by the risk allele at each locus showed the strongest association with the disease with an OR of 2.84 (95%CI = [2.03; 3.98]) that is significantly higher than the one observed for rs9469003 only. To dissect further this haplotypic association, the haplotype conditioning test implemented in Plink v1.06 was used and we found that none of the 6 SNPs by itself has an independent effect nor explains the haplotype association.

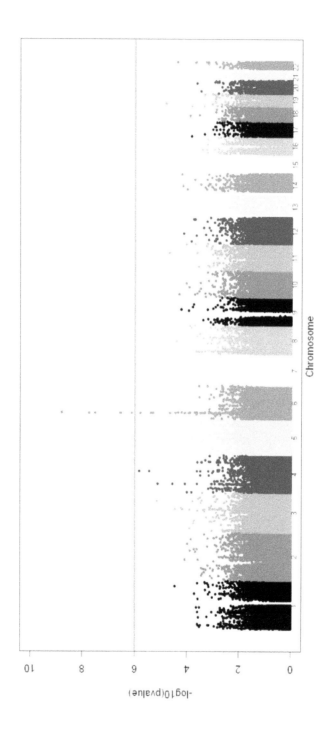

FIGURE 3: Results of the genome-wide association screen. The -log(p-value) of the PC-corrected Armitage-trend test implemented in Eigenstrat are plotted for the 268,818 autosomal SNPs that passed the QC with the different colors representing the different chromosome. The horizontal line represents the 10-6 p-value threshold that also corresponds to a False Discovery Rate q-value of 5%.

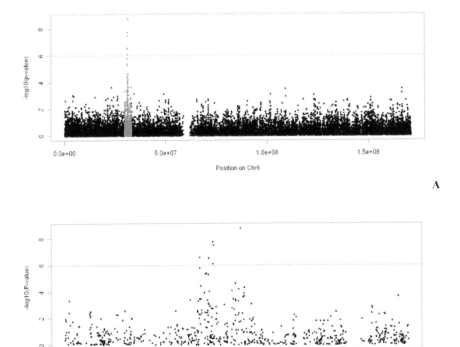

A

B

FIGURE 4: Results of the association test on chromosome 6. A The -log(p-value) of the PC-corrected Armitage-trend test implemented in Eigenstrat are plotted for the 18,1278 SNPs located on chromosome 6 as a function of their position on the chromosome in base pairs (Build 36). SNPs located in the HLA region are highlighted in red. B Detail of the association signals for the 706 SNPs located in the HLA region. The horizontal line represents the 10^{-6} p-value threshold.

TABLE 2: List of most significant SNPs associated with SJS/TEN

SNP	Chr	Position Build 36	RiskAllele (Other Allele)	Risk Allele Frequency in Cases	Risk Allele Frequency in Controls	PC corrected Trend test[b]	p-value[b]	q-value[c]	Odds-Ratio[d]	95% CI[d]	Pvalue of HW test	Annotation[a]
rs2844665	6	31114834	C (T)	0.72	0.62	26.46	2.69×10^{-7}	1.48×10^{-2}	1.54	1.30-1.82	0.77	C6orf205 flanking_3UTR
rs3815087	6	31201566	A (G)	0.31	0.21	26.32	2.89×10^{-7}	1.54×10^{-2}	1.53	1.29-1.80	0.08	PSORS1C1 UTR
rs3130931	6	31242867	C (T)	0.77	0.69	24.41	7.78×10^{-7}	3.34×10^{-2}	1.54	1.29-1.84	0.66	POU5F1 Intron
rs3130501	6	31244432	G (A)	0.83	0.74	31.78	1.73×10^{-8}	0.22×10^{-2}	1.74	1.43-2.13	0.59	POU5F1 Intron
rs3094188	6	31250224	A (C)	0.72	0.63	30.75	2.93×10^{-8}	0.26×10^{-2}	1.59	1.34-1.88	0.06	POU5F1 Flanking_5UTR
rs9469003	6	31515807	C (T)	0.24	0.15	36.41	1.60×10^{-9}	0.04×10^{-2}	1.73	1.44-2.08	0.003	HCP5 flanking_5UTR

a Annotations were obtained from the Illumina annotation file.
b Results obtained using the test implemented in Eigenstrat
c q-values (FDR adjusted p-values) were obtained by using the R package fdrtool [50].
d Results obtained using plink logistic regression test accounting for the top 2 PC.

Using a multinomial logistic regression approach, we found that the association with haplotype CACGAC was much stronger in the subgroup of patients with an allopurinol induced disease ($OR_{allopurinol}$ = 7.77, 95%CI = [4.66; 12.98]) than in other patients ($OR_{otherdrugs}$ = 1.92, 95%CI = [1.40;2.64]). The equality of odds-ratio ($OR_{allopurinol}$ = $OR_{otherdrugs}$) was strongly rejected (p = 6.56×10^{-7}). Results were the same in France and Germany a shown in Additional File 1 Tables S1 and S2.

To determine how much of the observed signal could be explained by the known association between allopurinol-induced SJS/TEN and HLA-B*5801, we took advantage of the availability of HLA-B two-digit resolution genotypes for 74 of the 424 patients [16]. Among the 74 HLA-B genotyped patients, 11 were carriers of an HLA-B58 allele and had an allopurinol-induced disease. The frequency of the CACGAC haplotype is increased in HLA-B58 carriers as compared to non HLA-B58 carriers (32.44% versus 10.44%, p-value of the test adjusted on the top 2 PC = 0.0052) but the linkage disequilibrium is not complete. Note however that given the small sample size of HLA-B typed individuals, the power was quite low to detect this difference (power of 60% at a type-one error rate level of 5%).

8.4 DISCUSSION

Discovering genes involved in severe cutaneous adverse reactions and especially SJS/TEN is a major challenge for pharmacogenetics as these reactions, when not fatal, are a sword of Damocles for those who already had the disease and fear to take any drug. Despite important efforts, only genes located in the HLA region have been identified so far. One possible explanation for this lack of success is the limited sample sizes of patients available. Indeed SJS/TEN is fortunately a very rare disease and in most studies, sample sizes rarely exceed one or two hundreds of patients, making it difficult to investigate more than a few candidate genes. Through a collaborative effort, the RegiSCAR group was able to collect detailed medical information and DNA of more than half a thousand of patients from Europe who were genotyped on Illumina 317 K chips. By comparing their genotypes to the ones of 1,881 controls genetically matched for

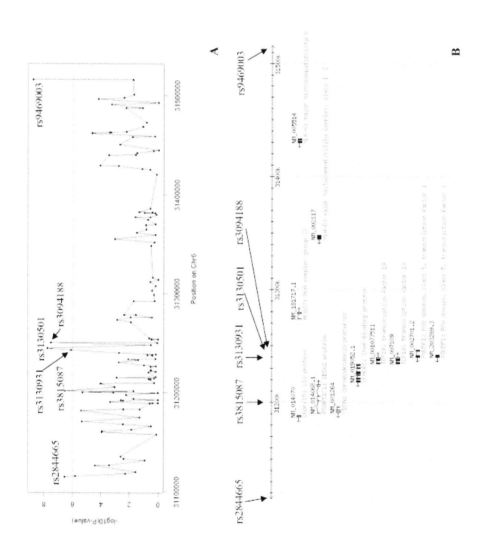

FIGURE 5: Association results in the 401 kb region encompassing the top six signals. A The -log(p-value) of the PC-corrected Armitage-trend test are plotted against the chromosome 6 positions for the 126 SNPs located in the region. B Listing of genes located in this region based on the HapMap web-browser.

the country of origin, we were able to study the SJS/TEN association at a genome-wide level on European samples. Apart from six SNPs located in the HLA region, no other locus was found to be associated with the disease at a high enough significance level to ensure it is not false positive due to the multiple tests performed. Sample sizes of half a thousand are not enough to ensure a good power to detect common variants with small effects (OR in the range between 1.1 and 1.3) similar to those identified in several common diseases such as diabetes or different cancers [29] where larger samples are easier to collect. However, given the 424 patients remaining after Quality Control, the study was powerful enough to detect common variants with modest effects (the power exceeded 80% to detect variants with an allele frequency above 15% conferring ORs above 1.7 under a multiplicative model). The fact that only the HLA region is detected, suggests that there might not be any other common variant that confer a substantial increase in disease risk and could thus be of interest as a predictive factor. These power computations might however be too optimistic as they are based on the assumption that the same genetic variants might be involved in cases of SJS/TEN associated to different drugs. If we consider that there might exist some levels of genetic heterogeneity depending on the drugs involved as it is the case for HLA associations, the power can be dramatically reduced. For example, considering only the 57 allopurinol-exposed cases, the power to detect a variant of frequency 15% conferring an OR of 2 is then only 5% at a nominal type-one error rate value of 10^{-6}. On the other hand, the power to detect a similar effect than the one observed at rs9469003 (i.e., risk allele of frequency of 15% with an associated OR of ~4 among allopurinol-induced SJS/TEN) is 99%, showing that the study was not underpowered to detect strong effects even if they were restricted to a small subset of patients.

It is true however that we might have missed some variants with important effects, especially those with minor allele frequencies below 5% that are not well covered by SNP-chips and it could thus be of interest to investigate the association with rare variants. If the "common disease-common variant hypothesis" was believed to explain the susceptibility to multifactorial diseases [30,31], results from GWAS have shown that common variants only explain a minor part of most common disease heritabilities. It has been suggested that rare variants could explain at least part of this "missing heritability" and indeed, rare variants have been found in several diseases [32-36]. Interestingly these rare variants are often functional variants with a direct impact on the protein functionality and they usually confer a much stronger increase in disease risk than common variants. They are also more likely to be affected by some moderate levels of negative selection [37,38]. For diseases such as adverse drug reactions, it is not unlikely that the genes and variants involved could be under selection. Indeed if most of the drugs have been introduced too recently to be directly responsible for selective pressures, they are often derived from nutrients that have been consumed by humans for a long time. Thus, genes involved in drug metabolism can potentially be involved in natural selection and this was confirmed by previous studies [39-41]. However, the implication of drug metabolism genes in SJS/TEN remains to be established.

Concerning the association with HLA detected in this sample, we confirmed that it is drug-specific with the strongest association found for the group of patients where allopurinol is suspected to be the cause of the disease. However the disease association was still detectable after exclusion of all patients exposed to allopurinol (whether considered as causing the reaction or not) (OR = 2.13, 95% CI = [1.41; 3.23]). This suggests that different HLA alleles are probably involved depending on the drug, and the 6 SNPs identified could be those that were in linkage disequilibrium (LD) with several of these HLA alleles. To further explore this hypothesis, we tried to impute HLA-B genotypes from the SNP data [42] but we were not successful because of the presence of multiple rare alleles at this locus such as HLA-B58 itself that is only present in one of the European families genotyped by de Bakker et al. [43] precluding the possibility to find appropriate tags for this allele using this sample. To overcome the problem, we tried to determine tag SNPs using our own HLA-B genotyped sample that was enriched in HLA-B58 but we did not find any SNP or group of SNPs that was correlated enough with the HLA-B58 allele (the maximum r^2 value was below 20%). This poor HLA-B tagging ability of the SNPs could certainly explain why the association found here in relation with allopurinol is much weaker than the one reported in studies on European samples using resolved HLA-B alleles [16,44]. However, even in these two latter studies, the association with HLA-B*5801 is much weaker than the one found in Han Chinese [15] or Thai populations [45] where all patients with allopurinol-induced SJS/TEN are carriers of this allele. It is thus possible that the HLA genetic determinants involved in SJS/TEN are not the HLA-B alleles themselves but some loci in LD with them. Further studies in this genomic region would certainly be necessary to better understand the mechanisms involved.

On a methodological point of view, this study illustrates how Reference Control Panels could be used to test for association when only cases are genotyped. Only patients were genotyped here and controls were selected from the CNG European Control Panel [22] based on their country of origin that was either France and Germany, since those were the two main countries where the cases come from. We also tried to use the whole panel of controls and correct for population stratification using Principal Component Analysis (PCA) and an adjustment on up to 20 PCs instead of the 2 used here. We found that including controls that were more genetically distant to the cases does not really improve the signal in the HLA region and instead could lead to false positive results in other genomic regions. This is somewhat in contradiction with the results recently reported by Zhuang et al. [46] and will probably need to be further investigated. We also used some alternative approaches where controls were genetically matched to cases using either a distance based on the identity by state as described by Guan et al. [47] or a distance derived from the top PCs of the PCA performed to identify axes of variations similar to the one proposed by Luca et al. [48]. Tests accounting for the matching units were then performed and we found that only the HLA region pointed out with significance levels in the same order of magnitude as the ones obtained here. These matching strategies might however be of interest as they could allow the analysis of more heterogeneous samples by including for example those non-European RegiSCAR cases that were excluded here and matching them to individuals from the CEPH-Human Genome Diversity Panel [49].

8.5 CONCLUSION

Our study confirms the involvement of genetic variants located in the HLA region in the susceptibility to SJS/TEN in European samples, especially in association with allopurinol. No difference is seen depending on the disease severity and no other locus reaches genome-wide association in this sample that is also the largest one collected so far.

REFERENCES

1. Lazarou J, Pomeranz BH, Corey PN: Incidence of adverse drug reactions in hospitalized patients: a meta-analysis of prospective studies. Jama 1998, 279(15):1200-1205.
2. Svensson CK, Cowen EW, Gaspari AA: Cutaneous drug reactions. Pharmacol Rev 2001, 53(3):357-379.
3. Wolverton SE: Update on cutaneous drug reactions. Adv Dermatol 1997, 13:65-84.
4. Roujeau JC, Stern RS: Severe adverse cutaneous reactions to drugs. N Engl J Med 1994, 331(19):1272-1285.
5. Auquier-Dunant A, Mockenhaupt M, Naldi L, Correia O, Schroder W, Roujeau JC: Correlations between clinical patterns and causes of erythema multiforme majus, Stevens-Johnson syndrome, and toxic epidermal necrolysis: results of an international prospective study. Arch Dermatol 2002, 138(8):1019-1024.
6. Rzany B, Mockenhaupt M, Baur S, Schroder W, Stocker U, Mueller J, Hollander N, Bruppacher R, Schopf E: Epidemiology of erythema exsudativum multiforme majus, Stevens-Johnson syndrome, and toxic epidermal necrolysis in Germany (1990-1992): structure and results of a population-based registry. J Clin Epidemiol 1996, 49(7):769-773.
7. Halevy S, Ghislain PD, Mockenhaupt M, Fagot JP, Bouwes JN, Sidoroff A, Naldi L, Dunant A, Viboud C, Roujeau JC: Allopurinol is the most common cause of Stevens-Johnson syndrome and toxic epidermal necrolysis in Europe and Israel. J Am Acad Dermatol 2008, 58(1):25-32.
8. Mockenhaupt M, Viboud C, Dunant A, Naldi L, Halevy S, Bouwes JN, Sidoroff A, Schneck J, Roujeau JC, Flahault A: Stevens-Johnson syndrome and toxic epidermal necrolysis: assessment of medication risks with emphasis on recently marketed drugs. The EuroSCAR-study. J Invest Dermatol 2008, 128(1):35-44.
9. Pirmohamed M, Park BK: Genetic susceptibility to adverse drug reactions. Trends Pharmacol Sci 2001, 22(6):298-305.

10. Melsom RD: Familial hypersensitivity to allopurinol with subsequent desensitization. Rheumatology (Oxford) 1999, 38(12):1301.

11. Pellicano R, Silvestris A, Iannantuono M, Ciavarella G, Lomuto M: Familial occurrence of fixed drug eruptions. Acta Derm Venereol 1992, 72(4):292-293.

12. Roujeau JC, Huynh TN, Bracq C, Guillaume JC, Revuz J, Touraine R: Genetic susceptibility to toxic epidermal necrolysis. Arch Dermatol 1987, 123(9):1171-1173.

13. Roujeau JC, Bracq C, Huyn NT, Chaussalet E, Raffin C, Duedari N: HLA phenotypes and bullous cutaneous reactions to drugs. Tissue Antigens 1986, 28(4):251-254.

14. Chung WH, Hung SI, Hong HS, Hsih MS, Yang LC, Ho HC, Wu JY, Chen YT: Medical genetics: a marker for Stevens-Johnson syndrome. Nature 2004, 428(6982):486.

15. Hung SI, Chung WH, Liou LB, Chu CC, Lin M, Huang HP, Lin YL, Lan JL, Yang LC, Hong HS, Chen MJ, Lai PC, Wu MS, Chu CY, Wang KH, Chen CH, Fann CS, Wu JY, Chen YT: HLA-B*5801 allele as a genetic marker for severe cutaneous adverse reactions caused by allopurinol. Proc Natl Acad Sci USA 2005, 102(11):4134-4139.

16. Lonjou C, Borot N, Sekula P, Ledger N, Thomas L, Halevy S, Naldi L, Bouwes-Bavinck JN, Sidoroff A, de Toma C, Schumacher M, Roujeau JC, Hovnanian A, Mockenhaupt M, RegiSCAR study group: A European study of HLA-B in Stevens-Johnson syndrome and toxic epidermal necrolysis related to five high-risk drugs. Pharmacogenetics and genomics 2008, 18(2):99-107.

17. Lonjou C, Thomas L, Borot N, Ledger N, de Toma C, LeLouet H, Graf E, Schumacher M, Hovnanian A, Mockenhaupt M, Roujeau JC, RegiSCAR Group: A marker for Stevens-Johnson syndrome ...: ethnicity matters. Pharmacogenomics J 2006, 6(4):265-268.

18. Pirmohamed M, Arbuckle JB, Bowman CE, Brunner M, Burns DK, Delrieu O, Dix LP, Twomey JA, Stern RS: Investigation into the multidimensional genetic basis of drug-induced Stevens-Johnson syndrome and toxic epidermal necrolysis. Pharmacogenomics 2007, 8(12):1661-1691.

19. Miller JW: Of race, ethnicity, and rash: the genetics of antiepileptic drug-induced skin reactions. Epilepsy Curr 2008, 8(5):120-121.

20. Abe R: Toxic epidermal necrolysis and Stevens-Johnson syndrome: soluble Fas ligand involvement in the pathomechanisms of these diseases. J Dermatol Sci 2008, 52(3):151-159.

21. Pirmohamed M: Genetic factors in the predisposition to drug-induced hypersensitivity reactions. Aaps J 2006, 8(1):E20-26.

22. Heath SC, Gut IG, Brennan P, McKay JD, Bencko V, Fabianova E, Foretova L, Georges M, Janout V, Kabesch M, et al.: Investigation of the fine structure of European populations with applications to disease association studies. Eur J Hum Genet 2008, 16(12):1413-1429.

23. Purcell S, Neale B, Todd-Brown K, Thomas L, Ferreira MA, Bender D, Maller J, Sklar P, de Bakker PI, Daly MJ, Sham PC: PLINK: a tool set for whole-genome association and population-based linkage analyses. American journal of human genetics 2007, 81(3):559-575.

24. Price AL, Patterson NJ, Plenge RM, Weinblatt ME, Shadick NA, Reich D: Principal components analysis corrects for stratification in genome-wide association studies. Nature genetics 2006, 38(8):904-909.

25. Patterson N, Price AL, Reich D: Population structure and eigenanalysis. PLoS genetics 2006, 2(12):e190.
26. Devlin B, Roeder K: Genomic control for association studies. Biometrics 1999, 55(4):997-1004.
27. StataCorp: Stata Statistical Software: Release 10. In College Station, TX. StataCorp LP; 2007.
28. Kazma R, Babron MC, Genin E: Genetic association and gene-environment interaction: a new method for overcoming the lack of exposure information in controls. Am J Epidemiol 173(2):225-235.
29. Manolio TA, Brooks LD, Collins FS: A HapMap harvest of insights into the genetics of common disease. J Clin Invest 2008, 118(5):1590-1605.
30. Reich DE, Lander ES: On the allelic spectrum of human disease. Trends Genet 2001, 17(9):502-510.
31. Lohmueller KE, Pearce CL, Pike M, Lander ES, Hirschhorn JN: Meta-analysis of genetic association studies supports a contribution of common variants to susceptibility to common disease. Nature genetics 2003, 33(2):177-182.
32. Bodmer W, Bonilla C: Common and rare variants in multifactorial susceptibility to common diseases. Nature genetics 2008, 40(6):695-701.
33. Manolio TA, Collins FS, Cox NJ, Goldstein DB, Hindorff LA, Hunter DJ, McCarthy MI, Ramos EM, Cardon LR, Chakravarti A, Cho JH, Guttmacher AE, Kong A, Kruglyak L, Mardis E, Rotimi CN, Slatkin M, Valle D, Whittemore AS, Boehnke M, Clark AG, Eichler EE, Gibson G, Haines JL, Mackay TF, McCarroll SA, Visscher PM: Finding the missing heritability of complex diseases. Nature 2009, 461(7265):747-753.
34. Gorlov IP, Gorlova OY, Sunyaev SR, Spitz MR, Amos CI: Shifting paradigm of association studies: value of rare single-nucleotide polymorphisms. American journal of human genetics 2008, 82(1):100-112.
35. McCarthy MI: Exploring the unknown: assumptions about allelic architecture and strategies for susceptibility variant discovery. Genome medicine 2009, 1(7):66.
36. Fearnhead NS, Winney B, Bodmer WF: Rare variant hypothesis for multifactorial inheritance: susceptibility to colorectal adenomas as a model. Cell cycle (Georgetown, Tex 2005, 4(4):521-525.
37. Boyko AR, Williamson SH, Indap AR, Degenhardt JD, Hernandez RD, Lohmueller KE, Adams MD, Schmidt S, Sninsky JJ, Sunyaev SR, White TJ, Nielsen R, Clark AG, Bustamante CD: Assessing the evolutionary impact of amino acid mutations in the human genome. PLoS genetics 2008, 4(5):e1000083.
38. Kryukov GV, Pennacchio LA, Sunyaev SR: Most rare missense alleles are deleterious in humans: implications for complex disease and association studies. American journal of human genetics 2007, 80(4):727-739.
39. Luca F, Bubba G, Basile M, Brdicka R, Michalodimitrakis E, Rickards O, Vershubsky G, Quintana-Murci L, Kozlov AI, Novelletto A: Multiple advantageous amino acid variants in the NAT2 gene in human populations. PloS one 2008, 3(9):e3136.
40. Patin E, Barreiro LB, Sabeti PC, Austerlitz F, Luca F, Sajantila A, Behar DM, Semino O, Sakuntabhai A, Guiso N, Gicquel B, McElreavey K, Harding RM, Heyer E, Quintana-Murci L: Deciphering the ancient and complex evolutionary history of human arylamine N-acetyltransferase genes. American journal of human genetics 2006, 78(3):423-436.

41. Wilson JF, Weale ME, Smith AC, Gratrix F, Fletcher B, Thomas MG, Bradman N, Goldstein DB: Population genetic structure of variable drug response. Nature genetics 2001, 29(3):265-269.

42. Leslie S, Donnelly P, McVean G: A statistical method for predicting classical HLA alleles from SNP data. American journal of human genetics 2008, 82(1):48-56.

43. de Bakker PI, McVean G, Sabeti PC, Miretti MM, Green T, Marchini J, Ke X, Monsuur AJ, Whittaker P, Delgado M, Morrison J, Richardson A, Walsh EC, Gao X, Galver L, Hart J, Hafler DA, Pericak-Vance M, Todd JA, Daly MJ, Trowsdale J, Wijmenga C, Vyse TJ, Beck S, Murray SS, Carrington M, Gregory S, Deloukas P, Rioux JD: A high-resolution HLA and SNP haplotype map for disease association studies in the extended human MHC. Nature genetics 2006, 38(10):1166-1172.

44. Kazeem GR, Cox C, Aponte J, Messenheimer J, Brazell C, Nelsen AC, Nelson MR, Foot E: High-resolution HLA genotyping and severe cutaneous adverse reactions in lamotrigine-treated patients. Pharmacogenetics and genomics 2009, 19(9):661-665.

45. Tassaneeyakul W, Jantararoungtong T, Chen P, Lin PY, Tiamkao S, Khunarkornsiri U, Chucherd P, Konyoung P, Vannaprasaht S, Choonhakarn C, Pisuttimarn P, Sangviroon A, Tassaneeyakul W: Strong association between HLA-B*5801 and allopurinol-induced Stevens-Johnson syndrome and toxic epidermal necrolysis in a Thai population. Pharmacogenetics and genomics 2009, 19(9):704-709.

46. Zhuang JJ, Zondervan K, Nyberg F, Harbron C, Jawaid A, Cardon LR, Barratt BJ, Morris AP: Optimizing the power of genome-wide association studies by using publicly available reference samples to expand the control group. Genetic epidemiology 2010. Advance Online

47. Guan W, Liang L, Boehnke M, Abecasis GR: Genotype-based matching to correct for population stratification in large-scale case-control genetic association studies. Genetic epidemiology 2009, 33(6):508-517.

48. Luca D, Ringquist S, Klei L, Lee AB, Gieger C, Wichmann HE, Schreiber S, Krawczak M, Lu Y, Styche A, Devlin B, Roeder K, Trucco M: On the use of general control samples for genome-wide association studies: genetic matching highlights causal variants. American journal of human genetics 2008, 82(2):453-463.

49. Cann HM, de Toma C, Cazes L, Legrand MF, Morel V, Piouffre L, Bodmer J, Bodmer WF, Bonne-Tamir B, Cambon-Thomsen A, Chen Z, Chu J, Carcassi C, Contu L, Du R, Excoffier L, Ferrara GB, Friedlaender JS, Groot H, Gurwitz D, Jenkins T, Herrera RJ, Huang X, Kidd J, Kidd KK, Langaney A, Lin AA, Mehdi SQ, Parham P, Piazza A, Pistillo MP, Qian Y, Shu Q, Xu J, Zhu S, Weber JL, Greely HT, Feldman MW, Thomas G, Dausset J, Cavalli-Sforza LL: A human genome diversity cell line panel. Science (New York, NY 2002, 296(5566):261-262.

50. Strimmer K: fdrtool: a versatile R package for estimating local and tail area-based false discovery rates. Bioinformatics (Oxford, England) 2008, 24(12):1461-1462.

There are several supplemental files that are not available in this version of the article. To view this additional information, please use the citation information cited on the first page of this chapter.

CHAPTER 9

GENOTYPING COMMON AND RARE VARIATION USING OVERLAPPING POOL SEQUENCING

DAN HE, NOAH ZAITLEN, BOGDAN PASANIUC, ELEAZAR ESKIN, AND ERAN HALPERIN

9.1 BACKGROUND

Recent advances in sequencing technologies have drastically reduced the cost of nucleotide sequencing [1,2] and are rapidly establishing themselves as very powerful tools for quantifying a growing list of cellular properties that include sequence variation, RNA expression levels, protein-DNA/RNA interaction sites, and chromatin methylation [3-8]. An expensive step in the sequencing process is sample preparation where time consuming procedures such as library preparation must be applied to each individual sample. This greatly reduces the utility of a sequencer for sequencing a small genomic region in many individuals because the cost of preparing each sample counteracts the efficiency of the sequencer. In fact the sequencing capacity in terms of the number of reads generated by the sequencer is often much higher than is necessary for the application. This raises the need for the development of multiplexing strategies that allow the processing of multiple samples per single sample preparation step at

This chapter was originally published under the Creative Commons Attribution License. He D, Zaitlen N, Pasaniuc B, Eskin E, and Halperin E. Genotyping Common and Rare Variation Using Overlapping Pool Sequencing. BMC Bioinformatics *12(Suppl 6),S2 (2011), doi:10.1186/1471-2105-12-S6-S2.*

the cost of requiring additional sequencing capacity. However, in several practical scenarios, the overall cost can be reduced. One such multiplexing scheme is the use of overlapping pools [9-11]. In this scheme subsets of samples are mixed together into pools followed by a single sample preparation for each pool. Typically in such a sample preparation, a barcoding technique is applied so each read generated from the pool will be able to be identified as originating from the pool. By combining the results of the sequencing with the information on which samples appeared in which pool, the mixed information from each pool can be "decoded" to obtain information on the sequence of each sample.

Multiplexing pools are practical for sequencing a short genomic region in many individuals. As sequence capacity increases, it is likely that this technique will become even more practical in the future. Sequencing capacity is constantly increasing and therefore it is plausible that multiplexing pools will benefit whole-genome sequencing in the future. We note that in Erlich et al [9] the use of multiplexing has been proven in the lab, showing that this methodology is not merely a theoretical exercise. The current techniques for overlapping pool sequencing [9-11] are based on group testing or compressed sensing schemes. Their main limitation is that they are only applicable to detect rare variants. If a variant is common in the population, it will be present in almost every pool, causing the above pooling schemes to fail in identifying which subset of the samples contain the common variant.

In this paper, we present an alternate scheme for sequencing using overlapping pools which, unlike all previous approaches, is able to quantify both rare and common variation. The key idea underlying our scheme is that we formulate the pooling problem within a likelihood framework that provides several advantages over previous methods. Our scheme is flexible and can be applied to a wide variety of applications. We demonstrate this by applying the scheme to two very different applications, each of which takes advantage of the likelihood framework within our approach and is difficult to solve using previously proposed combinatorial methods.

The first application we consider is obtaining highly accurate genotype information for a set of individuals. Currently, genotype microarrays are the most accurate method for measuring individual genetic variation at a base-pair level at variable locations across the genome (Single Nucleotide

Polymorphisms: SNPs). A typical array will collect up to 900,000 or more genotype calls at common SNPs across the genome. Using imputation techniques and a reference dataset such as the HapMap [12] or the 1,000 Genomes project, we can make predictions for the remaining common variants in the genome. While error rates of genotyping are usually less than .5% errors at imputed variants range from around 5% in Europeans, and it could be as high as 10-15% for non-European populations [12-14]. Imputation accuracy is particularly poor for rare SNPs and for SNPs in regions of low linkage disequilibrium. We introduce here a scheme for obtaining highly accurate genotype information on both common and rare SNPs by combining genotyping microarrays, imputation and sequencing in pools of samples. This application is possible because our likelihood framework allows us to integrate the information from the imputation into the procedure to help us "decode" the information obtained from each pool. Furthermore, our scheme allows us to utilize the variant frequency information obtained in each pool. Our results show that our algorithm is capable of calling rare SNPs with high accuracy, but in contrast to previous multiplexing methods, it can also call common SNPs with high accuracy, by combining the imputation data with the pooling scheme. In fact, the same experiment which can obtain genotype information for rare variants combined with imputation can obtain genotype information at the common variants. Importantly, the outcome of our approach results in genotype information on the common variation which is more accurate than what is collected using microarrays. This application is particularly practical because much of the follow up sequencing of populations will be done in the same cohorts in which genome-wide association studies were performed. For these individuals, genotyping at common SNPs using microarrays has already been performed and for many of these studies only the regions of interest are targeted for sequencing, or exome sequencing is being performed, which makes multiplexing pools a practical approach at present.

The second application we consider in this work is to rapidly screen for fusion genes in cancer samples. Fusion genes play an important role in cancers and are caused by genomic rearrangements in a tumor that create new genes consisting of several exons from one gene followed by several exons from a second gene. Our application considers the sequencing of

RNA obtained from cancer tumors with paired-end reads. The read pairs of interest are ones that span exon boundaries with each read of the pair coming from a different exon. The majority of such read pairs will map to the same gene when aligned to the reference genome and. However, read pairs from fusion genes will map to two exons from different genes. One potential approach in identifying fusion genes is to search for read pairs that contain reads mapping to different genes. The main drawback of this approach is that it leads to a very high level of false positives making it difficult to distinguish actual fusions from experimental artifacts. Our application will mix RNA from a large number of cancer tumors into overlapping pools and utilize the likelihood framework to decode which fusion genes come from which samples. In order to accomplish that, we extended the basic overlapping pool model to consider different levels of expression for each gene. This can be estimated from the data. Our decoding scheme is based on a likelihood formulation which presents novel computational challenges compared to previous approaches. Each possible configuration (genotype assignment or gene-fusion assignment) is assigned a likelihood and the goal of the algorithm is to identify the most likely decoding. We identify good solutions for the problem by formulating a related problem as a linear program which we can efficiently solve. We note that these results are just the first steps in applying this framework to multiplexing sequencing pools and it is likely that better optimization algorithms and better designs of pooling schemes can lead to more substantial savings.

9.2 RESULTS AND DISCUSSION

9.2.1 GENOTYPING USING OVERLAPPING POOLS AND IMPUTATION

We first report the results of applying our approach to genotyping individuals to obtain both common and rare variation using combining overlapping sequencing pools with genotyping and imputation. In this set of experiments, we utilize the 1958 Birth Cohort from the Wellcome Trust

Case Control Consortium [15] data which contains approximately 1,500 individuals. These individuals were genotyped at approximately 500,000 SNPs. For every 10th SNP, we set the values of the genotypes to missing and applied MACH [16] using the HapMap data [17], an imputation algorithm, on these SNPs to make predictions. Since the SNPs were genotyped in the dataset, we can evaluate the accuracy of the imputation. We filter out any SNP with minor allele frequency lower than 5% since rare variants are easily genotyped using overlapping sequencing pools and the goal of these experiments is to evaluate the methods ability to genotype common variants. We simulate applying our method by generating sequencing reads by generating reads consistent with the true values of the genotypes at the missing SNPs for each pool and then apply our method to make predictions of the genotypes incorporating the imputation information. We then measure the increase in accuracy of our prediction relative to the imputation information.

For our experiments we consider a total of 100 individuals mixed into 36 pools which is a reduction of the total number of sample preparations necessary by 1/3. We use a very high coverage of 150 per individual within a pool for our experiments under the assumption that the bottleneck is not the coverage, but the number of pools, each which requires a single sample preparation step. We assume a sequencing error of 0.005. We measure the accuracy of the predictions by comparing the predicted genotypes to the true genotypes and only call a prediction correct if the genotypes are correct for all 100 individuals. We note that this is a very high standard and only 1 of our 100 SNPs have a correct imputation prediction. Our method has very high accuracy significantly improving over imputation. Table 1 summarizes the results.

TABLE 1: Results of genotyping using overlapping sequence pools with imputation information

Parameter Values Num Pools	Individuals = 100	
	Imputation Accuracy	LP Accuracy
36	0.01	0.98
30	0.01	0.87

9.2.2 GENOTYPING USING OVERLAPPING POOLS WITHOUT IMPUTATION

We also apply our scheme to predict the genotypes without the imputation for rare variants such as those not found in the reference. The only difference in our methodology is that in the optimization problem, for the imputation vector we use a zero vector since we expect most individuals will not have the variant. This problem is actually much easier than the case of common alleles and we get perfect accuracy for the parameters above.

TABLE 2: Results of cancer fusion gene detection simulations

Parameter Values (Num Pools, Coverage, Error Rate)	# of Samples with Fusion		
	1	2	3
(10, 4 , 0.01)	0.980	0.760	0.340
(10, 12 , 0.01)	0.990	0.970	0.700
(10, 16, 0.01)	1.000	0.980	0.780
(10, 20, 0.01)	1.000	0.930	0.790
(10, 24, 0.01)	1.000	0.990	0.810
(10, 28, 0.01)	0.990	0.970	0.840
(4, 28, 0.01)	0.180	0.030	0.000
(6, 28, 0.01)	0.550	0.230	0.050
(8, 28, 0.01)	1.000	0.900	0.410

Each entry in the table is the fraction that the algorithm correctly identified the samples harboring the fusion gene.

9.2.3 CANCER FUSION GENE DETECTION

We evaluate our approaches ability to detect cancer fusion genes using a similar simulation framework. In this application, RNA from different tumors is is mixed into overlapping pools and sequenced. In each pool we search for reads which cross exon boundaries from different genes and are evidence of fusion genes. Counts of these fusion genes in each pool are then decoded to identify the samples which contain the fusion genes.

We simulate this process by generating reads in a similar fashion to the genotyping without imputation simulations. We assume that we have 100 cancer samples where either 1, 2 or 3 of the samples contain a specific fusion gene. We assume a sequence error rate of 1% and vary the coverage and number of pools in our experiments. A difficulty in this application is that each individual has a different level of expression for each gene. We simulate this by randomly selecting an expression level in the range such that the concentration of the fusion gene in the RNA will differ by up to a factor of 10. Table 2 shows the results of our cancer fusion gene detection simulation experiment. For each experiment, we report the fraction of the time that the algorithm identified correctly which samples contain the fusion gene.

9.3 CONCLUSIONS

In this paper we have described a flexible framework for overlapping pool sequencing and two applications of this framework. We presented some results showing the bounds on the performance of decoding rare and common variants (in Methods). We argue that due to information theoretic bounds, common variants are impossible to decode without the addition of external information; we propose to use imputation results as possible external information. In practice, it is often the case that such information is given, as many of the samples have already been genotyped through the massive effort of genome-wide association studies; in addition, current cost of genotyping has reduced considerably and is negligible compared to sequencing costs (especially when considering gene-targeted sequencing).

Our decoding framework is likelihood based framework and is general enough to account for different types of data and error models. Particularly, we demonstrate how our method can be extended to the case where there are different unknown concentrations of each variant in each sample as motivated by the cancer fusion gene detection example. We note that our approach to detect fusion genes using RNA sequences can only detect fusion genes that are expressed in the tumor since we are sequencing RNA, but this is the case for a significant subset of the total fusion genes [18].

We expect that with improved optimization algorithms and better designs of pooling schemes, we can achieve even more substantial savings.

9.4 METHODS

We consider the scenario in which a set of N individuals are to be sequenced for any application such as a disease association study, or fusion-gene detection. The most straightforward approach would be to barcode the individual and sequence them separately. When N is large, or when the desired coverage is high, this approach is infeasible due to budget constraints. A few methods have been suggested to tackle this problem using a set of overlapping pools [9-11]. These methods are based on the following generic idea. Let the sequences of the individuals be represented by a matrix $G = \{g_{ij}\}$ of dimension $N \times m$, where m is the length of the genome. $g_{ij} \in \{0,1,2\}$ is the number of occurrences of a genetic variant in position j of individual i - such variant could be a single nucleotide polymorphism (SNP), copy number variant (CNV), or a gene-fusion, as discussed in the introduction. The pooling based approach considers a $\{0, 1\}$ matrix A of dimension $T \times N$, representing a set of pools. Each row of A corresponds to a DNA pool; individual j participates in the i —th row if and only if $A_{ij} = 1$. Thus, the matrix A provides a compact description of the study design. When the study is performed, under an error-free model, the pooling results are given as Y $= AG$. In principle, one can now decode the matrix by finding a solution to the set of equations $AX = Y$. In reality though the pooling results are not as accurate, and therefore current methods are using a rounded version of Y; for every i,j, we define $c_{ij} = 1$ if $y_{ij} > 0$ and $c_{ij} = 0$ otherwise. Thus, if we replace the SUM operation by an OR operation then $AG = C$. Using this information, Erlich et al. [9],Prabhu et al. [10] and Shental et al. [11] provide a decoding algorithm which finds which individuals have $g_{ij} > 0$. For the simplicity of the exposition, we will assume from now on that only one variant is considered, and so Y, C, and G are column vectors of length N.

9.4.1 A LOWER BOUND ON DECODING ACCURACY

Unfortunately, by collapsing the data to a {0,1} matrix resolution is lost, and therefore there is no hope in decoding all genetic variants from the pools if the number of pools is not large. Note that for a given variant, there are 3^N possible genotype vectors. The number of possible column vectors C_j is 2^T. Therefore, in order to be able to decode all individual genotypes we need $2^T > 3^N$, or $T > N$. Even without rounding, the number of possible vectors B_j is at most $(2N)^T$, and therefore even in the error-free case we need $(2N)^T > 3^N$, or $T \geq \Omega(N/\log N)$. In practice, the pooling decoding methods work well when the allele frequency is low, under an error-free model. For a variant of allele frequency α, the number of possible genotype vectors is:

$$\binom{N}{\alpha N} \approx \left(\frac{1}{\alpha}\right)^{\alpha N}$$

and therefore, we get that in the case where the rounded solutions are provided (the matrix C), we need $(2N)^T \geq 3^N$, or $2^T \geq (1/\alpha)^{\alpha N}$, or $T \geq$ —$N\alpha \log \alpha$, and if we are using the full information given by the matrix B, we need $T \geq$ -$(N\alpha \log \alpha)/(\log N)$. Note that these are lower bounds, and it may theoretically be the case that a larger number of pools is required; however, it is easy to see that if A is chosen as a random matrix where each entry is 1 with probability 0.5, then the bounds given here are tight up to a constant factor (the proof is omitted from this version). Moreover, in [9-11], it is shown that using the matrix C one can decode low allele frequencies ($\alpha = O(1)/N$) then $T = O(\log N)$ suffices, which is consistent with the bounds we provide here. Since a random matrix provides a good decoding scheme in theory, we followed this intuition and generated a matrix A so that half of the entries in each row is 1 and the other half is 0. To obtain a better design matrix, we use local search; we repeatedly permute a random row and a random column and check to see if the Hamming distance between the permuted row/column and the other rows increased. If so, we keep the change, otherwise, we revert. After performing 1000 permutations, we result in a matrix whose rows and columns are farther apart, which improves our ability to decode.

9.4.2 INCORPORATING IMPUTATION INTO DECODING

As described above, from an information theoretical point of view, decoding the genotype vector is only possible when the allele frequency is low and therefore the genotype vector is sparse. For this reason, both Erlich et al. [9] and Shental et al. [11] make the connection between decoding and compressed sensing [19], where the requirement for the decoding success is based on the fact that the desired vector is sparse. We therefore suggest to incorporate imputation results into the decoding scheme; this allows us to overcome the information theoretical bound for the following reason. We can represent the true genotypes G as a sum of the (rounded) imputation predictions I, $i_{ij} \in \{0, 1, 2\}$, and a set of imputation residual errors R, $r_{ij} \in \{-2, 1, 0, 1, 2\}$, where G = I + R. Then, the observed data can be represented as the pools' results Y = AG, which is Y = AI + AR. Now, note that R is a sparse vector, and I is known; therefore, from a theoretical point of view, the above information theoretic lower bound does not hold on our case and there may be an algorithm that is able to decode the genotypes based on the sequencing and the imputation. In principle, we can solve for the imputation residual errors by solving the set of equations for AX = Y - AI. Once we obtain the residual vector, we can obtain the actual genotypes. In practice, as described below, we use the imputation dosage so our algorithm theoretically searches over the entire space, and not only over sparse vectors, but the search is pruned for vectors that are dense based on a likelihood model. As we show in the results section, this yields an improved imputation accuracy for high allele frequency SNPs.

9.4.3 POOLING USING READ COUNTS

Our approach differs from previous approaches in that we are considering the matrix Y and not C. As discussed above, this should allow us a gain of approximately log N factor in the number of pools needed, at least for higher allele frequencies. However, in order to do so, we need to explicitly model the sequencing errors. The error model may be different, depending on the application at hand. We will describe here the model we use for the detection of mutations (SNP calling). There are three main sources of noise that we include in the model:

1. There are slight differences in the concentration of each individual's DNA in each pool. This pooling noise is modeled as a normally distributed noise added to each non-zero element of A with mean 0 and variance σ_p. Thus, we set

$$\hat{A}_{ij} = A_{ij} + N(0, \sigma_p) \forall A_{ij} \neq 0$$

2. There is a variance in the coverage of any specific region in the genome. We denote by L the length of the sequenced genomic region; if the total number of bases sequenced is λLN, then we expect that each base will be covered by λ reads on average. λ is often termed the expected coverage. We will denote the number of reads covering individual i by r_{i1}, r_{i2} (corresponding to the two chromosomal copies). Then, r_{ij} is Poisson distributed, with mean mi. Prabhu et al. [10] showed that the mi are approximately drawn from a Gamma distribution with $\alpha = 6.3$ and $\beta = \lambda/\alpha$ for Illumina Solexa sequencers. We note that for a given variant it is easy to infer the value of m_i, since it is shared across all individuals in all pools. Thus, we have

$$m_i \sim \Gamma(\alpha, \beta), r_{ij} \sim \text{Poisson}(m_i)$$

3. The third source of error is sequencing error. The sequencing error rate depends on the location of the base in the read, but since the location of the base is uniformly distributed, we simply model the error rate by a constant probability ε for a substitution (1% is an acceptable estimate).

The above procedure produces a matrix $\hat{A}$ of noisy pools and a pair of vectors R^0, R^1 of noisy sequence reads; the number of sequence reads R_i^k

is generated by a Poisson distribution with an expectation that depends on the genotype g_i, and the coverage m_i, followed by a Binomial distribution to model the errors as explained above. R^0 corresponds to the reads with the major allele, while R^1 corresponds to the reads with the minor alleles. Note that even if $g_i = 0$, if $\varepsilon > 0$, then expected number of reads with the minor allele will be greater than 0 because of errors. The pooling results are given by (Y^0, Y^1), where $Y^k = \hat{A}R^k$.

9.4.4 A LIKELIHOOD MODEL

Given the pooling results Y, we need to find a decoding algorithm that estimates G from Y. To do so, we define a likelihood model which can evaluate each putative solution. Our likelihood model takes into account both the error model, as well as population genetics data and external information when available. We decompose the likelihood L(G; Pools) into several functions, and take their product as a composite likelihood.

9.4.5 HARDY-WEINBERG EQUILIBRIUM

We first note that the overall allele frequency p of the SNP can be estimated as the average across all pools. We can now compute the Hardy-Weinberg (HW) probability of the observed genotypes $\Pr^{HW}(G|p) = 2^{n_1} p^{n_2+n_1}(1 - p)^{n_0}$, where n_0, n_1, n_2 are the genotype counts in G. Using Bayes law we have

$$Pr^{HW}(Pools|G) = Pr^{HW}(G|Pools) \frac{\Pr(Pools)}{\Pr(G)}$$

Assuming no prior information, we observe that maximizing $\Pr^{HW}(\text{Pools} \mid G)$ is equivalent to maximizing $\Pr^{HW}(G \mid \text{Pools})$. We denote $f^{HW} = \Pr^{HW}(G|p) = 2^{n_1} p^{n_2+n_1}(1 - p)^{n_0}$.

9.4.6 LIKELIHOOD OF THE OBSERVED READS

We compute the probability of the observed reads in the pools given G based on the noise model. Note that the only unknown in the noise model is the concentrations of the individuals in the different pools. This is true since the coverage in any given region can be easily estimated. Assume that λ is the coverage. Then, the number of reads with the minor allele (or major) contributed by individual j in pool i are Poisson distributed with $\hat{A}_{ij}\lambda G_j$ (or $\hat{A}_{ij}\lambda(2 - G_j)$). Since the sum of Poisson distributions is Poisson distributed, we have that is Poisson distributed with a known expectation and thus we can write its likelihood. We denote this function by fnoise(G). In order to find the concentration values $\hat{A}_{ij}$ we need to use external information. One such possibility could be to genotype small set of SNPs across the population and use those as the ground truth in order to tune the values of $\hat{A}_{ij}$. These SNPs provide a set of linear equations for the values $\hat{A}$; for each pool we have one equation per SNP, and the number of variables is N. Therefore, genotyping as many as O(N) SNPs and using a least squares approach guarantees an accurate estimate of $\hat{A}_{ij}$.

9.4.7 LIKELIHOOD OF IMPUTED DATA

Due to the bounds given on the possibility for detection, it is clear that without external information we will not be able to do much better than detecting rare SNPs. One natural choice for external data could be the genotypes of the individuals using microarrays. Today's genotyping technology is extremely cheap compared to sequencing, and the genotyping of thousands of individuals is feasible within a given study. The genotype information, however, provides the information about less than a million SNPs and another million CNVs across the genomes, while many other genetic variants are left unmeasured. To cope with this, imputation methods have been developed, in which nearby SNPs are used to impute unmeasured variants using the linkage disequilibrium structure of the genome [16,20]. However, this process is inevitably noisy, especially when imputing SNPs of low allele frequencies or SNPs in regions of low linkage disequilibrium. Together with the pooling information we are able to provide a much more accurate calling of the imputed SNPs in all ranges of allele frequencies and linkage disequilibrium patterns. The output of the imputation method typically provides a distribution of the possible genotypes. For each individual j, we can assume that there is a given probability $h_1(0)$, $h_1(1)$, $h_1(2)$, where $h_1(j)$ is the probability that individual i has $G_i = j$. We can now use the imputation results for our likelihood model, by writing $f^{impute}(G) = Pr(G \mid imputation) = \Pi_{i=1}^{N} h_i(G_i)$.

9.4.8 A DECODING ALGORITHM USING LINEAR PROGRAMMING

We use a linear program to bound the possible errors of each of the pools. If the coverage for the SNP is λ, we have that the pools should roughly satisfy $\lambda \hat{A} G = Y$. We can therefore solve the following linear program:

$$LP(G') = \min \sum_{i=1}^{T} x_i + \beta |G_i - I_i|$$

$$s.t \quad \lambda \sum_{j=1}^{N} \hat{a}_{ij} G_j - Y_i \leq x_i, \forall i$$

$$\lambda \sum_{j=1}^{N} \hat{a}_{ij} G_j - Y_i \geq x_i, \forall i$$

$$0 \leq G_j \leq 2, j \in \{1, \dots N\}$$

The linear program provides a lower bound on the best possible l_1 distance between $\lambda \hat{A} G$ and Y as well as returning a solution G which is close to the imputation prediction I. β is a parameter that trades off the relative importance of being close to the imputation vector compared to being consistent with the pools. Note that if Y_j is distributed as Poisson with expectation μ_j for which $Y_j - \mu_j = x_j$, then

$$Pr(Y_j) = e^{\mu_j}\frac{\mu_j^{Y_j}}{Y_j!} = \frac{Y_j^{Y_i}e^{Y_i}}{Y_j!}e^{-xj}\left(1 - \frac{x_j}{y_j}\right)^{Y_j} \approx \frac{Y_j^{Y_i}e^{Y_i}}{Y_j!}e^{-2xj}$$

Therefore, we get that

$$f(v) \le e^{-2LP(G')}\prod_{j=1}^{T}\frac{Y_j^{Y_j}e^{Y_j}}{Y_j!}$$

9.4.9 APPLICATION TO GENE FUSION DETECTION

In order to detect gene fusions, we make several changes and extensions to the model presented above. The major additional complication in detection of fusion genes is that each sample may have a different expression level for a particular fusion gene. Even if we include the same amount of RNA from each tumor into each pool, the relative concentration of each gene will differ in each sample. However, this concentration is approximately constant across pools. Let e_{ij} be the normalized expression level of a particular variant j (in this case a fusion gene). Whether or not an individual i has the variant j is encoded as $G = \{g_{ij}\}$, $g_{ij} \in \{0, 1\}$. We define the matrix $H = \{e_{ij}g_{ij}\}$ as the concentrations of the samples and the results of the pools (assuming no noise) will then be $Y = AH$ instead of $Y = AG$ as in the genotyping application. In this application, we can also assume that the matrix G is sparse, but in order to perform the decoding, we must also estimate eij for the non-zero values of g_{ij}.

It is possible to estimate e_{ij} because they are constant across pools, however this introduces additional complexities in the optimization. We take advantage that fusion genes are very rare and most fusion genes are not shared across tumors. We constrain our optimization to allow for a maximum of k tumors to contain a given sample. We note that we only need to estimate the values of e_{ij} corresponding to non-zero elements of g_{ij}. To perform the optimization we enumerate over all possible genotype vectors and for each vector we estimate the corresponding e_{ij} values.

Since optimizing the likelihood function for each possible genotype vector is computationally impractical, we solve a linear program as a method to quickly eliminate poor solutions. Let A* be a matrix consisting of the only the columns of A corresponding to the non-zero entries of the genotype vector. If x is a vector which has a length the same as the number of non-zero elements in the genotype vector, the solution to A * x = Y will be an approximate estimate of the values for eij. We can incorporate errors by adding a vector of all 1s to A* and appending a term to x corresponding to the amount of errors expected in each pool. For the top 100 estimates obtained by using the pseudo-inverse, we then perform a grid search over the values of e_{ij} using the likelihood function described above.

REFERENCES

1. Margulies M, Egholm M, Altman WE, Attiya S, Bader JS, Bemben LA, Berka J, Braverman MS, Chen YJJ, Chen Z, Dewell SB, Du L, Fierro JM, Gomes XV, Godwin BC, He W, Helgesen S, Ho CH, Ho CH, Irzyk GP, Jando SC, Alenquer MLI, Jarvie TP, Jirage KB, Kim JBB, Knight JR, Lanza JR, Leamon JH, Lefkowitz SM, Lei M, Li J, Lohman KL, Lu H, Makhijani VB, McDade KE, McKenna MP, Myers EW, Nickerson E, Nobile JR, Plant R, Puc BP, Ronan MT, Roth GT, Sarkis GJ, Simons JF, Simpson JW, Srinivasan M, Tartaro KR, Tomasz A, Vogt KA, Volkmer GA, Wang SH, Wang Y, Weiner MP, Yu P, Begley RF, Rothberg JM: Genome sequencing in microfabricated high-density picolitre reactors. Nature 2005, 437(7057):376-80.
2. Harris TD, Buzby PR, Babcock H, Beer E, Bowers J, Braslavsky I, Causey M, Colonell J, Dimeo J, Efcavitch JW, Giladi E, Gill J, Healy J, Jarosz M, Lapen D, Moulton K, Quake SR, Steinmann K, Thayer E, Tyurina A, Ward R, Weiss H, Xie Z: Single-molecule DNA sequencing of a viral genome. Science 2008, 320(5872):106-9.
3. Wang Z, Gerstein M, Snyder M: RNA-Seq: a revolutionary tool for transcriptomics. Nat Rev Genet. 2009, 10(1):57-63.
4. Schuster SC: Next-generation sequencing transforms today's biology. Nat Methods 2008, 5(1):16-18.
5. Marioni JC, Mason CE, Mane SM, Stephens M, Gilad Y: RNA-seq: An assessment of technical reproducibility and comparison with gene expression arrays. [http://

genome.cshlp.org/content/18/9/1509.abstract] Genome Research 2008, 18(9):1509-1517.

6. Mortazavi A, Williams BA, McCue K, Schaeffer L, Wold B: Mapping and quantifying mammalian transcriptomes by RNA-Seq. Nat Methods. 2008, 5(7):621-628.

7. Johnson DS, Mortazavi A, Myers RM, Wold B: Genome-wide mapping of in vivo protein-DNA interactions. Science 2007, 316(5830):1497-1502.

8. Cokus SJ, Feng S, Zhang X, Chen Z, Merriman B, Haudenschild CD, Pradhan S, Nelson SF, Pellegrini M, Jacobsen SE: Shotgun bisulphite sequencing of the Arabidopsis genome reveals DNA methylation patterning. Nature 2008, 452(7184):215-219.

9. Erlich Y, Chang K, Gordon A, Ronen R, Navon O, Rooks M, Hannon GJ: DNA Sudoku-harnessing high-throughput sequencing for multiplexed specimen analysis. Genome Res 2009, 19(7):1243-53.

10. Prabhu S, Pe'er I: Overlapping pools for high-throughput targeted resequencing. Genome Res 2009, 19(7):1254-61.

11. Shental N, Amir A, Zuk O: Identification of rare alleles and their carriers using compressed se(que)nsing. Nucleic Acids Res 2010, 38(19):e179.

12. Integrating common and rare genetic variation in diverse human populations Nature 2010, 467(7311):52-58.

13. Marchini J, Howie B: Genotype imputation for genome-wide association studies. Nat Rev Genet 2010, 11(7):499-511.

14. Pasaniuc B, Avinery R, Gur T, Skibola CF, Bracci PM, Halperin E: A generic coalescent-based framework for the selection of a reference panel for imputation. Genet Epidemiol 2010, 34(8):773-782.

15. Consortium WTCC: Genome-wide association study of 14,000 cases of seven common diseases and 3,000 shared controls. Nature 2007, 447(7145):661-78.

16. Li Y, Abecasis G: Rapid Haplotype Reconstruction and Missing Genotype Inference. Am J Hum Genet 2006., S79(2290)

17. Consortium IH: A haplotype map of the human genome. Nature 2005, 437(7063):1299-320.

18. He D, Eskin E: Effective Algorithms for Fusion Gene Detection. In Algorithms in Bioinformatics, Volume 6293 of Lecture Notes in Computer Science Edited by Moulton V, Singh M, Springer Berlin / Heidelberg. 2010, :312-324.

19. Donoho D: Compressed Sensing. IEEE Transactions on Information Theory 2010, 52(4):1289-1306.

20. Marchini J, Howie B, Myers S, McVean G, Donnelly P: A new multipoint method for genome-wide association studies by imputation of genotypes. Nat Genet 2007, 39(7):906-13.

CHAPTER 10

LEARNING GENETIC EPISTASIS USING BAYESIAN NETWORK SCORING CRITERIA

XIA JIANG, RICHARD E. NEAPOLITAN, M. MICHAEL BARMADA, AND SHYAM VISWESWARAN

10.1 BACKGROUND

The advent of high-throughput genotyping technology has brought the promise of identifying genetic variations that underlie common diseases such as hypertension, diabetes mellitus, cancer and Alzheimer's disease. However, our knowledge of the genetic architecture of common diseases remains limited; this is in part due to the complex relationship between the genotype and the phenotype. One likely reason for this complex relationship arises from gene-gene and gene-environment interactions. So an important challenge in the analysis of high-throughput genetic data is the development of computational and statistical methods to identify gene-gene interactions. In this paper we apply Bayesian network scoring criteria to identifying gene-gene interactions from genome-wide association study (GWAS) data.

As background we review gene-gene interactions, GWAS, Bayesian networks, and modeling gene-gene interactions using Bayesian networks.

This chapter was originally published under the Creative Commons Attribution License. Jiang X, Neapolitan RE, Barmada MM, and Visweswaran S. Learning Genetic Epistasis Using Bayesian Network Scoring Criteria. BMC Bioinformatics *12,89 (2011), doi:10.1186/1471-2105-12-89.*

10.1.1 EPISTASIS

In Mendelian diseases, a genetic variant at a single locus may give rise to the disease [1]. However, in many common diseases, it is likely that manifestation of the disease is due to genetic variants at multiple loci, with each locus conferring modest risk of developing the disease. For example, there is evidence that gene-gene interactions may play an important role in the genetic basis of hypertension [2], sporadic breast cancer [3], and other common diseases [4]. The interaction between two or more genes to affect a phenotype such as disease susceptibility is called epistasis. Biologically, epistasis likely arises from physical interactions occurring at the molecular level. Statistically, epistasis refers to an interaction between multiple loci such that the net affect on phenotype cannot be predicted by simply combining the effects of the individual loci. Often, the individual loci exhibit weak marginal effects; sometimes they may exhibit none.

The ability to identify epistasis from genomic data is important in understanding the inheritance of many common diseases. For example, studying genetic interactions in cancer is essential to further our understanding of cancer mechanisms at the genetic level. It is known that cancerous cells often develop due to mutations at multiple loci, whose joint biological effects lead to uncontrolled growth. But many cancer-associated mutations and interactions among the mutated loci remain unknown. For example, highly penetrant cancer susceptibility genes, such as BRCA1 and BRCA2, are linked to breast cancer [5]. However, only about 5 to 10 percent of breast cancer can be explained by germ-line mutations in these single genes. "Most women with a family history of breast cancer do not carry germ-line mutations in the single highly penetrant cancer susceptibility genes, yet familial clusters continue to appear with each new generation" [6]. This kind of phenomenon is not yet well understood, and undiscovered mutations or undiscovered interactions among mutations are likely responsible.

Recently, machine-learning and data mining techniques have been developed to identify epistatic interactions in genomic data. Such methods include combinatorial methods, set association analysis, genetic programming, neural networks and random forests [7]. A well-known combinatorial method is Multifactor Dimensionality Reduction (MDR) [3,8-10]. MDR combines two or more variables into a single variable (hence leading to dimensionality reduction); this changes the representation space of the data and facilitates the detection of nonlinear interactions among the variables. MDR has been successfully applied to detect epistatic interactions in diseases such as sporadic breast cancer [3] and type II diabetes [8], typically in data sets containing at most a few hundred genetic loci.

10.1.2 GWAS

The most common genetic variation is the single nucleotide polymorphism (SNP) that results when a single nucleotide is replaced by another in the genomic sequence. In most cases a SNP is biallelic, that is it has only two possible values among A and G or C and T (the four DNA nucleotide bases). If the alleles are A and G, a diploid individual has the SNP genotype AA, GG, or AG. The less frequent (rare) allele must be present in 1% or more of the population for a site to qualify as a SNP [11]. The human genome contains many millions of SNPs. In what follows we will refer to SNPs as the loci investigated when searching for a correlation of some loci with a phenotype such as disease susceptibility.

The advent of high-throughput technologies has enabled genome-wide association studies (GWAS). A GWAS involves sampling in a population of individuals about 500,000 representative SNPs. Such studies provide researchers unprecedented opportunities to investigate the complex genetic basis of diseases. While the data in a GWAS have commonly been analyzed by investigating the association of each locus individually with the disease [12-16], there has been application of pathway analysis in some of these studies [15,16].

An important challenge in the analysis of genome-wide data sets is the identification of epistatic loci that interact in their association with disease. Many existing methods for epistasis learning such as combinatorial methods cannot handle a high-dimensional GWAS data set. For example, if we only investigated all 0, 1, 2, 3 and 4-SNP combinations when there are 500,000 SNPs, we would need to investigate 2.604×10^{21} combinations. Researchers are just beginning to develop new approaches for learning epistatic interactions using a GWAS data set [17-24]; however, the successful analysis of epistasis using high-dimensional data sets remains an open and vital problem. Cordell [25] provides a survey of methods currently used to detect gene-gene interactions that contribute to human genetic diseases. Most GWAS studies so far have been about "agnostic" discovery. Thomas [26] suggests combining data-driven approaches with hypothesis-driven, pathway-based analysis using hierarchical modeling strategies.

10.1.3 BAYESIAN NETWORKS

Bayesian networks [27-33] are increasingly being used for modeling and knowledge discovery in genetics and in genomics [34-41]. A Bayesian network (BN) is a probabilistic model that consists of a directed acyclic graph (DAG) G, whose nodes represent random variables, and a joint probability distribution P that satisfies the Markov condition with G. We say that (G,P) satisfies the Markov condition if each node (variable) in G is conditionally independent of the set of all its nondescendent nodes in G given the set of all its parent nodes. It is a theorem [31] that (G,P) satisfies the Markov condition (and therefore is a BN) if and only if P is equal to the product of the conditional distributions of all nodes given their parents in G, whenever these conditional distributions exist. That is, if the set of nodes is $\{X_1, X_2,...,X_n\}$, and PA_i is the set of parent nodes of X_1, then

$$P(x_1, x_2, ... x_n) = \prod_{j=1}^{n} P(x_j | PA_j)$$

BNs are often developed by first specifying a DAG that satisfies the Markov condition relative to our belief about the probability distribution, and then determining the conditional distributions for this DAG. One common way to specify the edges in the DAG is to include the edge X1 → X2 only if X1 is a direct cause of X2 [32]. Figure 1 shows an example of a BN. A BN can be used to compute conditional probabilities of interest using a BN inference algorithm [32]. For example, we can compute the

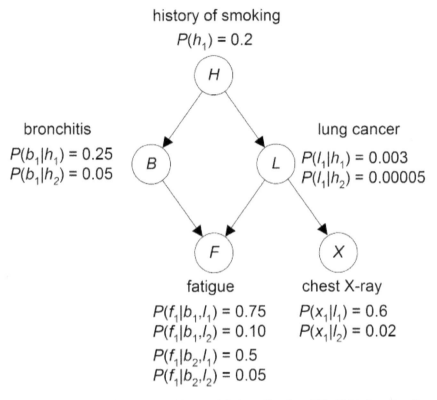

history of smoking
$P(h_1) = 0.2$

H

bronchitis

$P(b_1|h_1) = 0.25$
$P(b_1|h_2) = 0.05$

B

lung cancer

L

$P(l_1|h_1) = 0.003$
$P(l_1|h_2) = 0.00005$

F

X

fatigue

$P(f_1|b_1,l_1) = 0.75$
$P(f_1|b_1,l_2) = 0.10$
$P(f_1|b_2,l_1) = 0.5$
$P(f_1|b_2,l_2) = 0.05$

chest X-ray

$P(x_1|l_1) = 0.6$
$P(x_1|l_2) = 0.02$

FIGURE 1: An example BN. A BN that models lung disorders. This BN is intentionally simple to illustrate concepts; it is not intended to be clinically complete.

conditional probability that an individual has lung cancer and the conditional probability the individual has bronchitis given that the individual has a history of smoking and a positive chest X-ray.

Both the parameters and the structure of a BN can be learned from data. The Data consists of samples from some population, where each sample (called a data item) is a vector of values for all the random variables under consideration. Learning the structure of a BN is more challenging than learning the parameters of a specified BN structure, and a variety of techniques have been developed for structure learning. One method for structure learning, called constraint-based, employs statistical tests to identify DAG models that are consistent with the conditional independencies entailed by the data [42]. A second method, called score-based, employs heuristic search to find DAG models that maximize a desired scoring criterion [32]. Pierrier et al. [43] provide a detailed review of the methods for BN structure learning. Next we review scoring criteria since these criteria are the focus of this paper.

10.1.4 BN SCORING CRITERIA

We review several BN scoring criteria for scoring DAG models in the case where all variables are discrete since this is the case for the application we will consider. BN scoring criteria can be broadly divided into Bayesian and information-theoretic scoring criteria.

10.1.4.1 BAYESIAN SCORING CRITERIA

The Bayesian scoring criteria compute the posterior probability distribution, starting from a prior probability distribution on the possible DAG models, conditional on the Data. For a DAG G containing a set of discrete random variables $V = \{X_1, X_2,...,X_n\}$ and Data, the following Bayesian scoring criterion (or simply score) is derived under the assumption that all DAG models are equally likely a priori [44,45]:

$$\text{Score}_{\text{Bayes}}(G : \text{Data}) = P(\text{Data}|G)$$

$$= \prod_{i=1}^{n} \prod_{j=1}^{q_i} \frac{\Gamma\left(\sum_{k=1}^{r_i} a_{ijk}\right)}{\Gamma\left(\sum_{k=1}^{r_i} a_{ijk} + \sum_{k=1}^{r_i} s_{ijk}\right)} \prod_{k=1}^{r_i} \frac{\Gamma(a_{ijk} + s_{ijk})}{\Gamma(a_{ijk})}$$

$$(1)$$

where r_i is the number of states of X_i, q_i is the number of different values the parents of X_i in G can jointly assume, a_{ijk} is the prior belief concerning the number of times X_i took its kth value when the parents of X_i took their jth value, and s_{ijk} is the number of times in the data that X_i took its kth value when the parents of X_i took their jth value.

The Bayesian score given by Equation 1 assumes that our prior belief concerning each unknown parameter in each DAG model is represented by a Dirichlet distribution, where the hyperparameters aijk are the parameters for this distribution. Cooper and Herskovits [44] suggest setting the value of every hyperparameter a_{ijk} equal to 1, which assigns a prior uniform distribution to the value of each parameter (prior ignorance as to its value). Setting all hyperparameters to 1 yields the K2 score and is given by the following equation:

$$score_{K2}(G:Data) = \prod_{i=1}^{n}\prod_{j=1}^{q_i}\frac{\Gamma(r_i)}{\Gamma(r_i + \sum_{k=1}^{r_i} s_{ijk})}\prod_{k=1}^{r_i}\Gamma(1+s_{ikj})$$

The K2 score does not necessarily assign the same score to Markov equivalent DAG models. Two DAGs are Markov equivalent if they entail the same conditional independencies. For example, the DAGs X→Y and X ← Y are Markov equivalent. Heckerman et al. [45] show that if we determine the values of the hyperparameters from a single parameter α called the prior equivalent sample size then Markov equivalent DAGs obtain the same score. If we use a prior equivalent sample size α and want to represent a prior uniform distribution for each variable (not parameter) in the network, then for all i, j, and k we set $a_{ijk} = \alpha/r_iq_i$, where r_i is the number of states of the ith variable and q_i is the number of different values the parents of X_i can jointly assume. When we use a prior equivalent sample size α in the Bayesian score, the score is called the Bayesian Dirichlet equivalent (BDe) scoring criterion. When we also represent a prior uniform distribution for each variable, the score is called the Bayesian Dirichlet equivalent uniform (BDeu) scoring criterion and is given by the following equation:

$$score_{\alpha}(G:Data) = \prod_{i=1}^{n}\prod_{j=1}^{q_i}\frac{\Gamma(\alpha/q_i)}{\Gamma(\alpha/q_i + \sum_{k=1}^{r_i} s_{ijk})}\prod_{k=1}^{r_i}\frac{\Gamma(\alpha/r_iq_i + s_{ikj})}{\Gamma(\alpha/r_iq_i)}$$

The Bayesian score does not explicitly include a DAG penalty for network complexity. However, a DAG penalty is implicitly determined by the hyperparameters a_{ijk}. Silander et al. [46] show that if we use the BDeu score, then the DAG penalty decreases as α increases. The K2 score uses hyperparameters in a way that can be related to a prior equivalent sample size. When a node is modeled as having more parents, the K2 score effectively assigns a higher prior equivalent sample size to that node, which in turn decreases its DAG penalty.

10.1.4.2 MINIMUM DESCRIPTION LENGTH SCORING CRITERIA

The Minimum Description Length (MDL) Principle is an information-theoretic principle [47] which states that the best model is one that minimizes the sum of the encoding lengths of the data and the model itself. To apply this principle to scoring DAG models, we must determine the number of bits needed to encode a DAG G and the number of bits needed to encode the data given the DAG. Suzuki [48] developed the following MDL scoring criterion:

$$score_{Suz}(G:Data) = \sum_{i=1}^{n} \frac{d_i}{2} \log_2 m - m \sum_{i=1}^{n} \sum_{j=1}^{q_i} \sum_{k=1}^{r_i} P(x_{ik}, pa_{ij}) \log_2 \frac{P(x_{ik}, pa_{ij})}{P(x_{ik})P(pa_{ij})}$$

(2)

where n is the number of nodes in G, d_i is the number of parameters needed to represent the conditional probability distributions associated with the ith node in G, m is the number of data items, r_i is the number of states of X_i, x_{ik} is the kth state of X_i, q_i is the number of different values the parents of X_i can jointly assume, pa_{ij} is the jth value of the parents of X_i, and the probabilities are estimated from the Data. In Equation 2 the first sum is the DAG penalty, which is the number of bits sufficient to encode the DAG model, and the second term is the number of bits sufficient to encode the Data given the model.

Other MDL scores assign different DAG penalties and therefore differ in the first term in Equation 2, but encode the data the same. For example, the Akaike Information Criterion (AIC) score is an MDL scoring criterion that uses $\sum_{i=1}^{n} d_i$ as the DAG penalty. We will call this score $score_{AIC}$. In the DDAG Model section (acronym DDAG is defined in that section) we give an MDL score designed specifically for scoring BNs representing epistatic interactions.

10.1.4.3 MINIMUM MESSAGE LENGTH SCORING CRITERION

Another score based on information theory is the Minimum Message Length Score (MML) that is described in [30]. In the case of discrete variables it is equal to

$$score_{MML}(G:Data) = \sum_{i=1}^{n} d_i \left(\log_2 \frac{e^{3/2}\pi}{6} \right) - \log_2 score_{K2}(G:Data)$$

where d_1 is the number of parameters stored for the ith node in G and $score_{k2}$ is the K2 score mentioned previously.

To learn a DAG model from data, we can score all DAG models using one of the scores just discussed and then choose the highest scoring model. However, when the number of variables is not small, the number of candidate DAGs is forbiddingly large. Moreover, the BN structure learning problem has been shown to be NP-hard [49]. So heuristic algorithms have been developed to search over the space of DAGs during learning [32].

In the large sample limit, all the scoring criteria favor a model that most succinctly represents the generative distribution. However, for practical sized data sets, the results can be quite disparate. Silander et al. [46] provide a number of examples of learning models from various data sets showing that the choice of α in the BDeu scoring criterion can greatly affect how many edges exist in the selected model. For example, in the case of their Yeast data set (which contains 9 variables and 1484 data items), the number of edges in the selected model ranged from 0 to 36 as the value of α in the Bayesian scores ranged from 2×10^{-20} to 34,000. Although researchers have recommended various ways for choosing α and sometimes argued for the choice on philosophical/intuitive grounds [32], there is no agreed upon choice.

10.1.5 DETECTING EPISTASIS USING BNS

BNs have been applied to learning epistatic interactions from GWAS data sets. Han et al. [50] developed a Markov blanket-based method that uses a G^2 test instead of a BN scoring criterion. Verzilli et al. [51] represent the relationships among SNPs and a phenotype using a Markov network (MN), which is similar to a BN but contains undirected edges. They then use MCMC to do approximate model averaging to learn whether a particular edge is present. Both these methods model the relationships among SNPs besides the relationship between SNPs and a phenotype.

Jiang et al. [52] took a different approach. Since we are only concerned with discovering SNP-phenotype relationships, they used specialized BNs called DDAGs to model these relationships. DDAGs are discussed in the DDAG Model subsection of the Results section. They developed a combinatorial epistasis learning method called BNMBL that uses an MDL scoring criterion for scoring DDAGs. They compared BNMBL to MDR using the data sets developed in [10]. Each of these data sets was generated from a model that associates two SNPs with a disease and includes 18 unrelated SNPs. For each data set, BNMBL and MDR were used to score all 2-SNP models, and BNMBL learned significantly more correct models. In another study, Visweswaran et al. [53] employed a K2-based scoring criterion for scoring these same DAG models that also outperformed MDR.

In real data sets, we ordinarily do not know the number of SNPs that influence phenotype. BNMBL may not perform as well if we also scored models containing more than two SNPs. Although BNs are a promising tool for learning epistatic relationships from data, we cannot confidently use them in this domain until we determine which scoring criteria work best or even well when we try learning the correct model without knowledge of the number of SNPs in that model. We provide results of experiments investigating this performance in the Results section.

10.1.6 DIAGNOSTIC BNS CONTAINING SNP VARIABLES

BN diagnostic systems that contain SNP information have also been learned from data. For example, Sebastiani et al. [54] learned a BN that predicts stroke in individuals with sickle cell anemia, while Meng et al. [55] learned a BN that predicts rheumatoid arthritis. In these studies candidate SNPs were identified based on known metabolic pathways. This is in contrast to the agnostic search ordinarily used to analyze GWAS data sets (discussed above). For example, Sebastiani et al. [54] identified 80 candidate genes and analyzed 108 SNPs in these genes.

10.2 RESULTS

We first describe the BN model used to model SNP interactions associated with disease. Next, we develop a BN score tailored to this model and list the other BN scores that are evaluated. Finally, we provide the results of experiments that evaluate the various BN scores and MDR using simulated data and a real GWAS data set.

10.2.1 THE DDAG MODEL

We use BNs to model the relationships among SNPs and a phenotype such as disease susceptibility. Given a set of SNPs $\{S_1, S_2, ...,S_n\}$ and a disease D, we consider all DAGs in which node D has only incoming edges and no outgoing edges. Such DAGs have the causal interpretation that SNPs are either direct or indirect causes of disease. An example of a DAG for 9 SNPs is shown in Figure 2. This DAG does not represent the relationships among gene expression levels. Rather it represents the statistical dependencies involving the disease status and the alleles of the SNPs. Since we are only concerned with modeling the dependence of the disease on the SNPs and not the relationships among the SNPs, there is no need for edges between SNPs. So we need only consider DAGs where the only edges are

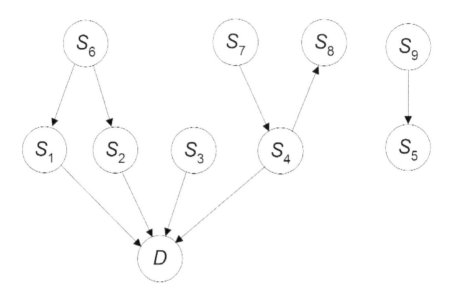

FIGURE 2: An example DAG. A DAG showing probabilistic relationships among SNPs and a disease D.

ones to D. An example of such a DAG is shown in Figure 3. We call such a model a direct DAG (DDAG).

The number of DAGs that can be constructed is forbiddingly large when the number of nodes is not small. For example, there are $\sim 4.2 \times 10^{18}$ possible DAGs for a domain with ten variables [56]. The space of DDAGs is much smaller: there are 2^n DDAGs, where n is the number of SNPs. So if we have ten SNPs, there are only 2^{10} DDAGs. Though the model space of DDAGs is much smaller that the space of DAGs, it still remains exponential in the number of variables. In the studies reported here, we search in the space of DDAGs.

10.2.2 THE BN MINIMUM BIT LENGTH (BNMBL) SCORE

An MDL score called BNMBL that is adapted to DDAGs is developed next. Each parameter (conditional probability) in a DAG model learned

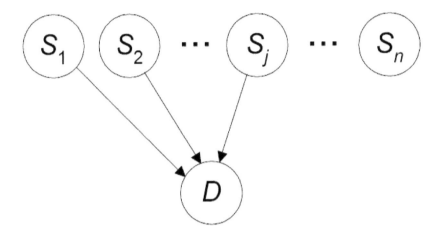

FIGURE 3: An example DDAG. A DDAG showing probabilistic relationships between SNPs and a disease D. A DDAG differs from the DAG in Figure 2 in that the relationships among the SNPs are not represented.

from data is a fraction with precision $1/m$, where m is the number of data items. Therefore, it requires $O(\log_2 m)$ bits to store each parameter. However, as explained in [57], the high order bits are not very useful. So we need use only $(1/2)\log_2 m$ bits and we arrive at the DAG penalty in Equation 2.

Suppose that k SNPs have edges into D in a given DDAG. Since each SNP has three possible values, there are 3^k joint states of the parents of D. The expected value of the number of data items, whose values for these k SNPs are the values in each joint state, is $m/3^k$. If we approximate the precision for each of D's parameters by this average, the penalty for each of these parameters is $(1/2)\log_2(m/3^k)$. Since the penalty for each parameter in a parent SNP is $(1/2)\log_2 m$, the total DAG penalty for a DDAG model is

$$\frac{3^k}{2}\log_2 \frac{m}{3^k} + \frac{2k}{2}\log_2 m \qquad (3)$$

The multiplier 2 appears in the second term because each SNP has three values. We need store only two of the three parameters corresponding to the SNP states, since the value of the remaining parameter is uniquely determined given the other two. No multiplier appears in the first term because the disease node has only two values. When we use this DAG penalty in an MDL score (Equation 2), we call the score scoreEpi.

10.2.3 BN SCORING CRITERIA EVALUATED

We evaluated the performance of MDR; three MDL scores: $score_{Epi}$, $score_{Suz}$, and $score_{AIC}$; two Bayesian scores: $score_{K2}$, and $score_{a}$; and the information-theoretic score $score_{MML}$. For $score_{a}$ we performed a sensitivity analysis over the following values of $\alpha = 1\ 3, 6, 9, 12, 15, 18, 21, 24, 30, 54, 162$. We evaluated two versions of each of the MDL scores. In the first version, all n SNPs in the domain are included in the model, though only k of them directly influence D and hence have edges to D in the DDAG. In this case the contribution of the SNP nodes to the DAG penalty is not included in the score because it is the same for all models. We call this version 1, and denote the score with the subscript 1 (scoreEpi1). In the second version, only the k SNPs that have edges to D are included in the model and the remaining n-k SNPs are excluded from the model. In this case, the contributions of the k SNP nodes to the penalty are included because models with different values of k have different penalties. We call this version 2, and denote the score with the subscript 2 (e.g., $score_{Epi2}$). The penalty term for $score_{Epi}$ that is given in Equation 3 is for version 2.

After describing the results obtained using simulated data, we show those for real data.

10.2.4 SIMULATED DATA RESULTS

We evaluated the scoring criteria using simulated data sets that were developed from 70 genetic models with different heritabilities, minor allele frequencies and penetrance values. Each model consists of a probabilistic relationship in which 2 SNPs combined are correlated with the disease, but

TABLE 1: Accuracies of scoring criteria

	Scoring Criterion	200	400	800	1600	Total
1	$score_{\alpha=15}$	4379	5426	6105	6614	22524
2	$score_{\alpha=12}$	4438	5421	6070	6590	22519
3	$score_{\alpha=18}$	4227	5389	6095	6625	22336
4	$score_{\alpha=9}$	4419	5349	5996	6546	22313
5	$score_{\alpha=21}$	3989	5286	6060	6602	21934
6	$score_{\alpha=6}$	4220	5165	5874	6442	21701
7	$score_{MML1}$	4049	5111	5881	6463	21504
8	$score_{\alpha=24}$	3749	5156	5991	6562	21448
9	$score_{MDR}$	4112	4954	5555	5982	20603
10	$score_{\alpha=3}$	3839	4814	5629	6277	20559
11	$score_{Epi2}$	3571	4791	5648	6297	20307
12	$score_{\alpha=30}$	3285	4779	5755	6415	20234
13	$score_{MML2}$	3768	4914	5754	5780	20216
14	$score_{Epi1}$	2344	5225	6065	6553	20187
15	$score_{Suz1}$	3489	4580	5521	6215	19805
16	$score_{\alpha=36}$	2810	4393	5464	6150	18817
17	$score_{\alpha=42}$	2310	4052	5158	5895	17415
18	$score_{K2}$	1850	3475	5095	6116	16536
19	$score_{Suz2}$	2245	3529	4684	5673	16131
20	$score_{\alpha=54}$	1651	3297	4492	5329	14769
21	$score_{AIC2}$	3364	3153	2812	2520	11847
22	$score_{AIC1}$	2497	1967	1462	1126	7052
23	$score_{\alpha=162}$	26	476	1300	2046	3848

The number of times out of 7000 data sets that each scoring criterion identified the correct model for sample sizes of 200, 400, 800, and 1600. The last column gives the total accuracy over all sample sizes. The scoring criteria are listed in descending order of total accuracy.

neither SNP is individually correlated. Each data set has sample size equal to 200, 400, 800, or 1600, and there are 7000 data sets of each size. More details of the datasets are given in the Methods section.

For each of the simulated data sets, we scored all 1-SNP, 2-SNP, 3-SNP, and 4-SNP DDAGs. The total number of DDAGs scored for each data set was therefore 6195. Since in a real setting we would not know the number of SNPs in the model generating the data, all models were treated equally in the learning process; that is, no preference was given to 2-SNP models.

We say that a method correctly learns the model generating the data if it scores the DDAG representing the generating model highest out of all 6195 models. Table 1 shows the number of times out of 7000 data sets that each BN scoring criterion correctly learned the generating model for each sample size. In this table, the scoring criteria are listed in descending order according to the total number of times the correct model was learned. Table 1 shows a number of interesting results. First, the AIC score performed reasonably well on small sample sizes, but its performances degraded at larger sample sizes. Unlike the other BN scores, the DAG penalty in the AIC score does not increase with the sample size. Second, the K2 score did not perform well, particularly at small sample sizes. However, the MML1 score, which can be interpreted as the K2 score with an added DAG penalty, performed much better. This indicates that the DAG penalty in the K2 score may be too small and the increased penalty assigned by the MML1 score is warranted. Third, MDR performed well overall but substantially worse than the best performing scores. Fourth, the best results were obtained with the BDeu score at moderate values of α. However, the results were very poor for large values of α, which assign very small DAG penalties.

TABLE 2: Statistical comparison of accuracies of scoring criteria

	Scoring Criterion	p-value
1	$score_{\alpha=15}$	NA
2	$score_{\alpha=12}$	0.996
3	$score_{\alpha=18}$	0.076
4	$score_{\alpha=9}$	0.046
5	$score_{\alpha=21}$	4.086×10^{-8}
6	$score_{\alpha=6}$	3.468×10^{-14}
7	$score_{MML1}$	1.200×10^{-20}

P-values obtained by comparing the accuracy of the highest ranking scoring criterion (score $\alpha = 15$) with the next six highest ranking scoring criteria using the McNemar chi-square test. Each p-value is obtained by comparing the accuracies for 28,000 data sets.

The ability of the highest ranking score (the BDeu score$_{\alpha=15}$) to identify the correct model was compared to that of the next six highest ranking scores using the McNemar chi-square test (see Table 2). In a fairly small interval around $\alpha = 15$ there is not a significant difference in performance. However, as we move away from $\alpha = 15$ the significance becomes dramatic, as is the significance relative to the highest scoring non-BDeu score (score$_{MML1}$).

BDeu scores with values of α in the range 12 - 18 performed significantly better than all other scores. If our goal is only to find a score that most often scores the correct model highest on low-dimensional simulated data sets like the ones analyzed here, then our results support the use of these BDeu scores. However, in practice, we are interested in the discovery of promising SNP-disease associations that may be investigated for biological plausibility. So perhaps more relevant than whether the correct model scores the highest is the recall of the correct model relative to the highest scoring model. The recall is given by:

$$recall(S,T) = \frac{\#(S \cap T)}{\#S}$$

where S is the set of SNPs in the correct model, T is the set of SNPs in the highest scoring model, and # returns the number of items in a set. The value of the recall is 0 if and only if the two sets do not intersect, while it is 1 if and only if all the SNPs in the correct model are in the highest scoring model. Therefore, recall is a measure of how well the SNPs in the correct model were are discovered. Recall does not measure, however, the extent to which the highest scoring model has additional SNPs that are not in the correct model (i.e., false positives).

Table 3 shows the recall for the various scoring criteria. The criteria are listed in descending order of total recall. Overall, these results are the reverse of those in Table 1. The BDeu scores with large values of α and the AIC scores appear at the top of the list. Part of the explanation for this is that these BDeu scores and AIC scores incorporate small DAG penalties, which results in larger models often scoring higher. A larger model has a greater chance of containing the two interacting SNPs. Not surprisingly scoreSuz1 and scoreSuz2, which have the largest DAG penalties of the MDL scores, appear at the bottom of the list. MDR again performed well but substantially worse than the best performing scores.

TABLE 3: Recall for scoring criteria

	Scoring Criterion	200	400	800	1600	Total
1	$score_{\alpha=162}$	5259	60433	6566	6890	24758
2	$score_{AIC2}$	5204	5969	6511	6849	24533
3	$score_{AIC1}$	5186	5960	6481	6830	24457
4	$score_{\alpha=54}$	5223	5941	6473	6813	24450
5	$score_{K2}$	5303	5962	6371	6747	24383
6	$score_{\alpha=42}$	5203	5902	6425	6794	24324
7	$score_{\alpha=36}$	5181	5866	6395	6768	24210
8	$score_{\alpha=30}$	5147	5816	6352	6754	24069
9	$score_{\alpha=24}$	5080	5767	6300	6725	23872
10	$score_{\alpha=21}$	5031	5733	6265	6704	23733
11	$score_{MDR}$	4870	5710	6324	6748	23652
12	$score_{\alpha=18}$	4973	5681	6230	6681	23565
13	$score_{\alpha=15}$	4902	5622	6183	6647	23354
14	$score_{Epi1}$	4984	5529	6105	6575	23193
15	$score_{\alpha=12}$	4786	5531	6119	6605	23041
16	$score_{\alpha=9}$	4649	5416	6026	6547	22638
17	$score_{\alpha=6}$	4383	5219	5901	6453	21956
18	$score_{MML1}$	4151	5159	5903	6473	21686
19	$score_{MML2}$	3881	4969	5780	6412	21042
20	$score_{Epi2}$	3895	4901	5715	6329	20840
21	$score_{\alpha=3}$	3953	4862	5652	6285	20752
22	$score_{Suz1}$	3618	4696	5595	6251	20160
23	$score_{Suz2}$	2500	3712	4811	5737	17760

The sum of the recall for each scoring criterion over 7000 data sets for sample sizes of 200, 400, 800, and 1600. The last column gives the total recall over all sample sizes. The scoring criteria are listed in descending order of total recall.

Perhaps the smaller DAG penalty is not the only reason that the BDeu scores with larger values of α performed best. It is possible that the BDeu scores with larger values of α can better detect the interacting SNPs than the BDeu scores with smaller values, but that the scores with larger values do poorly at scoring the correct model (the one with only the two interacting SNPs) highest because they too often pick a larger model containing those SNPs. To investigate this possibility, we investigated how well the scores discovered models 55-59 (See Supplementary Table one to [10]). These models have the weakest broad-sense heritability (0.01) and a minor allele frequency of 0.2, and are therefore the most difficult to detect.

Table 4 shows the number of times the correct hard-to-detect model scored highest for a representative set of the scores. Table 5 shows the p-values obtained when the highest ranking score (BDeu scoreα = 54) is compared to the next five highest ranking scores using the McNemar chi-square test. The BDeu score with large values of α performed significantly better than all other scores.

TABLE 4: Accuracies of scoring criteria on most difficult models

	Scoring Criterion	200	400	800	1600	Total
1	score$_{\alpha = 54}$	14	48	167	352	581
2	score$_{\alpha = 162}$	1	21	146	355	563
3	score$_{\alpha = 36}$	13	46	155	318	532
4	score$_{\alpha = 21}$	12	43	106	289	450
5	score$_{\alpha = 18}$	11	37	91	274	413
6	score$_{MDR}$	3	25	79	245	352
7	score$_{\alpha = 12}$	7	25	65	215	312
8	score$_{AIC2}$	16	33	80	138	267
9	score$_{\alpha = 9}$	5	20	48	186	259
10	Score$_{Epi1}$	4	16	47	179	246
11	score$_{MML1}$	2	7	23	140	172
12	score$_{\alpha = 3}$	3	6	13	86	108
13	score$_{Epi2}$	0	1	4	72	77
14	score$_{Suz1}$	0	1	2	41	44

The number of times out of 500 that each scoring criterion correctly learned the correct model in the case of the most difficult models (55-59) for sample sizes of 200, 400, 800, and 1600. The last column gives the total accuracy over all sample sizes. The scoring criteria are listed in descending order of accuracy.

TABLE 5: Statistical comparison of accuracies of scoring criteria on most difficult models

Scoring Criterion	p-value
$score_{\alpha = 54}$	NA
$score_{\alpha = 162}$	0.610
$score_{\alpha = 36}$	0.147
$score_{\alpha = 21}$	4.870×10^{-5}
$score_{\alpha = 18}$	1.080×10^{-7}
$score_{MDR}$	7.254×10^{-14}

P-values obtained by comparing the accuracy of the highest ranking scoring criterion (scoreα = 15) with the next five highest ranking scoring criteria using the McNemar chi-square test. Each p-value is obtained by comparing the accuracies for 2,000 data sets generated by the hardest-to-detect models.

The BDeu scores with large α values discovered the difficult models best, though they perform poorly on the average when all models were considered. An explanation for this phenomenon is that these scores can indeed find interacting SNPs better than scores with smaller values of α. However, when the interacting SNPs are fairly easy to identify, their larger DAG penalties makes it harder for them to identify the correct model relative to other scores. On the other hand, when the SNPs are hard to detect, their better detection capability more than compensates for their increased DAG penalty. Additional file 1 provides an illustrative example of this phenomenon. We hypothesize therefore that BDeu scores with larger values of α can better indentify interacting SNPs, even if they sometimes include extra SNPs in the highest scoring model.

10.2.5 GWAS DATA RESULTS

We evaluated the scoring criteria using a late onset Alzheimer's disease (LOAD) GWAS data set. LOAD is the most common form of dementia in the above 65-year-old age group. It is a progressive neurodegenerative disease that affects memory, thinking, and behavior. The only genetic risk factor for LOAD that has been consistently replicated involves the apolipoprotein E (APOE) gene. The ε4 APOE genotype increases the risk of development of LOAD, while the ε2 genotype is believed to have a protective effect.

The LOAD GWAS data set that we analyzed was collected and analyzed by Rieman et al. [16]. The data set contains records on 1411 participants (861 had LOAD and 550 did not), and consists of data on 312,316 SNPs and one binary genetic attribute representing the apolipoprotein E (APOE) gene carrier status. The original investigators found that SNPs on the GRB-associated binding protein 2 (GAB2) gene interacted with the APOE gene to determine the risk of developing LOAD. More details of this dataset are given in the Methods section.

To analyze this Alzheimer GWAS data set, for a representative subset of the scores listed in Table 1 we did the following. We pre-processed the data set by scoring all models in which APOE and one of the 312,316 SNPs are each parents of the disease node LOAD. The SNPs from the top 100 highest-scoring models were selected along with APOE. Using these 101 loci, we then scored all 1, 2, 3, and 4 parent models making a total of 4,254,726 models scored. We judged the effectiveness of each score according to how well it replicated the results obtained by the original investigators in [16] that the GAB2 gene is associated with LOAD. We did this by determining how many of the score's 25 highest-scoring models contained a GAB2 SNP. Table 6 shows the results. The number in each cell in Table 6 is the number of SNPs in the model, and the letter G appears to the right of that number if a GAB2 SNP appears in the model. The second to the last row in the table shows the total number of models in the top 25 that contain a GAB2 SNP. The last row in the table shows the total number of different GAB2 SNPs appearing in the top 25 models.

We included two new scores in this analysis. The first score is the BDeu score with α = 1000. We did this to test whether we can get good recall with arbitrarily high values of α. The second new score is an MDL score with no DAG penalty (labelled MDLn in the table). We did this to investigate the recall for the MDL score when we constrain the highest scoring model to be one containing four parent loci.

These results substantiate our hypothesis that larger values of α (54 and 162) can better detect the interacting SNPs. For each of the BDeu scores, the 25 highest-scoring models each contain 4 parent loci. However, when α equals 54 or 162, 19 and 18 respectively of the 25 highest-scoring models contain a GAB2 SNP, whereas for α equal to 12 only 7 of them contain a GAB2 SNP, and for α equal to 3 none of them do. The results for α equal to 1000 are not very good, indicating that we cannot obtain good results for arbitrarily large values of α. The MDL scores (MDLn, Suz1 and Epi2) all performed well, with the Suz1 score never selecting a model with more than 3 parent loci. This result indicates that the larger DAG penalty seems to have helped us hone in on the interacting SNPs. All the MDL scores detected the highest number of different GAB2 SNPs, namely 8. In comparison, MDR did not perform very well, having only 8 models of the top 25 containing GAB2 SNPs and none of the top 5 containing GAB2 SNPs.

10.3 DISCUSSION

We compared the performance of a number of BN scoring criteria when identifying interacting SNPs from simulated genetic data sets. Each data set contained 20 SNPs with two interacting SNPs and was generated from one of 70 different epistasis models. Jiang et al. [52] analyzed these same data sets using the BNMBL method and MDR (both of these methods are discussed in the Background section). However, that paper only investigated models with two interacting SNPs. So the 1-SNP, 3-SNP, and 4-SNP models were not competing and the learned model was restricted to be a 2-SNP model. In real applications we rarely would know how many SNPs are interacting. So this type of analysis is not as realistic as the one reported here.

Table 1 shows that the BDeu score with values of α between 12 and 18 was best at learning the correct model over all 28,000 simulated data sets. However, Table 3 shows that the BDeu score with large values of α (54 and 162) performed better at recall over all 28,000 data sets. Table 4 shows that these large values of α yield better detection of the models that are hardest to detect.

We evaluated the performance of a subset of the BN scores used in the simulated data analysis on a LOAD GWAS data set. The effectiveness of each score was judged according to how well it substantiated the previously obtained result that the GAB2 gene is associated with LOAD. As shown in Table 6, we obtained the best results with the BDeu score with large values of α. The various MDL scores also performed well.

Overall, our results are mixed. Although scores with moderate values of α performed better at actually scoring the correct model highest using simulated data sets, scores with larger values of α performed better at recall, at detecting models that are hardest to detect, and at substantiating previous results using a real data set. Our main goal is to develop a method that can discover SNPs associated with a disease from real data. Therefore, based on the results reported here, it seems that it is more promising to use the BDeu score with large values of α (54-162), rather than smaller values.

The MDL scores also performed well in the case of the real data set. An explanation for their poor performance with the simulated data sets is that their DAG penalties are either too large or too small. If we simply used an MDL score with no DAG penalty we should be able to discover interacting SNPs well (as indicated by Table 6). Once we determine candidate interactions using these scores, we can perform further data analysis of the interactions and also investigate the biological plausibility of the genotype-phenotype relationships. However, additional research is needed to further investigate a DAG penalty appropriate to this domain.

Another consideration which was not investigated here is the possible increase in false positives with increased detection capability. That is, although the BDeu score with large values of α performed best at recall and at identifying hard-to-detect models, perhaps these scores may also score some incorrect models higher, and at a given threshold might have more false positives. Further research is needed to investigate this matter.

Additional file 1 provides an illustrative example and some theoretical justification as to why a BDeu score with large values of α should perform well at discovering hard-to-detect SNP-phenotype relationships. However, further research, both of a theoretical and empirical nature, is needed to investigate the pattern of results reported here. In particular, additional simulated data sets containing data on a large number of SNPs (numbers appearing in real studies) should be analyzed to see if the BDeu score with large values of α or some other approach performs better in this more realistic setting.

10.4 CONCLUSIONS

Our results indicate that representing epistatic interactions using BNs and scoring them using a BN scoring criteria holds promise for identifying epistatic relationships. Furthermore, they show that the use of the BDeu score with large values of α (54-162) can yield the best results on some data sets. Compared to MDR and other BN scoring criteria, these BDeu scores performed substantially better at detecting the hardest-to-detect models using simulated data sets, and at confirming previous results using a real GWAS data set.

10.5 METHODS

10.5.1 SIMULATED DATA SETS

Each simulated data set was developed from one of 70 epistasis models described in Velez et al. [10] (see Supplementary Table one in [10] for details of the 70 models). These datasets are available at http://discovery. dartmouth.edu/epistatic_data/.

Each model represents a probabilistic relationship in which two SNPs together are correlated with the disease, but neither SNP is individually predictive of disease. The relationships represent various degrees of penetrance, heritability, and minor allele frequency. The models are distributed uniformly among seven broad-sense heritabilities ranging from 0.01 to 0.40 (0.01, 0.025, 0.05, 0.10, 0.20, 0.30, and 0.40) and two minor allele frequencies (0.2 and 0.4).

Data sets were generated with case-control ratio (ratio of individuals with the disease to those without the disease) of 1:1. To create one data set they fixed the model. Based on the model, they then generated data concerning the two SNPs that were related to the disease in the model, 18 other unrelated SNPs, and the disease. For each of the 70 models, 100 data sets were generated for a total of 7000 data sets. This procedure was followed for data set sizes equal to 200, 400, 800, and 1600.

10.5.2 GWAS DATA SET

Several LOAD GWA studies have been conducted. We utilized data from one such study [16] that contains data on 312,316 SNPs. In this study, Reiman et al. investigated the association of SNPs separately in APOE ε4 carriers and in APOE ε4 noncarriers. A discovery cohort and two replication cohorts were used in the study. Within the discovery subgroup consisting of APOE ε4 carriers, 10 of the 25 SNPs exhibiting the greatest association with LOAD (contingency test p-value 9×10^{-8} to 1×10^{-7}) were located in the GRB-associated binding protein 2 (GAB2) gene on chromosome 11q14.1. Associations with LOAD for 6 of these SNPs were confirmed in the two replication cohorts. Combined data from all three cohorts exhibited significant association between LOAD and all 10 GAB2 SNPs. These 10 SNPs were not significantly associated with LOAD in the APOE ε4 noncarriers.

10.5.3 IMPLEMENTATION

We implemented the methods for learning and scoring DDAGs using BN scoring criteria in the Java programming language. MDR v. mdr-2.0_beta_5 (available at http://www.epistasis.org) with its default settings (Cross-Validation Count = 10, Attribute Count Range = 1:4, Search Type = Exhaustive) was used to run MDR. All experiments were run on a 32-bit Server running Windows 2003 with a 2.33 GHz processor and 2.00 GB of RAM.

REFERENCES

1. Bateson W: Mendel's Principles of Heredity. New York; Cambridge University Press; 1909.
2. Moore JH, Williams SM: New strategies for identifying gene gene interactions in hypertension. Annals of Medicine 2002, 34:88-95. 2002
3. Ritchie MD, et al.: Multifactor-dimensionality reduction reveals high-order interactions among estrogen-metabolism genes in sporadic breast cancer. American Journal of Human Genetics 2001, 69:138-147.

4. Nagel RI: Epistasis and the genetics of human diseases. C R Biologies 2005, 328:606-615.

5. Armes BM, et al.: The histologic phenotypes of breast carcinoma occurring before age 40 years in women with and without BRCA1 or BRCA2 germline mutations. Cancer 2000, 83:2335-2345.

6. National Cancer Institute: Cancer Genomics [http://www.cancer.gov/cancertopics/understandingcancer/cancergenomics]

7. Heidema A, Boer J, Nagelkerke N, Mariman E, van der AD, Feskens E: The challenge for genetic epidemiologists: how to analyze large numbers of SNPs in relation to complex diseases. BMC Genetics 2006, 7:23. (21 April 2006)

8. Cho YM, Ritchie MD, Moore JH, Moon MK, et al.: Multifactor dimensionality reduction reveals a two-locus interaction associated with type 2 diabetes mellitus. Diabetologia 2004, 47:549-554.

9. Hahn LW, Ritchie MD, Moore JH: Multifactor dimensionality reduction software for detecting gene-gene and gene-environment interactions. Bioinformatics 2003, 19:376-382.

10. Velez DR, White BC, Motsinger AA, et al.: A balanced accuracy function for epistasis modeling in imbalanced data sets using multifactor dimensionality reduction. Genetic Epidemiology 2007, 31:306-315.

11. Brookes AJ: The essence of SNPs. Gene 1999, 234:177-186.

12. Herbert A, Gerry NP, McQueen MB: A common genetic variant is associated with adult and childhood obesity. Journal of Computational Biology 2006, 312:279-384.

13. Spinola M, Meyer P, Kammerer S, et al.: Association of the PDCD5 locus with long cancer risk and prognosis in Smokers. American Journal of Human Genetics 2001, 55:27-46.

14. Lambert JC, et al.: Genome-wide association study identifies variants at CLU and CR1 associated with Alzheimer's disease. Nature Genetics 2009, 41:1094-1099.

15. Coon KD, et al.: A high-density whole-genome association study reveals that APOE is the major susceptibility gene for sporadic late-onset Alzheimer's disease. Journal of Clinical Psychiatry 2007, 68:613-618.

16. Reiman EM, et al.: GAB2 alleles modify Alzheimer's risk in APOE carriers. Neuron 2007, 54:713-720.

17. Brinza D, He J, Zelkovsky A: Optimization methods for genotype data analysis in epidemiological studies. In Bioinformatics Algorithms: Techniques and Applications. Edited by Mandoiu I, Zelikovsky A. New York; Wiley; 2008::395-416.

18. Wu TT, Chen YF, Hastie T, Sobel E, Lange K: Genome-wide association analysis by lasso penalized logistic regression. Genome Analysis 2009, 25:714-721.

19. Wu J, Devlin B, Ringguist S, Trucco M, Roeder K: Screen and clean: A tool for identifying interactions in genome-wide association studies. Genetic Epidemiology 2010, 34:275-285.

20. Wongseree W, et al.: Detecting purely epistatic multi-locus interactions by an omnibus permutation test on ensembles of two-locus analyses. BMC Bioinformatics 2009, 10:294.

21. Zhang X, Pan F, Xie Y, Zou F, Wang W: COE: a general approach for efficient genome-wide two-locus epistasis test in disease association study. Journal of Computational Biology 2010, 17(3):401-415.

22. Meng Y, et al.: Two-stage approach for identifying single-nucleotide polymorphisms associated with rheumatoid arthritis using random forests and Bayesian networks. BMC Proc 2007, 1(Suppl 1):S56.

23. Wan X, et al.: Predictive rule inference for epistatic interaction detection in genome-wide association studies. Bioinformatics 2010, 26(1):30-37.

24. Logsdon BA, Hoffman GE, Mezey JG: A variational Bayes algorithm for fast and accurate multiple locus genome-wide association analysis. BMC Bioinformatics 2010, 11:58.

25. Cordell HJ: Detecting gene-gene interactions that underlie human diseases. Nat Rev Genetics 2009, 10(6):392-404.

26. Thomas D: Methods for investigating gene-environment interactions in candidate pathway and genome-wide association studies. Annu Rev Public Health 2010, 31:1-8.

27. Castillo E, Gutiérrez JM, Hadi AS: Expert Systems and Probabilistic Network Models. New York; Springer-Verlag; 2007.

28. Jensen FV: An Introduction to Bayesian Networks. New York; Springer-Verlag; 1997.

29. Jensen FV, Neilsen TD: Bayesian Networks and Decision Graphs. New York; Springer-Verlag; 2007.

30. Korb K, Nicholson AE: Bayesian Artificial Intelligence. Boca Raton, FL; Chapman & Hall/CRC; 2003.

31. Neapolitan RE: Probabilistic Reasoning in Expert Systems. New York; Wiley; 1990.

32. Neapolitan RE: Learning Bayesian Networks. Upper Saddle River, NJ; Prentice Hall; 2004.

33. Pearl J: Probabilistic Reasoning in Intelligent Systems. Burlington, MA; Morgan Kaufmann; 1988.

34. Fishelson M, Geiger D: Exact genetic linkage computations for general pedigrees. Bioinformatics 2002, 18(Suppl 1):189-198.

35. Fishelson M, Geiger D: Optimizing exact genetic linkage computation. Journal of Computational Biology 2004, 11:263-275.

36. Friedman N, Koller K: Being Bayesian about network structure: a Bayesian approach to structure discovery in Bayesian networks. Machine Learning 2003, 20:95-126.

37. Friedman N, Goldszmidt M, Wyner A: Data analysis with Bayesian networks: a bootstrap approach. In Proceedings of the Fifteenth Conference on Uncertainty in Artificial Intelligence. Edited by Laskey KB, Prade H. Burlington, MA; Morgan Kaufmann; 1999::196-205.

38. Friedman N, Linial M, Nachman I, Pe'er D: Using Bayesian networks to analyze expression data. Proceedings of the Fourth Annual International Conference on Computational Molecular Biology 2005, :127-135.

39. Friedman N, Ninio M, Pe'er I, Pupko T: A structural EM algorithm for phylogenetic inference. Journal of Computational Biology 2002, 9(2):331-353.

40. Neapolitan RE: Probabilistic Methods for Bioinformatics: with an Introduction to Bayesian networks. Burlington, MA: Morgan Kaufmann; 2009.

41. Segal E, Pe'er D, Regev A, Koller D, Friedman N: Learning module networks. Journal of Machine Learning Research 2005, 6:557-588.

42. Spirtes P, Glymour C, Scheines R: Causation, Prediction, and Search. second edition. New York; Springer-Verlag; Boston, MA; MIT Press; 1993.

43. Perrier E, Imoto S, Miyano S: Finding optimal Bayesian network given a superstructure. Journal of Machine Learning Research 2008, 9:2251-2286.

44. Cooper GF, Herskovits E: A Bayesian method for the induction of probabilistic networks from data. Machine Learning 1992, 9:309-347.

45. Heckerman D, Geiger D, Chickering D: Learning Bayesian Networks: The Combination of Knowledge and Statistical Data. Technical Report MSR-TR-94-09, Microsoft Research, Redmond, Washington; 1995.

46. Silander T, Kontkanen P, Myllymäki P: On sensitivity of the MAP Bayesian network structure to the equivalent sample size parameter. In Proceedings of the Twenty-Third Conference on Uncertainty in Artificial Intelligence. Edited by Parr R, van der Gaag L. Corvallis, Oregon; AUAI Press; 2002::360-367.

47. Rissanen J: Modeling by shortest data description. Automatica 1978, 14:465-471.

48. Suzuki J: Learning Bayesian belief networks based on the minimum description length principle: basic properties. IEICE Transactions on Fundamentals 1999, E82-A:2237-2245.

49. Chickering M: Learning Bayesian networks is NP-complete. In Learning from Data: Lecture Notes in Statistics. Edited by Fisher D, Lenz H. New York: Springer Verlag; 1996::121-130.

50. Han B, Park M, Chen X: A Markov blanket-based method for detecting causal SNPs in GWAS. BMC Bioinformatics 2010, 11(Suppl 3):S5.

51. Verzilli CJ, Stallard N, Whittaker JC: Bayesian graphical models for genomewide association studies. The American Journal of Human Genetics 2006, 79:100-112.

52. Jiang X, Barmada MM, Visweswaran S: Identifying genetic interactions from genome-wide data using Bayesian networks. Genet Epidemiol 2010, 34(6):575-581.

53. Visweswaran S, Wong AI, Barmada MM: A Bayesian method for identifying genetic interactions. Proceedings of the Fall Symposium of the American Medical Informatics Association 2009, :673-677.

54. Sebastiani P: Genetic dissection and prognostic modeling of overt stroke in sickle cell anemia. Nature Genetics 2005, 37:435-440.

55. Meng Y, et al.: Two-stage approach for identifying single-nucleotide polymorphisms associated with rheumatoid arthritis using random forests and Bayesian networks. BMC Proc 2007, 1(Suppl 1):S56.

56. Robinson RW: Counting unlabeled acyclic digraphs. In Lecture Notes in Mathematics. Volume 622. Edited by Little CHC. New York: Springer-Verlag; 1977::28-43.

57. Friedman N, Yakhini Z: On the sample complexity of learning Bayesian networks. Proceedings of the Twelfth Conference on Uncertainty in Artificial Intelligence 1996, :206-215.

There are several supplemental files, as well as one table, that are not available in this version of the article. To view this additional information, please use the citation information cited on the first page of this chapter.

CHAPTER 11

COMBINED ANALYSIS OF THREE GENOME-WIDE ASSOCIATION STUDIES ON vWF AND FVIII PLASMA LEVELS

GUILLEMETTE ANTONI, TIPHAINE OUDOT-MELLAKH, APOSTOLOS DIMITROMANOLAKIS, MARINE GERMAIN, WILLIAM COHEN, PHILIP WELLS, MARK LATHROP, FRANCE GAGNON, PIERRE-EMMANUEL MORANGE, AND DAVID-ALEXANDRE TREGOUET

11.1 BACKGROUND

Elevated plasma levels of factor VIII (FVIII) and von Willebrand factor (vWF), two key molecules of the coagulation cascade, are well-established risk factors for venous thrombosis (VT) [1-3]. More recent evidence shows that these plasma hemostatic proteins are also risk factors for other cardiovascular diseases (CVD) [4-8]. The broader role of FVIII and vWF is further supported by studies showing that genetic factors modulating the variability of these proteins are also associated with CVD. These include single nucleotide polymorphisms (SNPs) at the *BAI3* [9], *LDLR* [5,10], *VWF* [4] and *ABO* [11] genes, the latter being associated with other quantitative risk factors for CVD [12,13].

This chapter was originally published under the Creative Commons Attribution License. Antoni G, Oudot-Mellakh T, Dimitromanolakis A, Germain M, Cohen W, Wells P, Lathrop M, Gagnon F, Morange P-E, and Tregouet DA. Combined Analysis of Three Genome-Wide Association Studies on vWF and FVIII Plasma Levels. BMC Medical Genetics 12,102 (2011), doi:10.1186/1471-2350-12-102.

The estimated heritability of FVIII and vWF levels range between 40% and 60% [14,15] among which about 20% is attributable to the *ABO* locus. A genome wide association study (GWAS) within the CHARGE consortium [16] has recently identified five new genes, apart from their structural genes and *ABO*, consistently influencing vWF and/or FVIII plasma levels. These include *CLEC4M, SCARA5, STX2, STXBP5* and *TC2N*, collectively explaining ~10% of the variability of each two traits. These observations suggest that there are additional genetic factors remaining to be identified and contributing to the hidden heritability of these quantitative traits.

The increased power of selected samples has long been recognized in family-based studies but more recently the putative advantages of carefully selected samples for quantitative trait analysis of unrelated subjects has also been highlighted [17]. Therefore, we undertook the combined analysis of individual data from three GWAS performed in samples of VT patients and in extended families ascertained on VT and Factor V Leiden (FVL) to identify novel genetic factors implicated in the variation of plasma levels of FVIII and vWF.

11.2 METHODS

11.2.1 OVERALL STRATEGY

To achieve our primary goal of identifying new genetic factors that could influence vWF and/or FVIII plasma levels, we used data from three carefully selected independent GWAS. Great attention was drawn to the homogeneity across samples in terms of - ethnic background (most individuals were of French origin), exclusion criteria with respect to rare forms of inherited thrombophilia, objectively diagnosed VT, studied intermediate phenotypes (although some adjustments were done) and similar genotyping technologies (Illumina platform).

In the context of quantitative trait GWAS, individual genetic effect sizes are known to be small [18] and it is expected that a number of real associations do not reach genome-wide significance. Therefore, as part of our analytic strategy, we first tested for association in the individual studies, and results observed across samples were combined into a meta-analysis. We then focused on the consistency of associations across studies as our hypothesis was that real associations would more likely be consistently observed across studies given that each study samples were quite homogeneous with respect to the above-mentioned characteristics. Previously reported associations were also investigated using the above strategy.

As genetic variants associated to plasma levels of FVIII and vWF could be risk factors for VT, our secondary goal was to test the identified SNPs with VT using an in silico GWAS [19]. Analytic approaches and samples characteristics of the FVIII and vWF GWAS are described below.

11.2.2 FVL-FAMILIES SAMPLE

Five extended French-Canadian families were ascertained through single probands with idiopathic VT diagnosed at the Thrombosis Clinic of the Ottawa Hospital, and carrying the FVL mutation. VT cases secondary to cancer as well as rare forms of inherited VT (protein S, protein C, Anti-Thrombin deficiencies) were excluded. A pedigree was drawn from interviews with each potential probands. The largest families were invited to participate in the study, the family size and willingness to participate being the only criteria for the selection of the families (see Additional File 1, File S1 for the used questionnaire). The total number of family members was 255. Description of the extended families has been published elsewhere [9].

11.2.3 MARTHA SAMPLES

The MARseille THrombosis Association (MARTHA) project is composed of two independent samples of VT patients, named MARTHA08 (N = 1,006) and MARTHA10 (N = 586). MARTHA subjects are unrelated caucasians consecutively recruited at the Thrombophilia center of La Timone hospital (Marseille, France) between January 1994 and October 2005. All patients had a documented history of VT and free of well characterized genetic risk factors including AT, PC, or PS deficiency, homozygosity for FV Leiden or FII 20210A, and lupus anticoagulant. They were interviewed by a physician on their medical history, which emphasized manifestations of deep vein thrombosis and pulmonary embolism using a standardized questionnaire (see Additional file 2, File S2). The thrombotic events were confirmed by venography, Doppler ultrasound, spiral computed tomographic scanning angiography, and/or ventilation/perfusion lung scan. All the subjects were of European origin, with the majority being of French descent.

The main characteristics of the three samples are shown in Table 1.

TABLE 1: Main Characteristics of the Studied Samples

	FVL Families N = 253	MARTHA08 N = 972	MARTHA10 N = 570
Age (SD)	40.4 (17.9)	45.7 (14.9)	49.2 (15.7)
Sex (% female)	50.6%	70.8%	58.2%
Smoking (%)	24.4%	24.9%	22.71%
History of VT (%)	5.95%	100%	100%
PT G20210A carriers	0.40%	15.9%	10.6%
FV Leiden carriers	24.9%	26.6%	14.1%
ABO blood group (%)			
O	40.6%	22.9%	22.4%
A	57.8%	61.8%	59.3%
B	1.6%	10.3%	14.4%
AB	-	5%	3.9%
FVIII (SD) IU/dL	118.6 (38.51)	138.70 (55.34)	130.2 (46.35)
vWF (SD) IU/dL	130.3 (53.24)	152.33 (68.23)	152.9 (63.93)

11.2.4 IN SILICO GWAS STUDY ON VT

In a previously published GWAS on VT [19], 419 early age of onset and the idiopathic character of VT (ie without environemental risk factors) (< 50 years) VT cases were compared to 1,228 healthy controls at 291,872 SNPs. Cases were patients from four different French medical centers (Grenoble, Marseille, Montpellier, Paris) selected according to the same criteria as the MARTHA samples, except with the restriction on age of onset. Controls were French subjects selected from the SUVIMAX population [20].

11.2.5 MEASUREMENTS

In the French-Canadian (FVL) sample, plasma levels of FVIII activity were measured by a clotting assay on the BCS instrument (Siemens Diagnostics, Marburg Germany) and vWF antigen was measured with a commercially available ELISA kit from Diagnostica Stago. The interassay coefficients of variation for FVIII were ~ 1% and 6.1% for vWF.

In MARTHA subjects, plasma coagulant activity and vWF antigen were assayed in an automated coagulometer (STA-R; Diagnostica Stago, Asnières, France). The interassay coefficients of variation for FVIII and vWF were 6.96% and 2.27% respectively.

11.2.6 GENOTYPING

The French-Canadian sample was genotyped with the Illumina 660W-Quad Beadchip. The raw datafile contained data for 547,886 autosomal SNPs genotyped on 255 individuals. From these SNPs, 490,083 passed the quality control (QC) criteria of genotyping rate > 90% and more than 20 observations of the minor allele among all individuals. After removing the 88,390 SNPs that failed QC, the overall genotyping rate was 99.88%. The maximum missing rate per sample for all the 255 samples was 3.9%, with an average missing rate of 0.13%. The family structures had previously been checked using 1079 microsatellite markers and RELPAIR [9]. To further verify the correctness of the family structure, we used PREST [21] and computed IBD estimates for all the sample pairs, within and across pedigrees. PREST reported 14,949 Mendelian errors, which is equivalent to a very low Mendelian error rate of 0.012% among all genotypes. Genotypes showing Mendelian inconsistencies were excluded from the analysis. Finally, phenotypic and genotypic data were available on a total of 253 individuals.

The MARTHA08 study sample was typed in 2008 with the Illumina Human610-Quad Beadchip containing 567,589 autosomal SNPs while the MARTHA10 sample was recently typed (beginning of 2010) with the same Illumina Human660W-Quad Beadchip as in the FVL study sample. SNPs showing significant ($P < 10^{-5}$) deviation from Hardy-Weinberg equilibrium, with minor allele frequency (MAF) less than 1% or genotyping call rate $< 99\%$, in each study were filtered out. Individuals with genotyping success rates less than 95% were excluded from the analyses, as well as individuals demonstrating close relatedness as detected by pairwise clustering of identity by state distances (IBS) and multi-dimensional scaling (MDS) implemented in PLINK software [22]. Non-European ancestry was also investigated using the Eigenstrat program [23] leading to the final selection of 972 and 570 patients left for analysis in MARTHA08 and MARTH10, respectively. Plasma vWF levels were available in 834 and 537 MARTHA08 and MARTHA10 patients, respectively; corresponding numbers were 541 and 548 for plasma FVIII levels. A total of 442,728 SNPs were common to the three GWAS datasets (see Additional file 3, Figure S1).

11.2.7 STATISTICAL ANALYSIS

In the FVL families, association of SNPs with vWF and FVIII levels was tested by means of measured genotype linear association analysis as implemented in the SOLAR (version 4.0, http://solar.txbiomedgenetics.org/download.html) program. In MARTHA subjects, association was tested using linear model as implemented in the PLINK program [22].

In order to handle differences in phenotype distributions across studies (Figure 1), and any possible deviation from normality, plasma levels of vWF and FVIII were first normalized before any statistical analysis using the normal quantile transformation [24], separately in the French-Canadian sample, MARTHA08 and MARTHA10. This transformation assigns to each observed measurement the quantile value of the standard normal distribution that corresponds to the rank of this measurement in the original untransformed distribution. Transformed variables are then normally distributed making linear models applicable, and linear regression

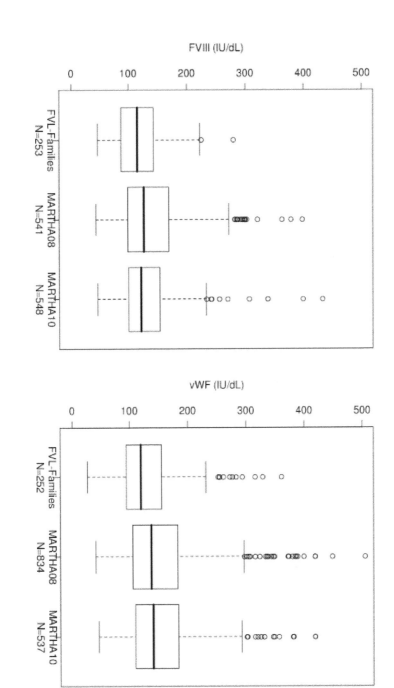

FIGURE 1: Box Plot Distribution of FVIII (left) and vWF (right) Plasma Levels in the Three GWAS Datasets.

coefficients comparable across studies. Association analyses were then carried out on the transformed variables assuming additive allele effects (0,1, 2 coding according to the number of minor alleles), and adjusting for age, sex and ABO blood group as tagged by the ABO rs8176746, rs8176704 and rs505922 [19]. When appropriate, haplotype association analyses were carried out in MARTHA samples using THESIAS software [25] to handle the correlation between SNPs, that is linkage disequilibrium (LD). This widely used software implements a stochastic-EM algorithm that simultaneously estimates the frequencies and the effect on the studied phenotype of each inferred haplotype. Haplotype - phenotype associations are then assessed by means of likelihood ratio tests.

Results obtained in each GWAS datasets were combined in a meta-analysis using the GWAMA program [26]http://www.sph.umich.edu/csg/abecasis/metal. Both fixed-effect and random-effect models- based analyses were conducted. Regression coefficients characterizing the minor allele effect of each SNP were then combined (after having checked that the minor allele was the same in the different populations) using the inverse-variance method to provide an overall allelic estimate. All reported P values were 2-sided.

11.3 RESULTS

A total of 442,728 QC-validated SNPs were common to the three GWAS and were tested through a meta-analysis for association with vWF and FVIII plasma levels. Quantile-quantile plots did not reveal any inflation from what was expected under the null hypothesis of no association (Figure 2), and no SNP reached the study-wide significance level of 1.12×10^{-7} that corresponds to the Bonferroni correction for the number of tested SNPs. Applying the less stringent Sidak correction corresponding to a significant threshold of $p = 1.16 \times 10^{-7}$ would not have modified this conclusion. We then further focused on genetic effects that were consistent across studies and with combined p-value of less than 10^{-5}. As fixed-effect and random-effect analyses provided similar results for most of the main associations (Tables 2 & 3), the following discussion is based on results obtained from the fixed-effect model analysis.

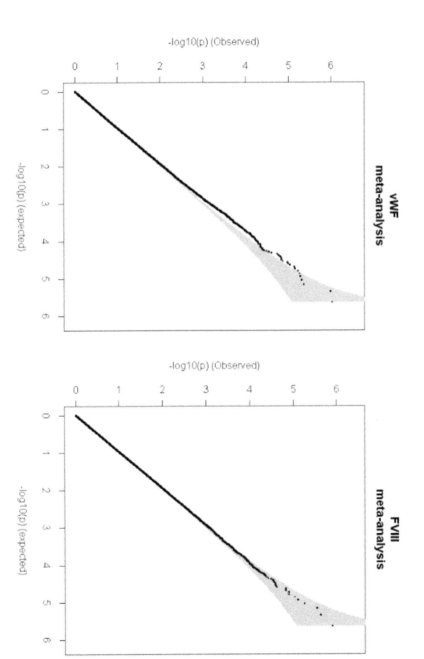

FIGURE 2: Quantile-Quantile Plots of the Association Results from the Meta-Analysis of the Three GWAS Datasets.

TABLE 2: Ten SNPs Showing Association with vWF levels Across the Three GWAS Datasets With Combined Significance P-value $< 10^{-5}$

Gene	SNP	Alleles*		MAF+	β (SE)	p	I²	phet	Random Effect		Fixed Effect	
									β (SE)	p	β (SE)	p
VPS8	rs4686760	A/G	FVL	0.47	-0.16 (0.08)	0.044						
			Martha08	0.46	-0.18 (0.04)	$4.11\ 10^{-5}$	0	0.549	0.15 (0.03)	$1.10\ 10^{-6}$	-0.15 (0.03)	$1.08\ 10^{-6}$
			Martha10	0.45	-0.11 (0.05)	0.047						
	rs13361927	G/A	FVL	0.15	0.44 (0.11)	$3.08\ 10^{-4}$						
			Martha08	0.06	0.28 (0.09)	0.003	0.53	0.119	-0.28 (0.09)	0.002	0.28 (0.06)	$4.51\ 10^{-6}$
EPB41L4A			Martha10	0.05	0.11 (0.11)	0.316						
	rs379440	A/G	FVL	0.12	0.46 (0.12)	$8.35\ 10^{-4}$						
			Martha08	0.04	0.31 (0.11)	0.004	0	0.502	-0.34 (0.07)	$9.99\ 10^{-7}$	0.34 (0.07)	$9.82\ 10^{-7}$
			Martha10	0.03	0.25 (0.14)	0.071						
ANKRD6	rs6454764	C/T	FVL	0.04	-0.01 (0.21)	0.977						
			Martha08	0.06	0.24 (0.09)	0.007	0.70	0.036	-0.29 (0.14)	0.035	0.31 (0.07)	$5.12\ 10^{-6}$
			Martha10	0.05	0.54 (0.12)	$8.97\ 10^{-6}$						
KRT18P24	rs1757948	T/G	FVL	0.27	0.34 (0.09)	$2.82\ 10^{-4}$						
			Martha08	0.27	0.1 (0.05)	0.030	0.62	0.071	-0.18 (0.06)	0.003	0.15 (0.03)	$7.37\ 10^{-6}$
			Martha10	0.30	0.15 (0.06)	0.009						

TABLE 2: *Cont.*

Gene	SNP	Alleles*		MAF+	β (SE)	p	I²	phet	Random Effect β (SE)	p	Fixed Effect β (SE)	p
	rs1438993	G/A	FVL	0.19	0.15 (0.1)	0.127						
			Martha08	0.28	0.18 (0.05)	1.11 10⁻⁴	0	0.666	-0.16 (0.03)	6.34 10⁻⁶	0.16 (0.03)	6.25 10⁻⁶
			Martha10	0.27	0.12 (0.06)	0.052						
desert	rs10745527	T/G	FVL	0.20	0.19 (0.1)	0.062						
			Martha08	0.28	0.18 (0.05)	1.63 10⁻⁴	0	0.663	-0.16 (0.03)	5.51 10⁻⁶	0.16 (0.03)	5.43 10⁻⁶
			Martha10	0.27	0.11 (0.06)	0.056						
	rs2579103	T/G	FVL	0.18	0.17 (0.11)	0.098						
			Martha08	0.26	0.19 (0.05)	8.24 10⁻⁵	0	0.533	-0.16 (0.04)	7.72 10⁻⁶	0.16 (0.04)	7.61 10⁻⁶
			Martha10	0.25	0.1 (0.06)	0.090						
CDH2	rs2298574	A/G	FVL	0.04	-0.02 (0.19)	0.905						
			Martha08	0.08	-0.34 (0.08)	2.77 10⁻⁵	0.19	0.290	0.26 (0.07)	1.81 10⁻⁴	-0.27 (0.06)	5.67 10⁻⁶
			Martha10	0.07	-0.24 (0.1)	0.022						
SAFB2	rs732505	G/A	FVL	0.05	0.32 (0.18)	0.080						
			Martha08	0.09	0.24 (0.08)	0.001	0	0.929	-0.25 (0.06)	9.50 10⁻⁶	0.25 (0.06)	9.38 10⁻⁶
			Martha10	0.08	0.25 (0.1)	0.013						

*Common/rare alleles
+ Allele frequency of the minor allele

TABLE 3: Six SNPs Showing Association with FVIII Activity Across the Three GWAS Datasets With Combined Significance P-value < 10^{-5}

Gene	SNP	Alleles*		MAF+	β (SE)	p	I²	phet	Random Effect		Fixed Effect	
									β (SE)	p	β (SE)	p
LBH	rs6708166	G/A	FVL	0.41	-0.12 (0.09)	0.156						
			Martha08	0.40	-0.23 (0.06)	8.98e-05	0	0.478	-0.17 (0.04)	1.32 10	-0.17 (0.04)	1.30 10⁻⁶
			Martha10	0.42	-0.15 (0.05)	0.007						
FAM46A	rs1321761	T/C	FVL	0.42	-0.20 (0.08)	0.014						
			Martha08	0.45	-0.10 (0.06)	0.074	0	0.451	-0.15 (0.04)	9.67 10⁻⁶	-0.15 (0.04)	9.54 10⁻⁶
			Martha10	0.47	-0.19 (0.05)	5.93e-04						
VAV2	rs12344583	A/G	FVL	0.17	0.28 (0.11)	0.012						
			Martha08	0.20	0.19 (0.07)	0.006	0	0.716	0.20 (0.04)	8.03 10⁻⁶	0.20 (0.04)	7.92 10⁻⁶
			Martha10	0.18	0.17 (0.07)	0.012						
STAB2	rs7306642	C/A	FVL	0.16	0.52 (0.12)	1.36e-05						
			Martha08	0.07	0.22 (0.11)	0.057	0.59	0.086	0.31 (0.10)	0.002	0.30 (0.06)	2.95 10⁻⁶
			Martha10	0.07	0.20 (0.1)	0.052						

TABLE 3: *Cont.*

Gene	SNP	Alleles*		MAF+	β (SE)	p	I²	phet	Random Effect β (SE)	p	Fixed Effect β (SE)	p
			FVL	0.53	0.09 (0.08)	0.293						
	rs1354492	G/A	Martha08	0.49	0.23 (0.05)	1.20e-05	0.39	0.192	0.16 (0.04)	5.47 10⁻⁶	0.16 (0.03)	2.41 10⁻⁶
ACCN1		Martha10	0.47	0.12 (0.05)	0.027							
			FVL	0.22	-0.29 (0.1)	0.004						
	rs12941510	G/A	Martha08	0.31	-0.17 (0.06)	0.002	0.12	0.321	-0.17 (0.04)	2.18 10⁻⁵	-0.17 (0.04)	5.67 10⁻⁶
		Martha10	0.33	-0.12 (0.06)	0.029							

*Common/rare alleles
+ Allele frequency of the minor allele

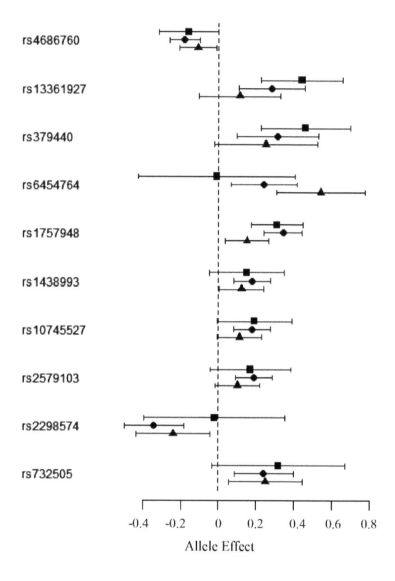

FIGURE 3: Forest plot representation of the ten SNPs that associated the most with vWF levels in the Three GWAS Datasets. Results observed in the FVL families, MARTHA08 and MARTHA10 studies are depicted by square, circle and triangle, respectively.

Ten SNPs covering seven different genes (Figure 3, Table 2) were associated with plasma vWF levels at $p < 10^{-5}$ with no strong evidence for heterogeneity across GWAS as the lowest Mantel-Haenszel observed p-value, $p = 0.036$, for the ANKDR6 rs645764 would not pass multiple testing correction for testing ten SNPs. The strongest association was observed for rs379440 ($P = 9.82 \ 10^{-6}$) mapping the *EPB41L4A* gene (Table 2). Another SNP at this locus was also associated with vWF, rs13361927 ($P = 4.51 \ 10^{-6}$), but its association was due to its complete LD with rs379440, with pairwise r^2 of 0.78, 0.69 and 0.62 in FVL, MARTHA08 and MARTHA10, respectively. Other vWF-associated SNPs included the *SAFB2* rs732505 ($P = 9.38 \ 10^{-6}$), *VPS8* rs4686760 ($P = 1.08 \ 10^{-6}$) and the *KRT18P24* rs1757948 ($P = 7.37 \ 10^{-6}$). The last three SNPs, rs1438993, rs10745527, rs2579103 (with P~ $6 \ 10^{-6}$), were located at the 12q21.33 locus with no known mapped gene and were in nearly complete association. Altogether, the independent signals derived from the rs4686760, rs379440, rs1757948, rs10745527 and rs732505 explained up to 5.7% and 3.8% of the variability of plasma vWF levels in MARTHA08 and MARTHA10, respectively, and 5.3% in the pooled MARTHA samples.

None of the ten vWF-associated SNPs were associated with plasma FVIII levels (all $p > 0.05$). However, six additional SNPs were specifically associated to FVIII levels with homogeneous effects (Mantel-Haenszel p-value > 0.05) across studies (Figure 4 - Table 3). The strongest effect ($P = 2.95 \ 10^{-6}$) was observed for rs7306642, a non synonymous Pro2039Thr variant within the STAB2 gene, which was one of the recently identified genes by the CHARGE consortium. However, our hit rs7306642 was not in LD with any of the two STAB2 SNPs recently identified, rs4981022 ($r^2 < 0.01$ in the three studies) and rs4981021 that served as a proxy for rs12229292 ($r^2 < 0.07$ in the three studies). Other FVIII-associated SNPs included the rs6708166 ($P = 1.30 \ 10^{-6}$) in the proximity of LBH, the rs1321761 ~ 300 kb apart from *FAM46A* ($P = 9.54 \ 10^{-6}$) and the intronic *VAV2* rs12344583 ($P = 7.92 \ 10^{-6}$) (Table 3). Lastly, two SNPs within the *ACCN1* gene, rs1354492 and rs12941510, were found modulating FVIII plasma levels, the A allele of the former being associated with increased FVIII levels ($\beta = +0.16$, $P = 2.42 \ 10^{-6}$) and the A allele of the latter being associated with decreased levels ($\beta = -0.17$, $P = 5.67 \ 10^{-6}$). These two SNPs were in complete negative LD generating three haplotypes, the sole

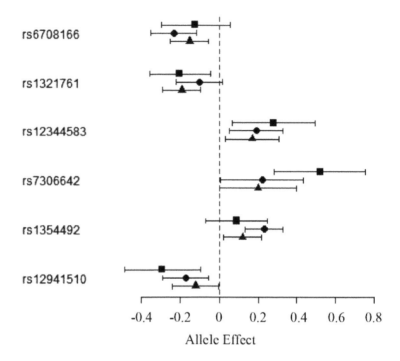

FIGURE 4: Forest plot representation of the six SNPs that associated the most with FVIII levels in the Three GWAS Datasets. Results observed in the FVL families, MARTHA08 and MARTHA10 studies are depicted by square, circle and triangle, respectively.

carrying the rs1354492-A allele being associated with highest levels (see Additional file 4, Table S1). Altogether, these five SNPs (i.e. rs6708166, rs1321761, rs12344583, rs7306642, rs1354492) explained 8.2% and 4.6% of the variability of FVIII levels in MARTHA08 and MARTHA10, respectively, and 6.3% in the combined MARTHA samples.

We then used our GWAS datasets to investigate SNPs that had previously been reported associated with vWF and/or FVIII [4,5,9,16]. As shown in Supplementary Table two, marginal associations ($P < 0.05$) with vWF levels at *STXBP5*, *VWF*, *STX2*, *TC2N* and *CLEC4M* were also observed in our study, the strongest ($P = 1.3 \ 10^{-4}$) being for SNP rs216335 at the structural *VWF* gene. All these associations were consistent (i.e the same allele was associated with a genetic effect in the same direction on the studied phenotype) with those previously reported. Together, these associations explained an additional 1.4% and 3.2% of the variance of plasma levels of vWF in MARTHA08 and MARTHA10, respectively. We did not observe any evidence for an effect of STAB2 rs4981022 or BAI3 rs9363864, while the effect of *SCARA5* rs2726953 was heterogeneous across the studies. For FVIII levels, we observed marginal associations of *SCARA5* rs9644133 (P = 0.009) and *VWF* rs1063856 (P = 0.020) that were consistent with those previously reported (Table 4), these two SNPs explaining 0.7% and 0.2% of FVIII variability in MARTHA08 and MARTHA10, respectively. No trend for association was observed for the previously reported associations with *STXBP5, STAB2* nor *LDLR* SNPs (Table 5).

We have recently observed that, among the newly identified vWF and/or FVIII genes by the CHARGE consortium, *TC2N* could also be associated with VT risk [27]. Therefore we investigated the effect of the SNPs identified in our meta-analysis on the risk of VT. Our working hypothesis was that SNPs associated with increased (decreased, resp.) plasma levels of these two molecules could be associated with increased (decreased, resp.) risk of disease. For this, we used the results of our previously published GWAS based on 419 VT patients and 1228 healthy subjects (in silico association) [19]. As indicated in Table 6, only two SNPs, *VPS8* rs4686760 and *ACCN1* rs12941510, showed some trend of association consistent with our hypothesis. The rs4686760-G allele found associated with decreased vWF levels was slightly less frequent in VT patients than in controls (0.441 vs 0.475, P = 0.101) and the rs12941510-A allele, associated with decreased FVIII levels, was also less frequent in cases than in controls (0.310 vs 0.350, P = 0.046). These associations can only be considered as suggestive as they would not pass correction for multiple testing. Nevertheless, the observed homogeneity of the allele frequencies of these two SNPs across all genotyped patients is noteworthy. Combining all the VT patients (n = 1946), and comparing to the healthy controls of the in silico GWAS, the association of rs4686760 with VT remained (0.454 vs 0.475, P = 0.108), and that of rs12941510 was strengthened (0.314 vs 0.348, P = 0.0056) (Table 7).

TABLE 4: Association of Previously Identified SNPs with vWF Levels in the three GWAS Datasets

Gene	SNP	Alleles*		MAF+	β (SE)	p	I²	phet	Random Effect		Fixed Effect	
									β (SE)	p	β (SE)	p
BAI3	rs9363864	A/G	FVL	0.42	0.04 (0.08)	0.618						
			Martha08	0.52	0.03 (0.04)	0.421	0	0.838	0.02 (0.03)	0.461	0.02 (0.03)	0.461
			Martha10	0.49	-0.002 (0.05)	0.973						
STXBP5	rs9390459	G/A	FVL	0.43	-0.08 (0.08)	0.366						
			Martha08	0.42	-0.06 (0.04)	0.197	0	0.545	-0.09 (0.03)	0.005	-0.09 (0.03)	0.005
			Martha10	0.43	-0.13 (0.05)	0.011						
SCARA5	rs10866867(1)	G/T	FVL	0.20	-0.08 (0.10)	0.446						
			Martha08	0.25	0.17 (0.05)	4.88e-04	0.71	0.03	0.05 (0.07)	0.466	0.09 (0.04)	0.015
			Martha10	0.25	0.01 (0.06)	0.830						
	rs216335(2)	G/A	FVL	0.06	-0.28 (0.19)	0.141						
			Martha08	0.08	-0.23 (0.08)	0.003	0	0.945	-0.23 (0.06)	$1.31\ 10^{-4}$	-0.23 (0.06)	$1.30\ 10^{-4}$
			Martha10	0.06	-0.21 (0.11)	0.059						

TABLE 4: *Cont.*

Gene	SNP	Alleles*		MAF+	β (SE)	p	I²	phet	Random Effect		Fixed Effect	
									β (SE)	p	β (SE)	p
VWF	rs1063856(3)	A/G	Martha08	0.37	0.08 (0.05)	0.094	0	0.889	0.09 (0.03)	0.006	0.09	0.006
			Martha10	0.38	0.11 (0.05)	0.041						
			FVL	0.48	-0.04 (0.08)	0.612						
			FVL	0.45	0.07 (0.08)	0.371						
	rs7306706	A/G	Martha08	0.45	0.02 (0.04)	0.634	0	0.754	0.01 (0.03)	0.664	0.01 (0.03)	0.664
			Martha10	0.46	0.03 (0.05)	0.604						
			FVL	0.30	-0.05 (0.09)	0.601						
STAB2	rs4981022	T/C	Martha08	0.30	0.02 (0.05)	0.652	0	0.541	-0.01 (0.03)	0.664	-0.01 (0.03)	0.664
			Martha10	0.28	-0.06 (0.06)	0.333						
			FVL	0.33	0.01 (0.09)	0.863						
STX2	rs4334059(4)	C/T	Martha08	0.37	0.08 (0.04)	0.067	0.01	0.363	0.1 (0.03)	0.004	0.1 (0.03)	0.003
			Martha10	0.36	0.15 (0.06)	0.008						

TABLE 4: *Cont.*

Gene	SNP	Alleles*		MAF+	β (SE)	p	I²	phet	Random Effect		Fixed Effect	
									β (SE)	p	β (SE)	p
TC2N	rs2402074(5)	G/A	FVL	0.52	0.05 (0.08)	0.548						
			Martha08	0.48	0.04 (0.04)	0.382	0	0.509	0.07 (0.03)	0.033	0.07 (0.03)	0.033
			Martha10	0.47	0.12 (0.05)	0.030						
CLEC4M	rs868875	A/G	FVL	0.22	-0.07 (0.1)	0.515						
			Martha08	0.32	-0.10 (0.05)	0.036	0	0.762	-0.08 (0.03)	0.026	-0.08 (0.03)	0.026
			Martha10	0.35	-0.05 (0.06)	0.424						

* *Common/rare alleles*

+ *Allele frequency of the minor allele*

(1) rs10866867 serves as proxy for rs2726953 (r2 = 0.92); (2) rs216335 serves as proxy for rs216318 (r2 = 1)

(3) rs1063856 serves as proxy for Rs1063857 (r2 = 1); (4) rs4334059 serves as proxy for rs7978987 (r2 = 1.0

(5) rs2402074 serves as proxy for rs10133762 (r2 = 0.96); No good proxy with r2 > 0.5 was available for the VWF rs4764478

TABLE 5: Association of Previously Identified SNPs with FVIII Activity in the three GWAS Datasets

Gene	SNP	Alleles*	MAF+		β (SE)	p	I²	phet	Random Effect β (SE)	p	Fixed Effect β (SE)	p
STXBP5	rs9390459	G/A		FVL	0.15 (0.08)	0.083						
			0.43	Martha08	-0.08 (0.06)	0.158	0.65	0.059	-0.02 (0.06)	0.795	-0.04 (0.03)	0.310
			0.42	Martha10	-0.07 (0.05)	0.199						
			0.43	FVL	-0.08 (0.1)	0.433						
SCARA5	rs9644133	C/T	0.24	Martha08	-0.16 (0.07)	0.029	0	0.753	-0.12 (0.05)	0.009	-0.12 (0.05)	0.009
			0.17	Martha10	-0.10 (0.07)	0.152						
			0.18	FVL	0.11 (0.08)	0.170						
VWF	rs1063856	A/G	0.45	Martha08	0.09 (0.06)	0.114	0	0.843	0.08 (0.03)	0.020	0.08 (0.03)	0.020
			0.37	Martha10	0.06 (0.05)	0.249						
			0.38	FVL	-0.13 (0.09)	0.146						
STAB2	rs4981021(1)	G/A	0.27	Martha08	-0.02 (0.06)	0.737	0	0.389	-0.02 (0.04)	0.521	-0.02 (0.04)	0.521
			0.32	Martha10	0.02 (0.06)	0.782						
			0.29	FVL	-0.03 (0.11)	0.816						
LDLR	rs2228671	C/T	0.14	Martha08	0.11 (0.09)	0.193	0.46	0.157	-0.01 (0.07)	0.890	-0.01 (0.05)	0.894
			0.11	Martha10	-0.13 (0.09)	0.161						
			0.10	FVL	-0.25 (0.09)	0.005						
	rs688	C/T	0.38	Martha08	0.06 (0.05)	0.235	0.79	0.010	-0.05 (0.08)	0.531	-0.02 (0.03)	0.652
			0.45	Martha10	-0.007 (0.05)	0.901						
			0.45									

* Common/rare alleles

+ Allele frequency of the minor allele

(1) rs4981021 serves as proxy for rs12229292 (r2 = 0.88)

TABLE 6: In Silico Association With Venous Thrombosis of the Identified vWF- and FVIII Associated SNPs

		Alleles*	Minor Allele Frequency		Cochran Armitage P-value
			Cases	Controls	
vWF associated SNPs					
VPS8	rs4686760	A/G	0.441	0.475	P = 0.101
EPB41L4A	rs13361927	G/A	0.065	0.062	P = 0.797
KRT18P24	rs1634352†	G/A	0.284	0.318	P = 0.055
12q21.33	rs1438933	G/A	0.256	0.294	P = 0.051
CDH2	rs2298574	A/G	0.084	0.093	P = 0.444
SAFB2	rs732505	G/A	0.061	0.064	P = 0.713
FVIII associated SNPs					
VAV2	rs12344583	A/G	0.217	0.193	P = 0.133
ACCN1	rs1354492	G/A	0.476	0.469	P = 0.740
ACCN1	rs12941510	G/A	0.310	0.350	P = 0.046

*Common/minor alleles, † serves as proxy for rs1757948 ($r^2 = 1$). No good proxy with $r^2 > 0.80$ was available for rs6708166 (LBH), rs1321761 (FAM46A) and rs7306642 (STAB2)

TABLE 7: Genotype Distributions of rs4686760 and rs12941510 Across VT Samples.

	rs4686760			
	AA	AG	GG	MAF (2)
MARTHA08	271	502	198	0.462
MARTHA10	173	281	115	0.449
GWAS patients	129	196	81	0.441
All VT patients	573	979	394	0.454
GWAS controls	354	581	292	0.475
	Test of association P = 0.108(1)			
	rs12941510			
	AA	AG	GG	MAF
MARTHA08	93	409	469	0.306
MARTHA10	67	243	259	0.331
GWAS patients	45	161	199	0.310
All VT patients	205	813	927	0.314
GWAS controls	139	576	512	0.348
	Test of association P = 0.0056			

(1) Cochran Armitage trend test, (2) Minor Allele Frequency

11.4 DISCUSSION

Theoretically, a sample size of 1,624 unrelated individuals should have a power of 95% to detect, at the significant level of 1.12 10^{-7}, the additive allele effect of a SNP explaining at least 3% if the variability of a quantitative trait [28]. This power would decrease to 86% and 66% for a SNP explaining 2.5% and 2%, respectively. Our meta-analysis of 1,624 carefully selected samples did not reveal any genome-wide significant association suggesting that the additional common SNPs tagged by current GWAS array and influencing vWF and FVIII plasma levels left to be identified would, if any, individually explain less than 2% of the variability of these two traits.

By lowering the statistical stringency to $p < 10^{-5}$ but focusing on the homogeneity of the effects observed in three independent samples, we identified several novel candidate genes that could contribute to modulate the variability of vWF and FVIII, and that deserve to be further studied. The novel candidate genes for vWF are *VPS8, EBP41L4A, KRT18P24, SAFB2* and a region on 12q21.3 where no known gene maps. Unfortunately, little is known about the biology of the associated proteins and their role in cardiovascular diseases. Among these, *VPS8* stands out. The rs4686760-G allele of the VPS8 gene, which was associated with decreased vWF levels, was also observed less frequently in VT cases than in healthy controls (0.45 vs 0.48) in the in silico GWAS, although this observation did not reach significance (P = 0.10). The vacuolar protein sorting 8 homolog gene (*VPS8*) is involved in protein traffic between the golgic appartus and the vacuaole [29] and could participate to the regulation of urokinase-type plasminogen activator [30], the latter known to be involved in thrombosis.

For FVIII levels, the candidate genes identified in our study were *LBH, FAM46A, VAV2, STAB2* and *ACCN1*. Both *LBH* and *VAV2* genes are thought to be involved in angiogenesis. The transcriptional cofactor limb-bud-and-heart (Lbh) was discovered as a small acidic nuclear protein highly conserved among species [31]. It has been demonstrated a dramatic suppression of VEGF mRNAs in cells that overexpress Lbh [32]. Vav2 is a guanine nucleotide exchange factor for Rho family proteins. The expression of a dominant negative form of Vav2 suppress the Vascular Endothelial-Protein Tyrosine Phosphatise (VE-PTP)-induced changes in endothelial cell morphology, such changes being implicated in regulation of angiogenesis [33].

Interestingly, we had previously shown that STAB2 was located within a linkage peak for vWF levels in our FVL extended families [9] while almost concomitantly STAB2 SNPs were found associated with both FVIII and vWF in the CHARGE consortium GWAS [16]. However, the non-synonymous rs7306642 (Pro2039Thr) found associated here with FVIII levels did not show a homogeneous effect on vWF levels across the three GWAS datasets (data not shown), and was in very low LD with others STAB2 SNPs found associated with these plasma levels. The substitution of a Proline by a Threonine at position 2039 is predicted to be damaging according to web resources http://genetics.bwh.harvard.edu/pph/index. html ; http://www.rostlab.org/services/SNAP . Investigating the effect of this substitution on VT risk would have been relevant but the corresponding SNP did not pass quality control in our in silico GWAS. These observations nevertheless suggest that an in-depth haplotype analysis of the STAB2 gene are required to gain better insight into which SNPs more likely influence plasma levels of FVIII and/or vWF.

ACCN1, encoding an amiloride-sensitive cation channel implicated in cell growth and migration [34], is another gene that deserves greater attention as its genetic variability was found here associated with both FVIII levels and VT risk. However, the SNP that seemed to modulate FVIII levels the most, rs1354492, was not the one that showed association with the disease. This could suggest that either different SNPs distinctly influence plasma levels and VT risk, or that the identified SNPs are in LD with unmeasured variant(s) that could simultaneously influence both phenotypes.

Our meta-analysis was also able to replicate several of the previously reported associations between SNPs and vWF/FVIII levels. Replicated associations include vWF-associated SNPs at *STXBP5, VWF, STX2, TC2N* and *CLEC4M* genes, and FVIII-associated SNPs within *SCARA5* and *VWF* genes. Other previously reported associations were not replicated, such as those involving *LDLR, BAI3*, and *STAB2* SNPs [5,9,16]. In addition to a lack of power, as previously discussed, this could be due to differential effects of SNP in normal range of plasma levels compared to the higher levels observed in VT patients. This could apply to the association of *BAI3* with vWF levels observed in healthy nuclear families [9] where plasma levels were lower than those observed in our VT samples. Conversely, this explanation does not completely hold for the LDLR SNPs that were found associated with FVIII activity in a population [5] where FVIII activity in healthy individuals were at higher levels than those observed in our VT patients. Besides, in these two studies, different methods from those we have used here were employed to measure vWF and FVIII activity, and this could also contribute to the discrepancies observed in our study.

11.5 CONCLUSIONS

In conclusion, a carefully planned meta-analysis of three independent samples gathering 1,624 individuals genotyped for more than 400,000 SNPs all over the genome replicated very recent findings but did not reveal any new genetic factors that could individually explain at least 2% of the plasma variability of vWF and FVIII levels.

REFERENCES

1. Koster T, Blann AD, Briet E, Vandenbroucke JP, Rosendaal FR: Role of clotting factor VIII in effect of von Willebrand factor on occurrence of deep-vein thrombosis. Lancet 1995, 345:152-155.
2. Kraaijenhagen RA, in't Anker PS, Koopman MM, Reitsma PH, Prins MH, van den Ende A, et al.: High plasma concentration of factor VIIIc is a major risk factor for venous thromboembolism. Thromb Haemost 2000, 83:5-9.
3. Tsai AW, Cushman M, Rosamond WD, Heckbert SR, Tracy RP, Aleksic N, et al.: Coagulation factors, inflammation markers, and venous thromboembolism: the longitudinal investigation of thromboembolism etiology (LITE). Am J Med 2002, 113:636-642.
4. van Schie MC, de Maat MP, Isaacs A, van Duin CM, Deckers JW, Dippel DW, et al.: Variation in the von Willebrand Factor gene is associated with VWF levels and with the risk of cardiovascular disease. Blood 2011, 117:1393-1399.
5. Martinelli N, Girelli D, Lunghi B, Pinotti M, Marchetti G, Malerba G, et al.: Polymorphisms at LDLR locus may be associated with coronary artery disease through modulation of coagulation factor VIII activity and independently from lipid profile. Blood 2010, 116:5688-5697.
6. Whincup PH, Danesh J, Walker M, Lennon L, Thomson A, Appleby P, et al.: von Willebrand factor and coronary heart disease: prospective study and meta-analysis. Eur Heart J 2002, 23:1764-1770.

7.   Folsom AR, Rosamond WD, Shahar E, Cooper LS, Aleksic N, Nieto FJ, et al.: Prospective study of markers of hemostatic function with risk of ischemic stroke. The Atherosclerosis Risk in Communities (ARIC) Study Investigators. Circulation 1999, 100:736-742.

8.   Cambronero F, Vilchez JA, Garcia-Honrubia A, Ruiz-Espejo F, Moreno V, Hernandez-Romero D, et al.: Plasma levels of von Willebrand factor are increased in patients with hypertrophic cardiomyopathy. Thromb Res 2010, 126:e46-50.

9.   Antoni G, Morange PE, Luo Y, Saut N, Burgos G, Heath S, et al.: A multi-stage multi-design strategy provides strong evidence that the BAI3 locus is associated with early-onset venous thromboembolism. J Thromb Haemost 2010, 8:2671-2679.

10.   Vormittag R, Bencur P, Ay C, Tengler T, Vukovich T, Quehenberger P, et al.: Low-density lipoprotein receptor-related protein 1 polymorphism 663 C > T affects clotting factor VIII activity and increases the risk of venous thromboembolism. J Thromb Haemost 2007, 5:497-4502.

11.   Carpeggiani C, Coceani M, Landi P, Michelassi C, L'Abbate A: ABO blood group alleles: A risk factor for coronary artery disease. An angiographic study. Atherosclerosis 2010, 211:461-466.

12.   Teupser D, Baber R, Ceglarek U, Scholz M, Illig T, Gieger C, et al.: Genetic regulation of serum phytosterol levels and risk of coronary artery disease. Circ Cardiovasc Genet 2010, 3:331-339.

13.   Barbalic M, Dupuis J, Dehghan A, Bis JC, Hoogeveen RC, Schnabel RB, et al.: Large-scale genomic studies reveal central role of ABO in sP-selectin and sICAM-1 levels. Hum Mol Genet 2010, 19:1863-1872.

14.   Souto JC, Almasy L, Borrell M, Gari M, Martinez E, et al.: Genetic determinants of hemostasis phenotypes in Spanish families. Circulation 2000, 101:1546-1551.

15.   Morange PE, Tregouet DA, Frere C, Saut N, Pellegrina L, Alessi MC, et al.: Biological and genetic factors influencing plasma factor VIII levels in a healthy family population: results from the Stanislas cohort. Br J Haematol 2005, 128:91-99.

16.   Smith NL, Chen M-H, Dehghan A, Strachan DP, Basu S, Soranzo N, et al.: Novel associations of multiple genetic loci with plasma levels of Factor VII, Factor VIII and von Willebrand Factor. The CHARGE (Cohorts for Heart and Aging Research in Genome Epidemiology) Consortium. Circulation 2010, 121:1392-1392.

17.   Abecasis GR, Cookson WO, Cardon LR: The power to detect linkage disequilibrium with quantitative traits in selected samples. Am J Hum Genet 2001, 68:1463-1474.

18.   Teslovich TM, Musunuru K, Smith AV, Edmondson AC, Stylianou IM, Koseki M, et al.: Biological, clinical and population relevance of 95 loci for blood lipids. Nature 2010, 466:707-713.

19.   Tregouet DA, Heath S, Saut N, Biron-Andreani C, Scheved JF, Pernod G, et al.: Common susceptibility alleles are unlikely to contribute as strongly as the FV and ABO loci to VTE risk: results from a GWAS approach. Blood 2009, 113:5298-5303.

20.   Hercberg S, Galan P, Preziosi P, Bertrais S, Mennen L, et al.: The SU.VI.MAX Study: a randomized, placebo-controlled trial of the health effects of antioxidant vitamins and minerals. Arch Intern Med 2004, 164:2335-2342.

21.   Sun L, Wilder K, McPeek MS: Enhanced pedigree error detection. Hum Hered 2002, 54:99-110.

22. Purcell S, Neale B, Todd-Brown K, Thomas L, Ferreira MA, Bender D, et al.: PLINK: a tool set for whole-genome association and population-based linkage analyses. Am J Hum Genet 2007, 81:559-575.

23. Price AL, Patterson NJ, Plenge RM, Weinblatt ME, Shadick NA, Reich D: Principal components analysis corrects for stratification in genome-wide association studies. Nat Genet 2006, 38:904-909.

24. Peng B, Yu RK, Dehoff KL, Amos CI: Normalizing a large number of quantitative traits using empirical normal quantile transformation. BMC Proc 2007, 1(Suppl 1):S156.

25. Tregouet DA, Garelle V: A new JAVA interface implementation of THESIAS: testing haplotype effects in association studies. Bioinformatics 2007, 23:1038-1039.

26. Magi R, Morris AP: GWAMA: software for genome-wide association meta-analysis. BMC Bioinformatics 2010, 11:288.

27. Morange PE, Saut N, Antoni G, Emmerich J, Tregouet DA: Impact on venous thrombosis risk of newly discovered gene variants associated with FVIII and VWF plasma levels. J Thromb Haemost 2011, 9:229-231.

28. Gauderman WJ, Morrison JM: Quanto 1.1: a computer program for power and sample sizee calculations for genetic-epidemiology studies. [http://hydra.usc.edu/gxe] 2006.

29. Chen YJ, Stevens TH: The VPS8 gene is required for localization and trafficking of the CPY sorting receptor in Saccharomyces cerevisiae. Eur J Cell Biol 1996, 70:289-297.

30. Agaphonov M, Romanova N, Sokolov S, Iline A, Kalebina T, et al.: Defect of vacuolar protein sorting stimulates proteolytic processing of human urokinase-type plasminogen activator in the yeast Hansenula polymorpha. FEMS Yeast Res 2005, 5:1029-1035.

31. Briegel KJ, Baldwin HS, Epstein JA, Joyner AL: Congenital heart disease reminiscent of partial trisomy 2p syndrome in mice transgenic for the transcription factor Lbh. Development 2005, 132:3305-3316.

32. Conen KL, Nishimori S, Provot S, Kronenberg HM: The transcriptional cofactor Lbh regulates angiogenesis and endochondral bone formation during fetal bone development. Dev Biol 2009, 333:348-358.

33. Mori M, Murata Y, Kotani T, Kusakari S, Ohnishi H, Saito Y: Promotion of cell spreading and migration by vascular endothelial-protein tyrosine phosphatase (VE-PTP) in cooperation with integrins. J Cell Physiol 2010, 224:195-204.

34. Vila-Carriles WH, Kovacs GG, Jovov B, Zhou ZH, Pahwa AK, Colby G, et al.: Surface expression of ASIC2 inhibits the amiloride-sensitive current and migration of glioma cells. J Biol Chem 2006, 281:19220-19232.

There are several supplemental files that are not available in this version of the article. To view this additional information, please use the citation information cited on the first page of this chapter.

PART IV

PROTEOMICS

CHAPTER 12

STATISTICAL METHODS FOR QUANTITATIVE MASS SPECTROMETRY PROTEOMIC EXPERIMENTS WITH LABELING

ANN L. OBERG AND DOUGLAS W. MAHONEY

12.1 BACKGROUND

In this manuscript we focus on statistical methods for quantitative mass spectrometry (MS) based proteomic experiments as they pertain to labeling protocols. Labeling of fragmented proteins (i.e., peptides) allows specimens to be labeled without altering the chemical properties of the peptides, mixed into a single aliquot and then subjected to MS simultaneously. The advantage of the labeling protocol is that specimens can be distinguished in the resulting data by leveraging known properties of the labels. For example, if stable isotopes are used, the known mass shift resulting from extra neutrons together with known naturally occurring distributions of isotopes in the atmosphere are used during the relative quantification step.

Several different labeling protocols have been developed. In iTRAQ labeling, each specimen is labeled with a different amine-specific isobaric

This chapter was originally published under the Creative Commons Attribution License. Oberg AL and Mahoney DW. Statistical Methods for Quantitative Mass Spectrometry Proteomic Experiments With Labeling. BMC Bioinformatics *13(Suppl 16)*,S7 (2012), doi:10.1186/1471-2105-13-S16-S7.

tag [1,2]. In $^{16}O/^{18}O$ labeling, one specimen is mixed with "light" water containing oxygen in its natural isotopic state (mostly ^{16}O) and a second specimen with "heavy" water containing mostly water molecules with the ^{18}O isotope that has two extra neutrons. With stable isotope labeling by amino acids in cell culture (SILAC) cells may be grown in "light" or "heavy" medium [3,4] or mice may be fed chow containing carbon in either the natural ("light") ^{12}C state or the ^{13}C ("heavy") isotopic state [5]. Similarly, with 15N labeling, cells may be grown in "light" or "heavy" medium [6,7].

Labeled protocols are appealing for multiple reasons. Mixing multiple specimens for simultaneous MS reduces the total MS machine time needed to perform an experiment. It also eliminates the between MS experiment variation for the specimens assayed together, thus reducing the variation in the study overall. We demonstrate here application of some fundamental experimental design principles, how to assess need for and success of normalization, and how to use statistical models to assess differential protein abundance for a study using data from multiple MS experiments.

There are three common objectives in high dimensional studies that produce data on a large number of endpoints such as global proteomics studies [8]. Class comparison involves comparing abundance levels between predefined groups. An example of this is comparing protein abundance levels between cancerous and benign tumors in order to gain biological insight into the mechanism of cancer. Class prediction involves development of a prediction rule consisting of a panel of biomarkers that are useful for classifying a new subject into pre-determined classes such as cancer or benign. Building on the cancer example, this process would combine multiple proteins present at differing abundance levels between cancer and benign tumors in this case, into a prediction rule that could be applied to a new subject with a tumor to determine whether the tumor was benign or cancerous. Class discovery involves use of abundance profiles to uncover yet unknown biological subtypes of disease. For example, in a proteomics study of high-grade serous ovarian cancers, the protein abundance data would be used to determine whether subtypes of serous cancer may exist that are currently unknown. The methods of this manuscript are focused on the class comparison objective.

In general we will use specimen to refer to the sample material labeled, tag to refer to the label applied to the specimen, experiment to refer to the set of specimens mixed and subjected to MS simultaneously, and study to refer to the collection of MS experiments used to test a particular hypothesis. We assume that protein and peptide identification has already been performed, and that a list of peptides, the associated proteins and abundance levels are available for analysis. Case studies will be used to demonstrate the principles discussed. The beginning portions of the "Assessing the need for and success of normalization" and " Estimation of model parameters and calculating significance" sections will likely be more accessible to statisticians than to non-statisticians; the case studies in those sections provide tangible examples of the concepts being discussed which will likely be more tangible to clinicians and practitioners of mass spectrometry.

12.2 METHODS

12.2.1 OVERVIEW

We utilize three 4-plex iTRAQ data sets as case studies throughout the manuscript. The iTRAQ 4-plex labeling protocol involves adding one of four amine specific isobaric labels which do not alter mass (e.g., 114, 115, 116, or 117) to each of four specimens for simultaneous mass analysis via tandem mass spectrometry. The four mixed specimens are not discernible in the first MS where the most abundant species in the chamber are chosen for relative quantification (see Figure 1). During the second MS, the isobaric tags are broken off and quantification is performed based on the relative abundance of these tags. An 8-plex iTRAQ protocol is also available. See the "Discussion" section for an example of how other labeling protocols may differ.

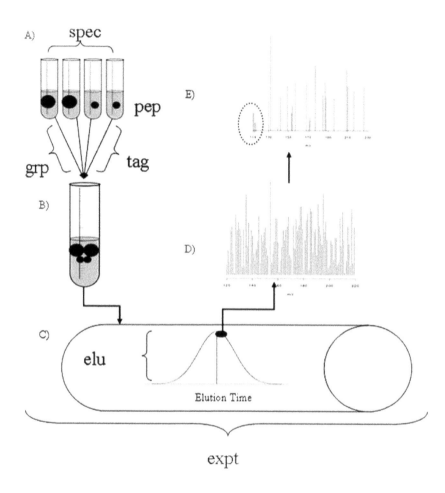

FIGURE 1: Cartoon depiction of the 4-plex iTRAQ labeling protocol for one MS experiment. A) Four specimens are each labeled with one of the four tags. The black dots indicate a given peptide that is present in different relative abundance according to size. B) The four specimens are then mixed into a single aliquot for simultaneous MS analysis. The resulting data constitute an MS experiment. C) Each peptide will take some amount of time to elute off of the LC column and so may be observed multiple times. D) In the first MS the top species according to abundance are chosen for a second MS. It is common for the top 3 or 5 to be chosen. E) During the second MS the iTRAQ tags are broken off and used for relative quantification (left in the dotted circle). It is these data that are used in downstream statistical analyses. The remaining peptides are fragmented further for identification purposes (right).

Here we provide a very brief explanation of each case study. Highly abundant proteins were removed in the GCM and prostate cancer studies, proteins were digested for all three studies, and fractionation was performed via strong cation exchange (SCX) in all three studies.

12.2.2 GIANT CELL MYOCARDITIS (GCM)

The study focused on three histologic subtypes of acute cardiomyopathy: 1) idiopathic dilated cardiomyopathy (DCM), 2) giant cell myocarditis (GCM) and 3) lymphocytic myocarditis (LM). These three subtypes present with similar clinical symptoms. However, GCM is much more lethal and requires a very different treatment strategy. Immediate objectives included comparing protein abundance profiles between these groups and long-term objectives included finding a protein present in blood useful as a diagnostic tool.

Six subjects of each subtype were included in the study. Though less than ideal (rationale will be discussed more in later sections), a pool of six normal healthy controls was used as a reference (N). Specimens were mass analyzed via capillary reverse-phase LC/MS/MS on a QSTAR quadripole time of flight mass spectrometer. Protein identification was performed via ProQuant. A total of six MS experiments were performed. Full experimental details are available elsewhere [9,10].

12.2.3 PROSTATE CANCER

This study used serum from prostate cancer patients to understand changes from pre- to post-androgen deprivation therapy (ADT) (n = 15 paired specimens) and to understand the differences between subjects experiencing ADT failure within a short (n = 10) versus long (n = 10) time-frame. Mass analysis was performed with an LTQ-Orbitrap Velos mass spectrometer. Final analyses are still being performed, so group membership is blinded for the current manuscript. A total of 13 MS experiments were performed. Two of the experiments were run a second time as indicated by an 'R' suffix (1R and 13R).

12.2.4 YEAST SPIKE-IN

A spike-in study was performed using yeast lysate to represent a complex background with the goal of understanding variance structure, systematic experimental biases and ability to detect fold changes of various magnitudes. Sixteen proteins with masses ranging from approximately 11 to 98 kDa were combined into two spike-in mixes; each protein was present in one mix at a "low" concentration and in the other mix at a "high" concentration. Each mix was then spiked into the yeast background at relative concentrations (fold changes) ranging from 1.0, 1.1, 1.2, 2.0, and 5.0. For each mix, two combinations of fold changes were performed: 1.0 : 1.5 : 1.0 : 5 and 1.1 : 1.0 : 2.0 : 1.2. Each of these was mass analyzed in duplicate for a total of eight MS experiments (2 mixes * 2 fold change layouts * 2 replicates). The yeast background was present at equal abundance (1.0 : 1.0 : 1.0 : 1.0) in all experiments. Mass analysis was performed on an LTQ Orbitrap. Full experimental details are available elsewhere [11]. These data are publicly available from http://ProteomeCommons.org/ Tranche webcite using the following hash search: YW9yck8PKhd5vyK-wUt0AIfVVllgXP9RoM0qTZDWQ05aNtae8uIHN/ 1Ird7APnNweSfqjV b9n5fT+oEyfqnOKZdRz3AUAAAAAAAB8Q==.

12.3 STATISTICAL EXPERIMENTAL DESIGN

12.3.1 OVERVIEW

The primary goals of statistical experimental design are to maximize information gain while minimizing resource expenditure and avoiding bias. Thoroughly considering the key aspects of replication, randomization and blocking prior to running an experiment ensures that enough of the necessary data is collected in a manner that ensures proper conclusions. In this "Statistical experimental design" section we first briefly describe the issues of bias and variability followed by discussion of the fundamental experimental design strategies to combat these issues.

12.3.2 BIAS

Bias is any trend in collection, analysis, interpretation, publication or review of data that can lead to conclusions that are systematically different from the truth. A confounded factor is one that is associated with both a real causal factor and the outcome of interest [12]. Bias and confounding may enter a study if samples in the comparative classes differ systematically on factors that affect the outcome. Dr. Ransohoff defines bias, describes ways to avoid it, and how to assess it and address it in various types of studies [13].

12.3.3 VARIABILITY

There are several levels of variability including technical, biological and institutional. Technical variability deals with reproducibility of an assay. Sample extraction, label, dye, technician, machine, reagent batch are all potential sources of assay variation and could alter the result produced in multiple assays of the same specimen. Biological variation is due to the difference between human subjects in a human study, mice in a mouse study, or Petri dishes/beakers of cell line in a cell line study. Institutional variation is due to differences between institutions and can be due to differences in patient populations seen, e.g. differences in disease severity or ethnicity, and differences in sample procurement protocols and implementation (even if identical on paper). These levels of variability all play a role in distinguishing signal from noise as well as in the generalizability of study conclusions. In general, technical variability is smaller than biological variability, which in turn is smaller than institutional variability. Generally, biological variability is the focus of most studies.

12.3.4 REPLICATION

One of the main threats to validity and generalizability of experiments where a large number of endpoints are measured on a small set of subjects is chance [13]. Replication is the tool that increases the precision of study conclusions and reduces the possibility that they are due to chance. There are several levels of replication that parallel the levels of variability. Technical replication involves repeated assays on the same biological replicate. This could involve one extraction of sample material undergoing sample preparation procedures as a unit but subjected to assay multiple times. It could also involve more than one extraction of sample material with each extraction then undergoing the sample preparation process on its own. Biological replication involves studying multiple members of the population being studied. For example, in a human study, each person in the study constitutes one biological replicate. If each human provides, say both cancer tumor tissue and normal tissue, then the pair of cancer-normal specimens constitutes one biological replicate. In an animal study, each animal constitutes one biological replicate. In a cell line study, each dish of cells grown up and subjected to treatment on its own constitutes a biological replicate. Institutional replication involves a study being performed at multiple institutions.

The optimal replication strategy depends on the goal of a study. A study with the goal of understanding and estimating sources of assay variability requires various types and levels of technical replication on a small number of biological replicates. Class comparison and class prediction studies have the goal of better understanding distinct classes of subjects. Study results are generally inferred back to population classes of subjects, making it ideal to maximize the precision of statements about those populations. Technical replication increases the information and precision about a specific subject while biological replication increases the information and precision about a population. Thus, the greatest information gain and increase in precision for inferences to the study population comes from allocating available resources to more biological replicates rather than technical replicates. The mathematics supporting this are demonstrated elsewhere [14].

In practice, it is wise to include technical replicates on a few of the biological replicates in high dimensional experiments, especially if the assay platform or protocol is new to the laboratory, for use in evaluating and reporting on reproducibility and quality. Institutional replication is often utilized in studies with validation as the goal.

12.3.5 BLOCKING

Statistical blocking is a tool that helps to guard against known potential biases and to minimize variance in a study. Blocking is sometimes referred to as matching in the context of sample selection, where for example, subjects are matched on gender or paired specimens are taken from the same subject. In the context of spectral acquisition, blocking is sometimes referred to as multiplexing. Specimens assayed within a block are more similar than specimens assayed between (in different) blocks. Use of this strategy in allocating specimens to MS experiments and tags is called a Randomized Block Design (RBD). MS experiment is a natural blocking factor in labeled work-flows and should be used as such. Labels or tags, day of MS assay, laboratory technicians, reagent batches, MS machines or LC columns are other examples of natural blocking factors. To protect against bias, avoid confounding and minimize variance about the question of interest, some specimens from each study group should be allocated to be assayed together within a block. This is the basis of the RBD and is demonstrated in the case study examples towards the end of this section on "Statistical experimental design". A labeled MS study with only one MS experiment will result in study groups being confounded with labels and very small sample sizes. It is good practice to utilize multiple MS experiments in order to avoid confounding of study groups and tag effects and reasonable sample sizes.

12.3.6 RANDOMIZATION

Randomization is a tool that protects a study from both known and unknown biases. This tool is utilized during both subject selection and during the allocation of specimens to sample processing order. Randomized selection of subjects generally ensures that potential biases which may influence the outcome are approximately balanced across the study groups and is discussed in greater detail elsewhere [15,16].

Randomized allocation of study specimens over assay run order generally ensures group membership is approximately balanced over run order, thereby eliminating the potential confounding of study group and run order. In a labeled workflow using MS experiment as a blocking factor, this allocation takes place in two steps. Consider the 4-plex iTRAQ workflow and a study with four groups of interest such as the GCM study. Thus, the number of groups is equal to the number of tags within each MS experiment block. The first step is to allocate one specimen from each study group to each block. To do this, a random number is generated for each biological replicate via a random number generator, such as the RAND function in excel. These numbers are then ranked within study group to determine which specimen is allocated to MS experiment 1, 2, etc. The second step is to allocate specimens to labels within a block. This can be done using the same random number, or a second random number could be generated, with the rank order of these random numbers determining the tag allocation.

Though a consistent tag bias affecting all proteins has not been demonstrated in iTRAQ data, there are likely protein-specific tag biases. Thus, it is wise to ensure tag and study group are not confounded. Check the randomization to be sure groups are approximately balanced over tag so that group and tag are not confounded. Alternatively, both MS experiment and tag can be used as blocking factors. This is especially wise in studies with very small sample sizes.

12.3.7 CASE STUDY: GCM DATA

Both MS experiment and labeling tag were used as blocking factors in this study. First, one specimen from each of the four study groups was allocated to an MS experiment. Second, within each MS experiment, the four specimens were randomly assigned to a tag so that the study groups were approximately balanced over tags. Both steps were accomplished using a random number generator. See Table 1 for the resulting allocation. Though the normal pool was included as a reference, it was randomly assigned to tag within a block in order to avoid confounding of tag and study group. As a result of the blocked randomization, any potential effects or biases due to tag can be distinguished from study group using a statistical model for differential abundance. This will be discussed in more detail in the "Differential abundance" section.

TABLE 1: Statistical experimental design of the GCM study demonstrating allocation of specimens to MS experiments and labeling tags.

Experiment	Tag			
	114	115	116	117
1	GCM1	DCM1	LM1	Normal Pool1
2	DCM2	Normal Pool2	GCM2	LM2
3	Normal Pool3	LM3	GCM3	DCM3
4	LM4	GCM4	Normal Pool4	DCM4
5	DCM5	GCM5	Normal Pool5	LM5
6	Normal Pool6	DCM6	LM6	GCM6

(Adapted with permission from [9]. Copyright 2008 American Chemical Society.) The abbreviations GCM, DCM, LM and N (normal control pool) denote the four groups under investigation as described in Section 2.2. The numbers denote biological replicates for GCM, DCM and LM, and technical replicate number for N. For example, GCM1 is the first sample in the GCM group. Experiment number also corresponds to run order.

The rationale for using a pool as a reference in a labeled design is based on the fact that the abundance measures are relative and the pool can be used as a normalizing factor of sorts to adjust for technical variation. With this strategy, abundance values are divided by the pool abundance values to create a "normalized" ratio. First, this strategy assumes the normalization factor is identical for each specimen within the MS experiment. However, normalization factors generally differ for each specimen due to slight but non-ignorable differences in sample handling from the time of extraction from the subject to mass analysis. Second, the resulting ratios are generally ill behaved and difficult to deal with in statistical analyses. This will be discussed further in the "Data quality and normalization" and "Differential abundance" sections. Third, this induces a correlation between observations, violating the independence assumption of statistical tests. Model-based methods for normalization are described in the "Data quality and normalization" section. Fourth, it is not possible to correctly perform statistical differential abundance between the six normal specimens in the pool and other study groups since biological variability cannot be estimated for the normal specimens. Statistical designs and the associated analysis methods were developed specifically to deal with relative measurements in the early 1900's[17,18], obviating the need for a reference sample in each MS experiment.

TABLE 2: Statistical experimental design of the prostate cancer study.

Experiment	Tag			
	114	115	116	117
1, 1R	Pre	Late	Early	Post
2	Post	Early	Late	Pre
3	Early	Post	Pre	Late
4	Post	Early	Late	Pre
5	Late	Pre	Post	Early
6	Late	Early	Post	Pre
7	Pre	Post	Early	Late
8	Pre	Pre	Post	Post
9	Early	Late	Pre	Post
10	Post	Post	Pre	Pre
11	Post	Pre	Late	Early
12	Early	Pre	Late	Post
13, 13R	Post	Late	Pre	Early

Statistical experimental design of the prostate cancer study demonstrating allocation of specimens to MS experiments (where an 'R' suffix indicates that experiment was re-run) and labeling tags. The abbreviations Pre, Post, Early and Late denote the four groups under investigation as described in Section 2.3, pre-ADT, post-ADT, ADT failure within a short time-frame, ADT failure within a long timeframe. The numbers denote biological replicates for each group. For example, Pre1 is the first sample in the pre-ADT group.

12.3.8 CASE STUDY: PROSTATE DATA

Two comparisons were of interest in the prostate cancer study. The first comparison was between pre- and post-ADT treatment protein profiles in paired specimens from each of 15 patients in order to understand proteins indicating early response to ADT. The second comparison was between ten subjects who failed ADT within 12 months (short) and ten subjects who failed after 30 months (long). In addition, for proteins found to be significantly differentially a in the pre- to post-ADT comparison, the investigator wished to assess behavior of those proteins in the short and long cohorts. Thus, it was important to keep paired pre and post specimens within the same MS experiment in order to minimize variability in that comparison. Second, it was important to allocate at least one short and one long specimen to the same MS experiment in order to minimize variability in that comparison. Third, it was important to observe most of the proteins in both sets of subjects. Thus, given the data-dependent acquisition process of global MS studies, it was important to include both pre/post specimens together with short-term/long-term in the same MS experiments.

The randomization plan accounted for these goals. Thirteen MS experiments were required to assay the 50 specimens and two technical replicates. First, one short-term and one long-term subject were randomly assigned to 10 of the 13 MS experiments, allocating all 20 of these specimens. Second, a pair of pre/post specimens was randomly assigned to those same 10 MS experiments, allocating 10 of the 20 pairs of specimens. Third, the remaining five pairs of specimens were randomly assigned to the remaining three MS experiments. Fourth, the four specimens assigned to each MS experiment were randomly assigned to tag, ensuring balance of the groups over tag. See Table 2 for the resulting allocation.

12.4 DATA QUALITY AND NORMALIZATION

12.4.1 OBTAINING THE DATA

Vendor software generally creates data reports in which abundance data has been divided by the abundance in one specimen or tag that is designated as the reference. This reference specimen may be a control or a pool, or represent one of the study groups of interest. However, ratios are generally ill behaved, and it is preferable to work with the individual abundance values in statistical analyses [14,19]. For example, when abundance values in the control are very small, the resulting ratios get incredibly large very quickly due to very small numbers in the denominator. In addition, such ratios are not immune to pipetting errors or differences in specimen processing.

Thus, it is preferable to work with data that have not been put into a ratio format. That is, we want the peptide level abundance values for each labeled specimen for use in statistical analyses. It is not always obvious how to obtain this data. In the ProteinPilot software with which we are familiar, individual reporter ion area under the curve values are contained in the Peptide Summary exports. These reports are generated by first opening the results file (*.group) in ProteinPilot and then clicking on Peptide Summary export on the left side of the page. The user is then prompted for a location to save the resulting .txt file. The desired data are near the last columns in the spreadsheet and are given variable names such as Area114, ..., Area117.

An a priori list of proteins does not exist for global MS studies. Rather, the goal is to catalogue as many proteins as possible in a specimen and obtain quantification information for them. A "divide-and-conquer" strategy is employed since MS instruments have a dynamic range of around 4-5 orders of magnitude while the human proteome spans over 12 [20]. A specimen undergoes many steps in this process including digestion to break proteins into peptides and fractionation to separate the specimen into less complex sub-samples via some chemical property such as charge state (saltiness) and/or hydrophobicity (ability to mix with water) [21,22]. As material is introduced into the mass spectrometer, generally only the most abundant species are selected for MS, e.g., the top three or five. Thus, the data acquisition is abundance-dependent. As a result, iTRAQ studies using multiple MS experiments typically have many proteins/peptides that are not observed in all MS experiments. Due to the dynamic range of the proteome, whether human or other species, approximately half of the species in a specimen are present at the level of detection. So even in technical replicate MS experiments there can be a large number of proteins which are not observed in both experiments.

The tandem MS is utilized in iTRAQ to choose a species in the first MS and then perform identification and quantification in the second MS, generally resulting in an observed abundance value for all for specimens within an experiment. Thus, there is generally not missing data for a given peptide within an MS experiment. This has implications for the normalization strategy. See Table 3 for an example of a typical data matrix.

TABLE 3: Snapshot of an iTRAQ data table.

MSMS Spectrum ID	Protein Accession	Peptide Sequence	114	115	116	117	Experiment Number
S4_F11.1140.1140.2	GPP1_YEAST	(F)EDAPAGIAAGK(A)	2813.568536	1595.741524	2475.724121	2458.306255	4
S1_F16.2850.2850.3	GPP1_YEAST	(K)GRNGLGFPINEQDPSK(S)	316.4418979	466.2738416	630.4750319	444.921289	1
S3_F16.2618.2618.3	GPP1_YEAST	(K)GRNGLGFPINEQDPSK(S)	869.2210037	544.1843783	1617.949095	665.3067241	3
S3_F16.2623.2623.3	GPP1_YEAST	(K)GRNGLGFPINEQDPSK(S)	1163.021548	925.1491063	1347.204837	1032.958433	3
S1_F13.1643.1643.2	GPP1_YEAST	(K)DDLLK(-)	10607.97083	8544.75492	10953.83841	9005.777375	1
S1_F13.1513.1513.2	GPP1_YEAST	(K)DDLLK(-)	1748.258583	2893.388823	1861.30691	2715.653088	1
S1_F13.1507.1507.2	GPP1_YEAST	(K)DDLLK(-)	606.7841803	919.8748238	1144.338397	1025.119065	1
S1_F13.1643.1643.2	GPP1_YEAST	(K)DDLLK(-)	10607.97083	8544.75492	10953.83841	9005.777375	1
S2_F13.1291.1291.2	GPP1_YEAST	(K)DDLLK(-)	2618.558021	1367.979923	2947.928581	2321.749983	2
S2_F13.1291.1291.2	GPP1_YEAST	(K)DDLLK(-)	2618.558021	1367.979923	2947.928581	2321.749983	2
S3_F13.1582.1582.2	GPP1_YEAST	(K)DDLLK(-)	1849.138156	2882.532646	3456.336093	3333.133633	3
S3_F14.1374.1374.2	GPP1_YEAST	(K)DDLLK(-)	88.57809719	39.54738544	113.4348917	128.6087568	3
S3_F13.1360.1360.2	GPP1_YEAST	(K)DDLLK(-)	5897.197655	8115.16893	5413.842313	6349.146183	3
S3_F13.1360.1360.2	GPP1_YEAST	(K)DDLLK(-)	5897.197655	8115.16893	5413.842313	6349.146183	3
S3_F13.1357.1357.2	GPP1_YEAST	(K)DDLLK(-)	3232.418762	6524.148517	5246.904457	5391.07817	3
S3_F13.1582.1582.2	GPP1_YEAST	(K)DDLLK(-)	1849.138156	2882.532646	3456.336093	3333.133633	3
S3_F14.1374.1374.2	GPP1_YEAST	(K)DDLLK(-)	88.5780919	39.54738544	113.4348917	128.6087568	3
S3_F13.1360.1360.2	GPP1_YEAST	(K)DDLLK(-)	88.57809719	39.54738544	113.4348917	128.6087568	3
S3_F13.1357.1357.2	GPP1_YEAST	(K)DDLLK(-)	3232.418762	6524.148517	5246.904457	5391.07817	3
S4_F13.1399.1399.2	GPP1_YEAST	(K)DDLLK(-)	3195.952185	3638.020997	6349.053364	6973.840279	4
S4_F13.1395.1395.2	GPP1_YEAST	(K)DDLLK(-)	2404.371623	2571.938103	4057.845902	3907.827732	4
S4_F13.1395.1395.2	GPP1_YEAST	(K)DDLLK(-)	3195.952185	3638.020997	6349.053364	6973.840279	4
S4_F13.1399.1399.2	GPP1_YEAST	(K)DDLLK(-)	2404.371623	2571.938103	4057.845902	3907.827732	4

12.4.2 ASSESSING THE NEED FOR AND SUCCESS OF NORMALIZATION

Observed abundance values produced by global mass spectrometry machines are relative rather than absolute. In addition, experimental effects between MS runs have been demonstrated in several proteomic workflows [23]. Even in labeled work-flows which reduce between MS experiment variability, abundance values are subject to other experimental factors such as sample handling from the time the specimen was extracted from the subject, pipetting errors or other potential sources of bias [24]. Thus, data must generally be normalized prior to performing comparisons between groups of interest.

Normalization via standard curves is problematic in these experiments that catalogue and quantify hundreds to thousands of proteins in a single assay. However, normalization methods have been developed utilizing the entire data distributions. These make some specific assumptions about the data. Most algorithms assume: 1) only a small portion of the proteins are differentially abundant between groups of interest, 2) the fold change distribution of differentially abundant proteins is symmetric about 1.0, 3) data must be available on a sufficient number of proteins with abundance levels distributed throughout the dynamic range to estimate global biases without over-fitting [25]. For example, quantile [26,27] and cyclic loess normalization [28-30] are examples of normalization algorithms developed for one- and two-color gene expression arrays that make these assumptions. The iterative ANOVA model [9] described in the "Data quality and normalization" section is an example of such a normalization algorithm which can be applied to both labeled and label-free proteomics abundance data.

There are several visualization tools which are useful for assessing data quality, the need for normalization and the success of normalization. These include peptide or protein coverage plots, box-and-whisker plots (box plots for briefness), and minus versus average (MVA or MA) plots. We define these and provide some examples of each in subsequent paragraphs.

Peptide and protein coverage plots are useful for understanding the magnitude of missing data in a data set, and how many peptides/proteins were detected in multiple MS experiments. They can highlight systematic effects present in the data for further investigation. The axes indicate MS experiment number versus some rank order of the peptide or protein ID. The sort order of the peptides can be by average abundance, by number of experiments it was observed in, or other. A line is placed on the plot if the peptide was detected in that experiment, white space if it was not detected. A peptide that was detected in all MS experiments in a study would show as a solid line across the entire plot.

Box plots provide a visual summary of a distribution. The bottom, mid and top lines of the box represent the 25th, 50th (median) and 75th percentiles of the distribution. A "whisker" extends above the box to 1.5 times the inter quartile range (i.e., the distance from the 75th percentile to the 25th percentile) or to the maximum value in the distribution, whichever is smallest. Similarly, a whisker extends below the box the same distance or to the minimum value, whichever is largest. If points exist beyond these whiskers, they are represented by dots. There is one box-and-whisker for each specimen in the study. Global biases which affect all peptides are indicated by shifts up or down in the box-and-whiskers. Usually such a shift is not expected due to the disease, i.e., a global increase or decrease in protein concentration in the biological subject is not expected. The sort order of the boxes can be chosen strategically. For example, sorting by MS experiment first and then by tag would help the eye identify global experiment effects whereas sorting by tag first and then experiment would help the eye identify global tag effects. Changes in dynamic range are evident from compression or expansion of the box and whiskers. If normalization has effectively removed global biases, the box plots of post-normalization data should demonstrate similar per-specimen box and whiskers. They typically demonstrate less variability than in the pre-normalization plots as well, as evidenced by reduced height of the box and whiskers.

Minus versus average (MVA) plots are useful for assessing whether bias is a function of mean abundance. Nonlinear bias of this type is common in gene expression data from both single and multi-channel arrays [30,31]. Traditional MVA plots demonstrate agreement in the global distributions (or lack thereof) for two specimens, have the average of the two on the horizontal (x) axis and the difference between the two on the vertical (y) axis, and a point for each peptide or protein that is observed in both specimens. If two replicates yielded identical results, all points would lie

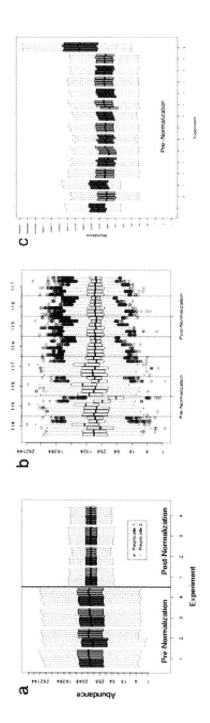

FIGURE 2: Box plots of global protein abundance distribution. Box plots for A) the yeast spike in study, B) the GCM study and C) the prostate cancer study. The log2 scale raw mass spectrometric signal abundance (labeled on the raw scale) is plotted as a function of MS experiment number. Sort order in A is first by MS experiment and then by tag, and boxes are grouped in sets of 8 where the first four (dark grey) are the first technical replicate MS experiment and the second four (light grey) are the second technical replicate MS experiment for a given spike-in combination. Sort order in B was chosen to be first by tag and then by MS experiment in order to help assess whether a systematic tag bias was present. Sort order in C is first by MS experiment and then by tag. (Panel B is reproduced with permission from [9]. Copyright 2008 American Chemical Society.)

on the $y = 0$ horizontal line (indicated on the plots for reference). Residual MVA plots are advantageous because they allow one plot for every specimen (rather than all pair-wise combinations) and demonstrate visually how a specimen is similar to or different from the average of the others. Here, the horizontal axis is the average over all specimens instead of the average of two specimens and the vertical axis is the difference between that specimen and the average over all specimens.

12.4.3 CASE STUDY: YEAST DATA

Pre-normalization box plots of peptide abundance values from the yeast study demonstrate that, even in a well-controlled experiment where all but 16 proteins are present at $1.0 : 1.0 : 1.0 : 1.0$ ratios, between MS experiment and tag effects exist (see Figure 2a, left panel). Post-normalization box plots (see Figure 2a, right panel) demonstrate that the global distributions have similar percentiles and the variability has been reduced, both indicators of successful normalization.

MVA plots in the yeast study demonstrate a small amount of global shift in abundance (see Figure 3), more between MS experiments than within as would be expected. The fact that the smoother is shifted away from the $y = 0$ line indicates global bias. The curvature in the smoother indicates the bias may be abundance-dependent. If normalization has been effective at removing global biases, the smoothers on post-normalization MVA plots should overlay the $y = 0$ line. This is nearly true in these data. Some nonlinearity remains post-normalization. However, these are in a region where there are very few data points as demonstrated by the smoothed histogram at the bottom of the plot. Completely removing this bias would be viewed as over-fitting the data. Most experimental biases we have seen in iTRAQ data have been mostly linear in nature, but this should be evaluated on a per-study basis.

The abundance-dependent data acquisition process is evident in a protein coverage plot for the yeast data through the gradation of shading; there are fewer proteins present on the left at low abundance levels than on the right at high abundance levels (see Figure 4a). It is also evident that a larger portion (relative to the other case studies) of proteins were observed in most of the MS experiments in this well controlled spike-in study.

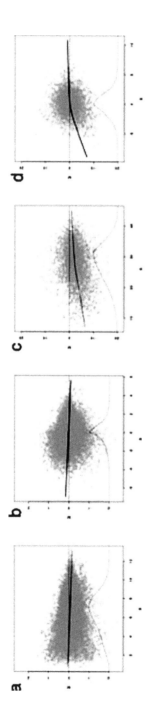

FIGURE 3: MVA plots. Pre- (panel A) and post-normalization (panel B) within-experiment MVA plots. Pre- (panel C) and post-normalization (panel D) between-experiment MVA plots. The vertical axis is difference between the intensities in two specimens on the log2 scale and the horizontal axis is the average of the two intensities on the log2 scale (note the different in axes labels between the top and bottom plots); there is one point for each peptide observed in both specimens. A locally weighted moving average smoother is indicated to demonstrate the average bias curve as a function of average abundance. A smoothed histogram is included at the bottom of the plots to demonstrate the number of data points represented directly above that area in the plots.

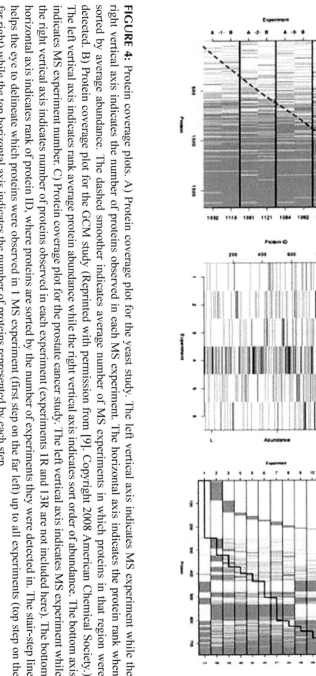

FIGURE 4: Protein coverage plots. A) Protein coverage plot for the yeast study. The left vertical axis indicates MS experiment while the right vertical axis indicates the number of proteins observed in each MS experiment. The horizontal axis indicates the protein rank when sorted by average abundance. The dashed smoother indicates average number of MS experiments in which proteins in that region were detected. B) Protein coverage plot for the GCM study (Reprinted with permission from [9]. Copyright 2008 American Chemical Society.) The left vertical axis indicates rank average protein abundance while the right vertical axis indicates sort order of abundance. The bottom axis indicates MS experiment number. C) Protein coverage plot for the prostate cancer study. The left vertical axis indicates MS experiment while the right vertical axis indicates number of proteins observed in each experiment (experiments 1R and 13R are not included here). The bottom horizontal axis indicates rank of protein ID, where proteins are sorted by the number of experiments they were detected in. The stair-step line helps the eye to delineate which proteins were observed in 1 MS experiment (first step on the far left) up to all experiments (top step on the far right) while the top horizontal axis indicates the number of proteins represented by each step.

12.4.4 CASE STUDY: GCM DATA

The coverage plot from the GCM study demonstrates that many more pep-tides were detected in experiment 4 than the other experiments (see Figure 4b). In discussing the results with the researchers, we learned that experiments 1-3 had been performed within a short time-frame, experiment 4 was performed approximately two months later followed by another gap in time before experiments 5 and 6 were performed. Pre- and post-normalization box plots (see Figure 2b) demonstrate linear biases have been removed and variability reduced through normalization.

12.4.5 CASE STUDY: PROSTATE DATA

Protein coverage plots from the prostate study (See Figure 4c) indicate a systematic difference between experiments (1, 9-13) and (2-8) as demonstrated by the blocks of proteins present in all of one set of experiments or the other. Upon discussion with laboratory personnel including the mass spectrometry expert and the bioinformatics expert, we determined that a change in the protein identification labels had occurred in between the eighth and ninth MS experiments (experiment 1 was actually run between numbers 8 and 9). This change resulted in protein names represented two different ways for a subset of proteins. Once the naming conventions were applied similarly across all experiments, these "blocks" of proteins were no longer evident.

Box plots from this study demonstrate that the distributions for experiments 1, 2 and 13R (recall the 'R' suffix indicates a repeated MS experiment) were shifted up relative to the other experiments in the box plot (see Figure 2c). In talking with the mass spectrometry expert, there was no known explanation for the shifts in experiment 1 and 2, and review of the spectra deemed the data to be of good quality. Through the discussion we determined that a machine setting had been changed prior to experiment 13R resulting in a nearly 10 fold increase in abundance and far fewer proteins observed compared to other experiments, thus the data was rendered not useable. Experiment 1R was done due to questionable quality of Experiment 1. Thus, the MS experiments used statistical analysis were 1R, 2-13.

12.4.6 BUILDING THE NORMALIZATION MODEL

Vendor software generally applies a normalization factor within an MS experiment which results in equal median fold changes between the chosen reference specimen and the remaining specimens. This is not adequate with the abundance-dependent data acquisition process [32]. Here, we describe how to build a model for normalization.

We use the observed data, y, to indicate the true abundance. However, the observed values are influenced by multiple factors. There are both known biological and experimental factors as well as unknown factors which can be put into a statistical model. Biological factors include study group, subject or specimen, protein and peptide. Experimental factors include MS experiment, tag and elution time (see Figure 1). On the raw scale effects are generally considered to be multiplicative. Thus, the model can be written as

$$y_{ijkpm} = expt_i \times tag_j \times spec_{ij} \times grp_k \times prot_p \times pep_{kpm} \times err_{ijkpm}$$

where, y_{ijkpm} is the observed abundance value, $expt_i$ indicates the ith MS experiment, tag_j indicates the jth labeling tag, $spec_{ij}$ indicates the ijth specimen (which is also the $expt_i \times tag_j$ interaction), grp_k indicates the kth study group, $prot_p$ indicates the pth protein observed in the ith MS experiment, pep_{kpm} indicates the mth peptide observed for the pth protein in the ith experiment and err_{ijkpm} indicates random, unspecified error. Note that subscripts may be helpful for some readers. For others, it is important simply to understand the conceptual framework of representing known effects in the model to explain sources of variability in the data. A complete discussion of model terms and the rationale for each can be found elsewhere [33].

The most common and simplest statistical models are based upon additive rather than multiplicative effects. Since it is generally easier to transform data to obtain the proper scale for the mean and then worry about how to model the variance in that framework, the data are generally transformed to the log scale. Log2 is commonly used since it is easy to interpret in your head with differences of 1, 2, 3, etc. corresponding to fold changes of 2, 4, 8, etc., respectively (powers of 2). On the additive scale then, this model can be written as

$$\log_2(y_{ijkpm}) = \text{expt}_i + \text{tag}_j + \text{spec}_{ij} + \text{grp}_k + \text{prot}_p + \text{pep}_{kpm} + \varepsilon_{ijkpm}$$

where the ε_{ijkpm} are assumed to identically and independently distributed according to a Gaussian distribution. This is the basis of the analysis of variance (ANOVA) model, explaining the sources of variation. Experimental factors are not of interest specifically, but should be accounted for in order to minimize variability and ensure accurate conclusions. Conceptually, including terms such as MS experiment in the statistical model performs group comparisons within an experiment, and then averages these comparisons over all experiments in the study to achieve a unified result based on all available data. It is this concept that allows multiple MS experiments to be combined for unified analysis.

The experimental effects serve as the normalization portion of the model, and the biological effects serve to test the hypotheses of interest. The experimental effects in labeled MS studies include MS experiment and label. These effects should be chosen based on the study at hand, and may also include others such as LC column or laboratory technician in larger studies. Biological effects will be discussed further in a subsequent section.

The experimental effects are global terms, and are assumed to affect all proteins and peptides similarly. Thus, they should be estimated using all available data. However, due to the size of data sets generated from these experiments it is generally not possible with current computing infrastructure to fit the entire model at once. Thus, the model is broken into normalization and differential abundance pieces which are each fit separately. If good study design is utilized, then normalization and group effects are close to independent, allowing these to be estimated in two separate models to achieve the desired results. Due to the abundance-dependent data acquisition process, peptide must be included in the normalization model in order to estimate the normalization parameters properly [9,32]. Code to implement this via SAS is available from the authors. See the "Discussion" section for potential extensions to the normalization model.

12.4.7 CASE STUDY: GCM DATA

The GCM study had six MS experiments and four iTRAQ tags. Thus, experiment and tag are two known experimental effects to be included into the normalization model. Specimen is included as well to obtain a specimen-specific normalization. Thus, the normalization model on the additive scale is $\log(y_{ijkpm}) = \text{expt}_i + \text{tag}_j + \text{spec}_{ij} + \text{pep}_{kpm} + \varepsilon_{ijkpm}$ where model terms are as defined in the previous section. With the 2,637 unique peptides observed in this study, the matrix is too large to invert and as a result, even this normalization model must be fit iteratively as is generally the case with these studies. The normalized data are then the residuals from the normalization model, $y\_\text{norm}_{ijkpm} = \log(y_{ijkpm}) - [\text{expt}_i + \hat{\text{tag}}_j + \hat{\text{spec}}_{ij}]$ where the hat indicates estimated parameter values. The pep_{kpm} term is not subtracted off since it is a biological effect and is included in the normalization only to appropriately line up the distributions between specimens. The normalization models for the other case studies contained the same terms.

We have investigated the utility of accounting for the abundance-dependent data acquisition, and therefore non-random missing data by incorporating a censoring mechanism into the normalization and differential abundance models [34]. iTRAQ-like data with either peptide competition alone or peptide competition plus a machine threshold for inducing missing data were simulated with MS experiment effects ranging from 0.5 to 2.0 and study group differences of 0.5, 1.0, 1.5, 2.0 and 2.5, all on the log2 scale. Incorporating a censoring mechanism into the modeling process reduces the bias in MS experiment effect estimation but does not reduce the variability in estimates (see Figure 5). However, due to the balance of study groups over MS experiments and tags in a properly designed study, the MS experiment effects cancel out in the class comparison calculation, resulting in essentially no difference in estimation of study group effects

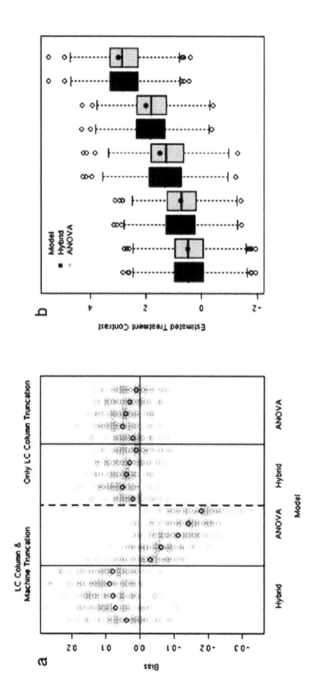

FIGURE 5: Bias in parameter estimates. Bias in MS experiment study group comparison estimates under two different mechanisms described in the text giving rise to missing data between MS experiments using either the ANOVA model normalization or a hybrid model incorporating censoring. A) Bias (vertical axis) is the difference between estimated and true MS experiment effects. The horizontal axis indexes varying MS experimental effects and analysis methods. B) Box and whisker plots of estimated study group differences. The dot indicates the true simulated difference.

under the two models. Note that this does not imply that normalization is not necessary; it is still required to account for and therefore remove variability and improve reliability of treatment comparisons.

12.5 DIFFERENTIAL ABUNDANCE

12.5.1 OVERVIEW

Statistical models can be used to assess which peptides or proteins are significantly differentially abundant between study groups. The models are flexible, can accommodate nearly any experimental design, and consider the magnitude of signal relative to the variation in the data in order to determine whether the signal is appreciably larger than random noise. These methods have been shown to be the most powerful for hypothesis testing and enable estimates of fold change based on all available data. They are more straightforward than many ad hoc methods and result in simple summary statistics for each protein or peptide.

12.5.2 BUILDING THE DIFFERENTIAL ABUNDANCE MODEL

We pick up the modeling discussion we began in the previous section where we discussed and demonstrated estimation and removal of the experimental effects. Now we turn our attention to the biological effects in the model. Differential abundance models are generally fit on a per-protein basis due to computational limitations. Thus, the differential abundance model reduces to $y\_norm_{ijkpm} = grp_k + pep_{kpm} + \varepsilon_{ijkpm}$. The hypothesis test of grpk is of greatest interest, as this is a measure of the difference in abundance between the two groups relative to the noise in the data. Research has shown use of all peptide information associated with a protein without summarization in a statistical model is more efficient than ad hoc summaries or decision rules [35].

12.5.3 VARIANCE STRUCTURE

It is important to understand the variance structure or precision in your data as this has implications for the statistical models and estimation strategies used. We and others have found that precision is generally a function of mean abundance in iTRAQ data [11,36-40]. This varying precision is not evident in standard residual plots, but is evident in per-MS experiment plots. The variance structure will likely depend on the MS technology used. Thus, this should be examined for each study to determine the structure and appropriate modeling approaches in light of this (See the "Estimation of model parameters and calculating significance" section).

12.5.4 CASE STUDY: YEAST DATA

We demonstrate the mean-variance relationship graphically. The within MS experiment coefficient of variation (CV), which corresponds to the standard deviation on the raw scale, plotted versus the mean abundance demonstrates that precision increases as abundance increases (see Figure 6). We have observed this relationship in several iTRAQ data sets produced from human and yeast specimens on Orbitrap and TOF mass spectrometers. It is important to look at your data to understand the correct modeling procedure to use.

12.5.5 ESTIMATION OF MODEL PARAMETERS AND CALCULATING SIGNIFICANCE

When variance or precision is constant, ordinary least squares (OLS) are used to estimate model parameters. However, as shown in the previous section, precision can be abundance-dependent in iTRAQ data. Thus, other means must be used for parameter estimation. Including MS scan, i.e., elution time, in the model to account for varying precision results in a saturated model. Thus, weighted least squares (WLS) is used to estimate model parameters. In WLS, each abundance value is given a weight that is inversely proportional to the precision. As a result, peptides measured

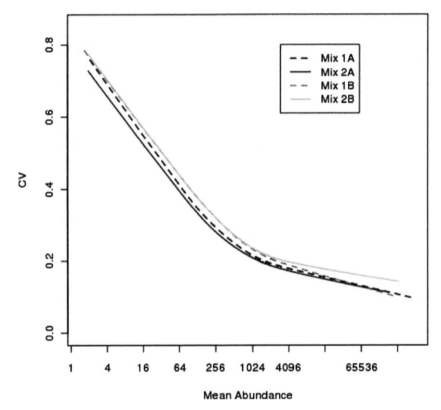

FIGURE 6: CV as a function of protein abundance. Within experiment peptide coefficient of variation (CV) on the vertical axis versus average abundance on the horizontal axis for the yeast data. The line is a moving average smoother indicating average CV as a function of mean abundance.

with more precision are given more weight in the analysis, whereas those measured with less precision are given less weight. The weight can be estimated theoretically using the relationship between the Gaussian and Lognormal distributions. Alternatively, it can be estimated empirically. We have chosen to use an empirical estimate, assigning each peptide the value of the moving average smoother at its abundance value on a CV plot such as that in Figure 6. In these data, this weighting accounts for the variability due to differences in elution time.

It is not computationally feasible to estimate all parameters within the biological model simultaneously. Thus, in practice, differential abundance models are fit on a per-protein or per-peptide basis depending on the goals of the study at hand. We focus on per-protein level models here. In biological terms, fitting models on a per-protein basis allows estimation of the amount of random variability for each protein separately rather than forcing it to be the same across all proteins.

Peptides mapped to multiple proteins are not included in differential abundance models. Shared peptides, peptides that are present in more than one protein, are common in shotgun proteomic experiments. These shared peptides have been found to be beneficial to determine the presence of a protein [41]. However, these same shared peptides can become problematic in estimating relative abundance of a protein. A simple example is demonstrated in Figure 7 containing two specimens, each of which contain two proteins which are represented by solid or dotted line circles. The true relative ratios for Specimen A to Specimen B are 3:1 and 1:1 for proteins ABC and DEF, respectively, and peptide 4 is shared between both proteins. If the shared peptide is ignored, the fold change difference between Sample A and B for protein ABC is simply $(3+3+3)/(1+1+1) = 9/3 = 3$ and for DEF is $(1+1)/(1+1) = 2/2 = 1$ which match the true fold changes. However, after the identification process Peptide 4 will be assigned a total abundance of 4 in Specimen A and 2 in Specimen B, and these abundance values will be attributed to both proteins in the resulting output. The resulting fold change estimates for ABC and DEF now become $(3+3+3+4)/(1+1+1+2) = 14/5 = 2.8$ and $(1+1+4)/(1+1+2) = 64 = 1.5$, respectively. Thus, both estimates of fold change for the proteins are biased away from their true values as a result of including the shared peptide. For this reason, when doing quantitative analyses, peptides that appear in more than one protein are excluded from analysis.

Due to the large number of proteins being examined in global mass spectrometry studies, stringent criteria must be used to determine significance of a peptide. One strategy is to use the Bonferroni correction which involves computing a significance threshold based on the number of proteins being tested as 0.05/(the number of proteins being tested). This is generally accepted to be too stringent and frequently results in no significant

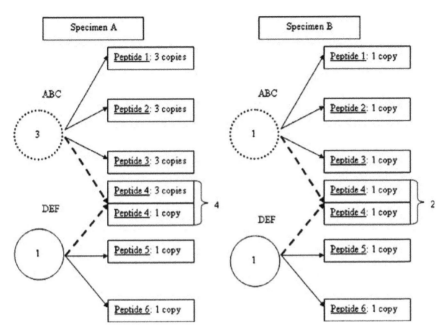

FIGURE 7: Cartoon illustration of the impact of including peptides mapped to multiple proteins in relative quantification. As described in the text, the inclusion of including peptides which are mapped to multiple proteins results in biased estimates of fold changes. Therefore, these peptides are generally included in the normalization step, but excluded from the relative quantification step.

proteins. The distribution of p-values can be used to compute an expected false discovery rate (FDR) [42,43]. These numbers, called q-values, give an indication of the level of significance in the study. An FDR value is the number of genes among those declared to be significant which are expected to be falsely declared significant. A study resulting in a uniform distribution of p-values (which would be expected by chance under the null hypothesis of no differences between the study groups) will have large FDR values. However, a study with a skewed distribution of p-values having a spike near zero will have smaller FDR values.

12.5.6 VISUALIZING AND INTERPRETING SIGNIFICANCE AND FOLD CHANGES

Digesting the volumes of data resulting from a high dimensional study can be challenging. Here we present some visualization and computational tools we have found helpful for drawing biological conclusions.

12.5.7 CASE STUDY: GCM DATA

Recall the primary goal of the GCM study was to compare abundance for proteins between four types of subjects, GCM, DM, LM and normal controls. We focus on the GCM versus DM comparison as an example. Note that due to the fact that the normal controls were pooled prior to mass analysis, it is not possible to properly estimate biological variability within this group. The differential abundance model was fit in SAS [44] with the following commands:

```
proc mixed data=abundance;

by protein_id;

class dx_grp;

model logYnorm=dx_grp;

/*This performs all pair wise comparisons between diagnostic groups*/

lsmeans dx_grp/pdiff;

ods output diffs=dx_grp_contrasts;

ods output tests3=overallFtest;

run;
```

A few lines of the output listing are shown in Table 4. The "Accession" column is the protein name. The "Comparison" column indicates which groups are being compared and which group is in the numerator for the fold change estimate. The "Estimate" column is the model estimate of the difference between GCM and DM on the log2 scale. The "Standard error" column contains the standard error of this estimate, and is an indicator of the precision associated with the comparison. The "Fold Change" column is 2 raised to the power in the "Estimate" column, so $2^{-2.068}$ in the first row of the table. 95% confidence interval limits for the fold change are the next two columns and the p-value is contained in the last column.

TABLE 4: Differential abundance output.

Accession	Comparison	Estimate	Standard Error	Fold Change	Lower 95th CI	Upper 95th CI	P-value
hCP1788782	GCM/DCM	-2.068	0.1272	0.238	0.186	0.306	2.09E-27
hCP1887960	GCM/DCM	1.894	0.08586	3.717	3.142	4.399	2.65E-18
hCP1780445	GCM/DCM	1.145	0.08317	2.211	1.878	2.602	9.99E-17
1OPH_A	GCM/DCM	-2.764	0.2218	0.147	0.095	0.227	1.27E-16
AAH78670.1	GCM/DCM	2.156	0.1805	4.458	3.130	6.350	1.51E-15
AAF29581.1	GCM/DCM	-3.013	0.266	0.124	0.074	0.207	5.60E-13

This table shows sample differential abundance output from the top 5 proteins when ranked by p-value in the GCM study. Columns are explained in the text.

A volcano plot helps to understand the level of significance and magnitude of changes observed in the study as a whole (see Figure 8). The fold change on the log2 scale is placed on the horizontal axis (sometimes labeled on the log2 scale, sometimes labeled on the fold-change scale) and p-value on the -log10 scale is placed on the vertical axis. Points on the plot tend to look like lava spewing from a volcano, hence the name. Points nearest the far right and left hand sides of the plot have the largest fold changes while those along the top of the plot are the most statistically significant. Thus, these may help one to use both fold change and significance in determining which proteins to carry forward for further study based on both statistical and biological criteria.

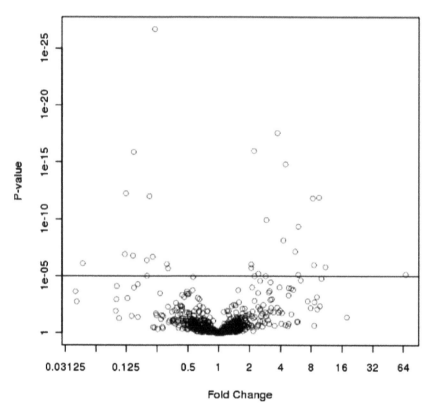

FIGURE 8: Volcano plot. A volcano plot from the GCM study demonstrating magnitude and significance of the protein comparisons between the GCM and DM groups. The vertical axis indicates -log10(p-value). The horizontal axis indicates log2 fold change, here labeled on the fold change scale.

While plots of p-values and FDR rates cannot help to distinguish true and false positive test results, they are useful for understanding the likelihood of real change. If there are no differences between the two groups, a uniform distribution of p-values would be expected. The presence of the spike for small p-values indicates that there are more significant differences than would be expected by chance (see Figure 9a). An FDR value (or q-value) for a given protein, indicates the expected number of false positive tests if the p-value for that protein is used as the significance cutoff (see Figure 9b). Figures 9c and 9d can help determine an acceptable

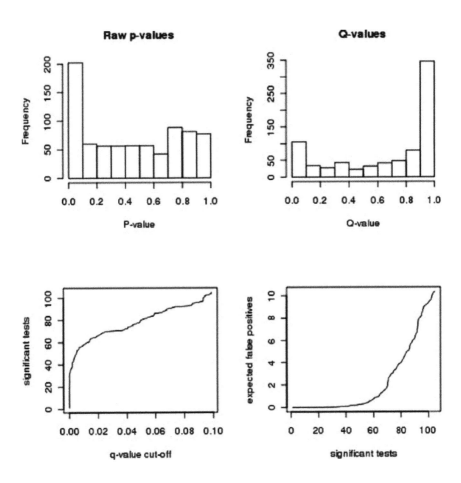

FIGURE 9: Visualization of statistical significance in the GCM study. A) Histogram of the p-values. B) Histogram of the q-values (FDR values). C) Number of tests declared to be significant (vertical axis) as a function of the FDR cut-off used (horizontal axis). D) Expected number of false positive tests (vertical axis) as a function of the number of significant tests (horizontal axis).

significance threshold in light of the number of expected false discoveries. In this particular example, a q-value threshold of 2% would result in approximately 60 expected false positive tests (see Figure 9c). On the other hand, if approximately the top 70 proteins are declared significant, one of these is expected to be a false positive (see Figure 9d).

Summary statistics such as estimates of fold change and p-values are useful. However, it is wise to also look at the data being summarized. A dot plot is useful for visualizing the behavior of the peptides within a given protein, and understanding the underling variability (see Figure 10). At least one study group is statistically significantly different from the other groups in this example peptide dot plot, but there is still a lot of variability in the underlying peptide distributions. There is substantial overlap in the abundance distribution between study groups, indicating this peptide may not be a good biomarker of disease. This particular peptide was detected in all six MS experiments; this is not the case for all peptides.

12.6 DISCUSSION

In this work, the primary focus has been on the iTRAQ labeling protocol, but the basic statistical principles highlighted here are directly applicable to other experiments which utilize different labeling protocols. What does vary between labeling protocols is the mathematical model governing the labeling process which ultimately dictates the analytical methods used to quantify relative abundance information from the raw data. Thus, each labeling protocol will require different analytical methods. For example, in the case of $^{16}O/^{18}O$ stable isotope labeling, all peptides mixed in heavy water would be shifted two Daltons to the right of those mixed in light water (^{18}O has two extra neutrons, thus is 2 Daltons heavier) and peak picking algorithms would be used to identify these provided that 100% of the oxygen atoms were fully exchanged. However, due to less than pure ^{18}O water, naturally occurring isotopes, and a probabilistic model governing the oxygen exchange rates, some of the labeled mixture will have 0, 1 or 2 extra neutrons. Regression modeling strategies can be used to tease apart just how much came from the light and heavy samples, respectively [45,46]. Coupled with sound statistical practices, a full understanding of

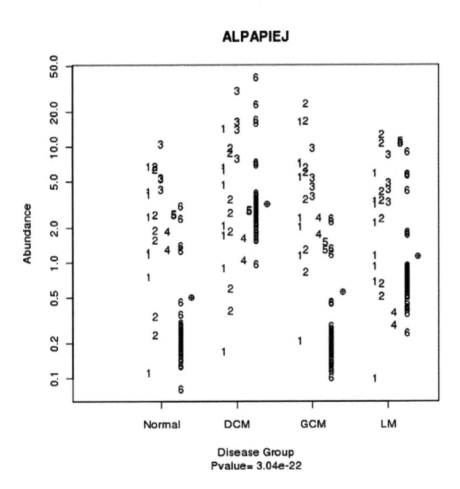

FIGURE 10: Dot plot for peptide with sequence ALPAPIEJ in the GCM study. The vertical axis indicates abundance on the log2 scale. The horizontal axis indicates study group. Numbers in the plot indicate the MS experiment in which the peptide was detected. The circles with + inside to the right of the points for a given study group indicate the mean for that study group. While this peptide has a small p-value, it appears that observations in run 6 are driving the significance. Relying on p-value alone isn't enough; one needs to look at data for a complete interpretation.

the labeling protocol being used and the necessary analytical steps to follow will maximize the information content of the experiment.

There is evidence that the variance is a function of mean abundance as discussed in the "Differential abundance" section. The analytical strategy demonstrated herein utilized that information in the differential abundance models by using WLS as the estimation technique. However, the normalization models were estimated via OLS which does not account for the varying levels of precision. Ideally both of these models would incorporate the weighting. This poses computational challenges since the entire model, normalization plus differential abundance, cannot be fit at once with current computing resources. Incorporation of the weighting into both steps would require iterating between estimation of normalization parameters and differential abundance parameters and is work that requires further investigation.

The models described herein are considered "fixed" effect models. It may be desired to utilize a "mixed" effect model in which some effects are considered fixed while others are considered to be random. Likely random effects are subject and peptide. Designating subject as a random effect would broaden the scope of inference from only the subjects selected for the current study to the population of subjects the sample represents. Designating peptide as a random effect acknowledges that due to the data-dependent acquisition process, the same peptides may not be observed every time. Use of global experimental factors as random effects in the normalization model is currently problematic due to computational limitations and the fact that iterative estimation processes are not yet worked out for random effects. Fixed effect models have been shown to have greater sensitivity than mixed effect models, and therefore more desirable in discovery studies whereas properties of the mixed effect models make them more attractive for studies validating results [35].

12.7 CONCLUSIONS

Use of replication, randomization and blocking in the process of experimental design for labeled MS studies can avoid confounding of experimental and biological effects and minimize variability. A statistical model can be used to account for experimental and biological sources of variation to describe the observed data and produce unified estimates of changes between study groups along with associated measures of uncertainty.

REFERENCES

1. Ross PL, Huang YN, Marchese JN, Williamson B, Parker K, Hattan S, Khainovski N, Pillai S, Dey S, Daniels S, et al.: Multiplexed protein quantitation in Saccharomyces cerevisiae using amine-reactive isobaric tagging reagents. Mol Cell Proteomics 2004, 3(12):1154-1169.

2. Pierce A, Unwin RD, Evans CA, Griffiths S, Carney L, Zhang L, Jaworska E, Lee CF, Blinco D, Okoniewski MJ, et al.: Eight-channel iTRAQ enables comparison of the activity of six leukemogenic tyrosine kinases. Mol Cell Proteomics 2008, 7(5):853-863.

3. Mann M: Functional and quantitative proteomics using SILAC. Nat Rev Mol Cell Biol 2006, 7(12):952-958.

4. Ong SE, Blagoev B, Kratchmarova I, Kristensen DB: Stable isotope labeling by amino acids in cell culture, SILAC, as a simple and accurate approach to expression proteomics. Mol Cell Proteomics 2002, 1:376-386.

5. Kruger M, Moser M, Ussar S, Thievessen I, Luber CA, Forner F, Schmidt S, Zanivan S, Fassler R, Mann M: SILAC mouse for quantitative proteomics uncovers kindlin-3 as an essential factor for red blood cell function. Cell 2008, 134(2):353-364.

6. Oda Y, Huang K, Cross FR, Cowburn D, Chait BT: Accurate quantitation of protein expression and site-specific phosphorylation. Proc Natl Acad Sci USA 1999, 96(12):6591-6596.

7. Pratt JM, Robertson DH, Gaskell SJ, Riba-Garcia I, Hubbard SJ, Sidhu K, Oliver SG, Butler P, Hayes A, Petty J, et al.: Stable isotope labelling in vivo as an aid to protein identification in peptide mass fingerprinting. Proteomics 2002, 2(2):157-163.

8. Dobbin K, Simon R: Comparison of microarray designs for class comparison and class discovery. Bioinformatics 2002, 18:1438-1445.

9. Oberg AL, Mahoney DW, Eckel-Passow JE, Malone CJ, Wolfinger RD, Hill EG, Cooper LT, Onuma OK, Spiro C, Therneau TM, et al.: Statistical analysis of relative labeled mass spectrometry data from complex samples using ANOVA. J Proteome Res 2008, 7(1):225-233.

10. Cooper LT, Onuma OK, Sagar S, Oberg AL, Mahoney DW, Asmann YW, Liu P: Genomic and proteomic analysis of myocarditis and dilated cardiomyopathy. Volume 6. Heart Failure Clin Elsevier, Inc.; 2010::75-85.

11. Mahoney DW, Therneau TM, Heppelmann CJ, Higgins L, Benson LM, Zenka RM, Japtap P, Nelsestuen GL, Bergen HR, Oberg AL: Relative quantification: characterization of bias, variability and fold changes in mass spectrometry data from iTRAQ labeled peptides. Journal of Proteome Research 2011, 10(9):4325-4333.

12. Potter JD: At the interfaces of epidemiology, genetics and genomics. Nature reviews 2001, 2(2):142-147.

13. Ransohoff DF: Bias as a threat to the validity of cancer molecular-marker research. Nature Reviews Cancer 2005, 5:142-149.
14. Oberg AL, Vitek O: Statistical design of quantitative mass spectrometry-based proteomic experiments. J Proteome Res 2009, 8(5):2144-2156.
15. Ransohoff DF: Rules of evidence for cancer molecular-marker discovery and validation. Nat Rev Cancer 2004, 4(4):309-314.
16. Ransohoff DF, Gourlay ML: Sources of bias in specimens for research about molecular markers for cancer. J Clin Oncol 2010, 28(4):698-704.
17. Fisher RA: Statistical Methods for Research Workers. London: Oliver and Boyd; 1932.
18. Fisher RA: The Design of Experiments. Oliver and Boyd. Edinburgh; 1937.
19. Kerr MK, Martin M, Churchill GA: Analysis of variance for gene expression microarray data. J Comput Biol 2000, 7(6):819-837.
20. Anderson NL, Anderson NG: The human plasma proteome: history, character, and diagnostic prospects. Mol Cell Proteomics 2002, 1(11):845-867.
21. Steen H, Mann M: The ABC's (and XYZ's) of peptide sequencing. Nat Rev Mol Cell Biol 2004, 5(9):699-711.
22. Eckel-Passow JE, Oberg AL, Therneau TM, Bergen HR: An insight into high-resolution mass-spectrometry data. Biostatistics 2009, 10(3):481-500.
23. Prakash A, Piening B, Whiteaker J, Zhang H, Shaffer SA, Martin D, Hohmann L, Cooke K, Olson JM, Hansen S, et al.: Assessing bias in experiment design for large scale mass spectrometry-based quantitative proteomics. Mol Cell Proteomics 2007, 6(10):1741-1748.
24. Applied biosystems: Using pro group reports 2004.
25. Cunningham JM, Oberg AL, Borralho PM, Kren BT, French AJ, Wang L, Bot BM, Morlan BW, Silverstein KA, Staggs R, et al.: Evaluation of a new high-dimensional miRNA profiling platform. BMC medical genomics 2009, 2:57.
26. Astrand M: Normalizing oligonucleotide arrays. Clinical Science 2001. OpenURL
27. Bolstad BM, Irizarry RA, Astrand M, Speed TP: A comparison of normalization methods for high density oligonucleotide array data based on variance and bias. Bioinformatics 2003, 19(2):185-193.
28. Dudoit S, Yang YH, Callow MJ, Speed TP: Statistical methods for identifying differentially expressed genes in replicated cDNA microarray experiments. Statistica Sinica 2002, 12:111-139. OpenURL
29. Ballman KV, Grill DE, Oberg AL, Therneau TM: Faster cyclic loess: normalizing RNA arrays via linear models. Bioinformatics 2004, 20(16):2778-2786.
30. Eckel JE, Gennings C, Therneau TM, Burgoon LD, Boverhof DR, Zacharewski TR: Normalization of two-channel microarray experiments: a semiparametric approach. Bioinformatics 2005, 21(7):1078-1083.
31. Bolstad BM: Probe level quantile normalization of high density oligonucleotide array data. 2001.
32. Wang P, Tang H, Zhang H, Whiteaker J, Paulovich AG, Mcintosh M: Normalization regarding non-random missing values in high-throughput mass spectrometry data. Pacific Symposium of Biocomputing 2006, 11:315-326.

33. Hill EG, Schwacke JH, Comte-Walters S, Slate EH, Oberg AL, Eckel-Passow JE, Therneau TM, Schey KL: A statistical model for iTRAQ data analysis. J Proteome Res 2008, 7(8):3091-3101.

34. Mahoney DW, Oberg AL, Malone CJ, Therneau TM, Bergen HR: Use of censored regression models for relative quantification in global mass spectrometry data. Poster log#220, US HUPO 5th Annual Conference, San Diego, CA 2009.

35. Clough T, Key M, Ott I, Ragg S, Schadow G, Vitek O: Protein quantification in label-free LC-MS experiments. J Proteome Res 2009, 8(11):5275-5284.

36. Van PT, Schmid AK, King NL, Kaur A, Pan M, Whitehead K, Koide T, Facciotti MT, Goo YA, Deutsch EW, et al.: Halobacterium salinarum NRC-1 PeptideAtlas: toward strategies for targeted proteomics and improved proteome coverage. J Proteome Res 2008, 7(9):3755-3764.

37. Gan CS, Chong PK, Pham TK, Wright PC: Technical, experimental, and biological variations in isobaric tags for relative and absolute quantitation (iTRAQ). Journal of Proteome Research 2007, 6:821-827.

38. Song X, Bandow J, Sherman J, Baker JD, Brown PW, McDowell MT, Molloy MP: iTRAQ experimental design for plasma biomarker discovery. Journal of Proteome Research 2008, 7:2952-2958.

39. Zhang Y, Askenazi M, Jiang J, Luckey CJ, Griffin DJ, Marto JA: A robust error model for iTRAQ quantification reveals divergent signaling between oncogenic FLT3 mutants in acute myeloid leukemia. Molecular & Cellular Proteomics 2010, 9:780-790.

40. Karp NA, Huber W, Sadowski PG, Charles PD, Hester SV, Lilley KS: Addressing accuracy and precision issues in iTRAQ quantitation. Molecular & Cellular Proteomics 2010, 9(9):1885-1897.

41. Gerster S, Qeli E, Ahrens CH, Buhlmann P: Protein and gene model inference based on statistical modeling in k-partite graphs. PNAS 2010, 107(27):12101-12106.

42. Benjamini Y, Hochberg Y: Controlling the false discovery rate: a practical and powerful approach to multiple testing. Journal of the Royal Statistical Society B 1995, 57:289-300. OpenURL

43. Storey JD: A direct approach to false discovery rates. Journal of the Royal Statistical Society, Series B 2002, 64:479-498.

44. SAS Institute I: SAS®/STAT User's Guide. In Version 9 2005. Cary NC: SAS Institute Inc.;

45. Eckel-Passow JE, Oberg AL, Therneau TM, Mason CJ, Mahoney DW, Johnson KL, Olson JE, Bergen HR: Regression analysis for comparing protein samples with 16O/18O stable-isotope labeled mass spectrometry. Bioinformatics 2006, 22(22):2739-2745.

46. Eckel Passow JE, Mahoney DW, Oberg AL, Zenka RM, Johnson KL, Nair KS, Kudva YC, Bergen HR, Therneau TM: Bi-linear regression for 18O quantification: modeling across the elution profile. Journal of Proteomics & Bioinformatics 2010, 3(12):314-320.

CHAPTER 13

Mrcquant: AN ACCURATE LC-MS RELATIVE ISOTOPIC QUANTIFICATION ALGORITHM ON TOF INSTRUMENTS

WILLIAM E. HASKINS, KONSTANTINOS PETRITIS, AND JIANQIU ZHANG

13.1 BACKGROUND

The large-scale identification, characterization and quantification of proteins in biological samples by liquid chromatography-mass spectrometry (LC-MS) and liquid chromatography-tandem mass spectrometry (LC-MS/MS)-based proteomic methods play a crucial role in biomedical research [1,2]. For example, in biomarker discovery studies, a common aim is to elucidate a set of proteins that can be used to reliably differentiate diseased and normal samples by abundance measurements. Precision and accuracy are critical for confident protein biomarker discovery and validation. In "bottom-up" approaches, proteins are cleaved by sequence-specific proteases such as trypsin prior to analysis. A protein fold change can be inferred from the relative abundance of peptides across samples, where peptide identification and quantification can be accomplished in separate steps [3]. In this paper, we consider the problem of relative isotopic quantification

This chapter was originally published under the Creative Commons Attribution License. Haskins WE, Petritis K and Zhang J. MRCQuant: An Accurate LC-MS Relative Isotopic Quantification Algorithm on TOF Instruments. BMC Bioinformatics 12,74 (2011), doi:10.1186/1471-2105-12-74.

of peptides in LC-MS based on time-of-flight (TOF) instruments. It is assumed herein that a list of candidate peptides has been compiled a priori, and that we are interested in measuring the relative abundance of their isotopes (natural or labeled).

The measurement of peptide abundance is complicated by the fact that a peptide forms both LC and MS peaks during its LC elution interval. To quantify a peptide, it requires the integration of its complete LC peaks, which is sometimes impossible due to strong interference from other peptide species or contaminants. However, relative quantification is still possible for the uncorrupted segments of LC peaks with slightly different isotopic compositions. Relative isotope abundance measurement is particularly important in chemical and metabolic labeling experiments for the quantification of differential expression of isotopically-labeled peptide pairs and their corresponding proteins. In "label-free" LC-MS peptide detection, measurement of relative natural isotope abundance is employed for peptide detection. In both cases, there exist several significant challenges: 1. The determination of LC peak boundaries to exclude noisy scans; 2. Background noise suppression in LC peaks; 3. Interference detection and removal; and 4. Mass drift correction. To achieve accurate relative quantification, these issues have to be addressed. Current software packages have not addressed these issues effectively. QUIL [4] and ProteinQuant [5] determine LC peak boundaries by the apex and the full-width-half-maximum (FWHM) of a peak, i.e., it is assumed that for a given LC elution peak, the distance between its starting point and its apex is the FWHM of the peak. This assumption is problematic when elution peaks (especially low abundance ones) are asymmetrical and jagged. Some software packages use an intensity threshold or local minima to determine the boundaries of LC peaks. The main problem of these methods is: one is never sure whether noise or interference-corrupted scans are included within the peak boundaries, which could greatly degrade quantification accuracy. Among popular software packages, msInspect [6] and SuperHirn [7] use thresholds, ASAPRatio [8] and MapQuant [9] use peak apex and FWHM. Recently, MaxQuant [10] uses local minima for LC peak detection after Extracted-Ion-Chromatogram (XIC) smoothing. See [2] for a comprehensive review of software tools currently available for LC-MS quantification.

On the problem of background noise suppression, almost all current software packages use Savitzky-Golay or other types of filters [6,8,10] to smooth XICs. However, through our own observation, elution process variations share similar frequency characteristics with that of instrument and Poisson noise (see [Additional file 1] for a detailed description of this phenomenon). Applying filters will distort elution process variations which adversely affect quantification accuracy.

For interference detection and removal, most software packages deconvolute peptide peaks and only consider peak centroids. Although this procedure decouples peptides with similar masses to a degree, it is susceptible to thermal noise, which can cause errors in the calculation of peak centroids. In addition, this procedure cannot provide interference detection, which is critical for accurate quantification.

Also, automatic mass drift correction is not implemented in these software packages, and users are generally expected to supply mass calibration information. This requirement introduces another source of variability, since the accurate determination of mass drift over all m/z ranges is a challenging problem. These issues become more severe when peptide abundance is low. Consequently, they have been bottlenecks in quantitative proteomic studies. For example, it is observed that whenever the signal intensity is low, the measurement of isotopically-labeled peptide pairs tends to be erroneous [11]. If we can computationally improve the coverage of accurate quantification, the chance for protein biomarker discovery will improve accordingly.

We limit the scope of this paper to TOFMS instruments where the Gaussian additive thermal noise model is appropriate [12,13], (note that this is different from the Poisson plus multinomial noise model for the XICs). In contrast, in FTMS, the assumption of Gaussian additive noise does not hold which is noted in [12] as the phenomenon of increased noise in XICs.

In this paper, we propose a Maximum Ratio Combining (MRC) based Quantification (MRCQuant) algorithm to address current issues in quantification. MRCQuant was developed based on the observation that peptide species register identical MS peak signals (scaled and noise corrupted) in different MS scans and m/z locations. Sometimes, the registered peaks have high Signal-to-Noise ratios (SNRs), while in other occasions, the peaks are noisy with low SNRs. While quantification at high SNRs is very accurate, quantification at low SNRs is problematic due to noise. We can extract the Maximum Likelihood estimate of peptide MS signals from MS peaks at high SNRs using MRC, hence referred to as MS templates. Note that these templates are extracted directly from experiment, and are not "predefined", thus they can capture slight variations in the shape and center locations of MS peaks caused by different environmental factors and instrument designs. Subsequently, extracted MS templates can be used as references when quantifying low SNR peaks. This method can effectively remove background noise without filtering out elution process variations. In addition, extracted MS templates can be compared to MS peaks for interference detection and removal. After interference and noise removal, accurate quantification can be performed.

MRCQuant provides measurements of isotopic abundance for each peptide of interest at all charge states and all isotope positions of interest. The output of the algorithm can be further processed to infer relative protein abundance in labeled experiments, or the results can be used for peptide detection based on isotope pattern in LC-MS data. The peptide list of interest can be compiled from peptides identified from multiple LC-MS/MS runs or from LC-MS peak detection algorithms such as msInspect [6].

13.1.1 DEFINITIONS

Before we describe the MRCQuant algorithm, we first define several key terminologies that we use throughout the paper.

1. Maximum Ratio Combining (MRC) is an averaging method that has been widely applied in Telecommunications [14] for estimating the actual transmitted signal from multiple copies received through Additive White Gaussian Noise (AWGN) channels. MRC assigns averaging weights proportional to the square root of SNRs of received copies. MRC is mathematically derived based on the Maximum Likelihood principle. MRC provides an estimation of the transmitted signal with the highest SNR possible among all averaging methods. Given a peptide, we consider its MS peaks in

multiple MS scans as copies of its real MS signal, which can be optimally estimated through MRC.

2. A reference template, not specific to any particular peptide, is defined as an estimation of the general MS peak shape in an LC-MS experiment. Such a peak shape is usually determined by instrument characteristics and environmental factors. Slight variations could exist between a reference template and particular peptide peak. This template can be translated and adjusted (in width) to different mass/charge (m/z) locations. (See support information for details of template translation). A reference template is described by its center m/z and its m/z -intensity pair values. Reference templates can either be extracted from LC-MS datasets at high SNRs, or can be theoretically predicted based on instrument resolution and characteristics. There may be several reference templates at different m/z values in an LC-MS dataset.

3. A peptide template is defined as an estimation of the MS peak signal registered by a specific peptide in one experiment. Comparing to reference templates, peptide templates are better estimations of MS peak signals for individual peptides. Peptide templates are generally extracted from MS peaks registered at the highest (most abundant) isotope and charge state position of peptides, where SNRs are high. Each peptide has its own template.

13.2 METHODS

13.2.1 MRCQUANT ALGORITHM

Here we describe the MRCQuant algorithm for relative peptide isotope quantification on LC-MS. The input of the algorithm includes an LC-MS dataset and a list of peptides to be quantified annotated by their monoisotopic mass and/or amino acid sequence. The mass annotation can be obtained through an LC-MS peptide identification algorithm like msInspect. The output of the algorithm is a matrix of abundance measurements, with a maximum of P columns, where P is the total number of peptides to be quantified, and whose rows are indexed by cs * maxcs + iso, where cs ∈ [1, maxcs] represents charge state, maxcs is the maximum number of charge states considered, and iso represents the isotope position. For a given peptide, we need to first detect its LC peaks. A peptide at a given mass forms a series of 2 D peptide peaks at different isotope and charge state positions. These 2 D peaks form LC and MS peaks if they are viewed from the elution time and m/z dimension. To establish the connection between a group of 2 D peaks to a specific peptide mass, we need to verify that: 1. their LC peaks at different isotope and charge state positions should be the same; and 2. their MS peaks match a reference template translated to their expected m/z locations. After LC peak identification, we need to accurately detect LC peak boundaries and perform quantification. To accomplish these goals, the proposed algorithm performs the following: 1. Extracts or theoretically predicts reference templates. 2. For each peptide of interest, performs LC peak detection at its highest isotope and charge state position using a reference template. 3.

Extracts peptide templates based on the MRC principle, which are used for accurate LC peak boundary detection and interference/noise removal at lower SNRs. Finally, quantification is performed based on peptide templates. The goal of the algorithm is to record accurate relative ion counts at all charge states and isotope positions.

A flow diagram of the entire process is shown in Figure 1, which is explained in detail below.

13.2.2 GENERATION OF REFERENCE TEMPLATES

Reference templates can either be extracted from experiments directly, or obtained through theoretical prediction. Theoretically predicted templates can adopt different peak shapes according to different instrument characteristics (resolution for example). Mass drifts can be accounted by shifting the center of theoretically predicted templates according to mass calibrations. Next, we discuss in detail how to extract reference templates from LC-MS data at high SNRs. Given an input peptide list, we select a subset of peptide that register uncorrupted MS peaks, from which we

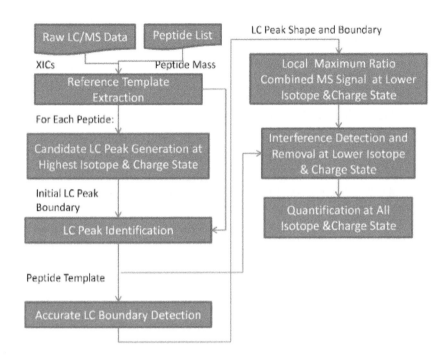

FIGURE 1: Flow diagram of the MRCQuant algorithm.

extract a number of reference templates centered at different m/z values. The underlying assumption is that MS peaks registered by the same instrument should be similar (except that MS peaks are scaled both in m/z and intensity). Thus, it is possible to use the estimated MS peak signals at high SNRs as reference templates for initial LC peak detection. Note that slight deviation of actual MS signals to reference templates is allowed since the templates are used for LC peak detection but not quantification. The number of templates can be selected by the user, and 4 templates have been used in our simulations with good results. Later, when quantifying a peptide at a given m/z value, we will not use a reference template, instead, we will use a peptide template for accurate quantification at low SNRs. This ensures that the template with the closest m/z value will be selected for LC peak boundary detection, interference detection and removal.

To extract the list of reference templates, we go through the following process for each peptide in the input list:

1. Determine the XICs of the peptide of interest at all charge states and isotope positions.

2. Determine the LC elution interval for the peptide of interest. To achieve this, we apply a high threshold at half maximum of the most intense (base) LC peak among all XICs. On the XIC with the tallest LC peak, all intervals above the threshold are considered as possible LC elution intervals. Then at the charge state of the base peak, we further check the correlations between the LC peaks on possible intervals at the two highest isotope positions (usually ^{12}C and ^{13}C). The interval corresponding to the peptide of interest should have a high correlation; otherwise the LC peaks must have been registered by other peptides, or have been corrupted by interference signals. The correlation is checked by R-squared statistics [15], and we apply a stringent threshold (> 0.9). We accept LC intervals with correlations higher than the threshold. If none of the intervals pass the threshold, we move on to the next peptide for possible template extraction. If multiple intervals have high correlations, which indicates that multiple peptides with similar mass occur on the same XIC, then we reject all intervals and move on to the next peptide since we can not detect the peptide interval unambiguously. This iterative procedure ensures that 1. We select a correct and unambiguous elution interval for the peptide of interest, and 2. The MS signal has not been corrupted by interference or noise.

3. If the elution interval is accepted, we determine the range of m/z values that the reference template spans (defined as the MS window of the template). The size of the MS window is determined by instrument resolution.

4. We average all MS peaks within the MS window and the accepted elution interval based on the MRC principle. The resulted MRC signal is an estimation of the MS peak signal registered by the peptide, and it can be used as a reference template.

After performing the above steps for each peptide, a list of reference templates has been obtained for LC peak detection. The details of XIC extraction, determination of MS windows, and the theoretical derivation of reference templates can be found in [Additional file 1].

13.2.3 LC PEAK DETECTION

After obtaining a list of reference templates, the algorithm moves on to accurately detect and quantify the LC peak for each peptide of interest. Given a peptide, we start LC peak detection by inspecting its XICs. Usually, several LC peaks above the background noise level exist on an XIC, where, one is generated by the peptide of interest and the rest belong to others. We need to correctly identify the LC peak and its boundaries so that noise signals are not included in relative quantification. We perform the following processing steps:

13.2.4 CANDIDATE LC PEAK GENERATION

The goal of this step is to detect high intensity intervals (LC peak candidates) on XICs of the peptide of interest for further investigation. Ideally, we should perform such detection at the most abundant charge state and isotope position where the LC peak has the highest SNR possible. Given peptide sequence information or mass, it is possible to predict its isotopic pattern [16], and its most abundant isotope position (base position). On the other hand, it is difficult to predict the most abundant charge state, and an exhaustive search must be conducted. We perform the following processing steps at all charge states:

1. Given a peptide's mass (m) at a charge state (z), determine its theoretical m/z values at different isotope positions.
2. At the m/z value of its base peak, estimate its MS window and generate the XIC.
3. Apply an intensity threshold at 3 times the estimated background noise standard deviation to identify LC peak candidates.

4. Determine the FWHM elution intervals of LC peak candidates by applying thresholds at half maximum of these LC peak candidates. These FWHM boundaries are set as initial LC peak boundaries. In this way, we only include MS scans with relatively high SNRs.

5. Check the correlation between LC peaks at the most intense and the second most intense isotope positions within the initial boundaries of each LC peak candidate. The correlation is checked using R-squared statistics, and all candidates with R statistics greater than 0.9 will be accepted. In this way, all intervals with good correlations between two isotopes will be selected.

6. If the maximum R-statistic is less than 0.9, then the LC peak candidate with the maximum R statistics will be selected. This corresponds to the case when correlations between isotope elution profiles are poor due to noise or interference, and the peptide of interest may or may not exist. In such cases, we select the best candidate for further verification in the MS dimension.

At the end of this process, a list of k LC peak candidates, each denoted by its start and end scan, is generated at each charge state. The charge state with the highest total ion count within initial LC peak boundaries will be selected.

Next, one of these LC peak candidates will be identified as the initial LC peak.

13.2.5 LC PEAK IDENTIFICATION

From previous steps, we find k LC peak candidates, but generally only one of them is generated by the peptide of interest, which can be further identified by matching a reference template to the MS peaks within the elution interval of each candidate:

1. We select the closest reference template to the peptide of interest in m/z values, which ensures the best match between the template and local MS peaks. We then translate the template to the local m/z

value of the peptide of interest. Details of template translation can be found in [Additional file 1].

2. For each LC peak candidate, estimate its local MS peak signal by averaging all MS peaks (using MRC) within its initial boundaries. By employing MRC, noise in individual MS peaks will be maximally suppressed.

3. The estimated local MS signal will be compared to the selected reference template. The LC peak candidate with the best matched local MS signal will be identified as the final LC peak.

4. If none of the local MS signals match with the reference template well (with R statistics < 0.4), then LC peak detection failed for the given peptide. This could happen when a peptide identification algorithm wrongly reports the center mass of the peptide, which leads to a mismatch between the reference template and the local MS signal. Although it is possible to correct such wrongly reported mass, however, it is beyond the scope of this paper.

At the end of this processing step, an LC peak has been identified for the peptide of interest with initial boundaries detected using a high intensity threshold at half the maximum of the LC peak.

We do not assume specific LC peak shapes (e.g. Gaussian), and the algorithm can be applied in various LC conditions (e.g., different reverse-phase gradients). If reference templates are extracted from an LC-MS experiment directly, then they will be centered at their theoretical m/z values plus the mass drift of the experiment. Thus, mass drift will be automatically accounted when applying such reference templates for LC peak detection. If a theoretical reference template is used, then its center needs to be shifted according to user provided mass calibration information.

13.2.6 PEPTIDE TEMPLATE EXTRACTION

For a peptide of interest, its identified LC peak has an initial elution interval that covers the intensity region above half of the LC peak maximum, and it is obtained at the highest charge state and isotope position. These conditions ensure that the MRC signal associated with the identified LC peak is estimated at a high SNR, and it can be treated as the peptide template of interest. Such a template captures accurate MS peak shape information, which can be used for LC peak boundary detection and quantification.

13.2.7 ACCURATE LC PEAK BOUNDARY DETECTION

The initial LC peak boundaries are obtained by applying a high intensity threshold, and many MS scans that belong to the peptide of interest are excluded. We need to accurately extend the boundaries so that all MS scans of the peptide will be accounted. If the boundaries exclude a significant segment of the LC peak, then quantification will be less accurate since combining fewer scans cannot suppress noise sufficiently. If the boundaries are extended too far to include scans that contain interference and noise, then quantification accuracy will also be reduced.

The problem of LC peak boundary detection can be translated to the problem of detecting of all scans that contain the peptide template. It can be further formulated as a hypothesis testing problem:

H0: A given MS scan only contains noise;
H1: The scan contains noise plus the peptide template.

We test the hypothesis by comparing the peptide template to the MS peak signal in a given scan. If the R-statistic is greater than a threshold (0.5), then H1 is accepted.

We start this hypothesis testing procedure from the initial LC peak starting scan to extend the head of the LC peak. Then we apply the same procedure to the tail end of the peak. Whenever encountering a scan that does not contain the template, the extension process will be terminated.

Accurate boundary detection plays a critical role in quantification accuracy. For example, in Figure 2, we plot the 2 D peaks of a peptide at ^{12}C and ^{13}C positions in charge state 2. The peptide signal actually resides from scan 200 to 211. In scan 194 - 199, an interfering peptide with similar m/z produces MS peaks at the ^{12}C position. However, inspecting the peaks at ^{13}C, it is evident that interference peaks do not exist in scans 194 - 199.

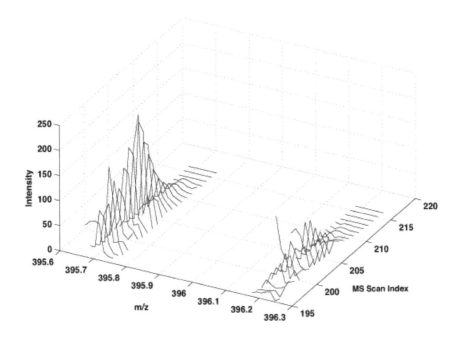

FIGURE 2: Example of peptide 2 D peaks with interference at C^{12}. Comparison of C^{12} and C^{13} peaks reveals interference at C^{12} in scans 194 - 199.

If the interfering scans are included, the resulted relative quantification accuracy will be greatly degraded. In Figure 3, we compare different boundary detection methods. The threshold method includes all scans from the interfering peptide. The FWHM method includes a few interfering peptide scans and excludes a few scans that belong to the peptide of interest. In contrast, the proposed method accurately detected the boundary from scan 200-211.

13.2.8 QUANTIFICATION

For a given peptide, we have obtained its peptide template and LC peak boundaries after LC peak detection. Based on these inputs, we can accurately

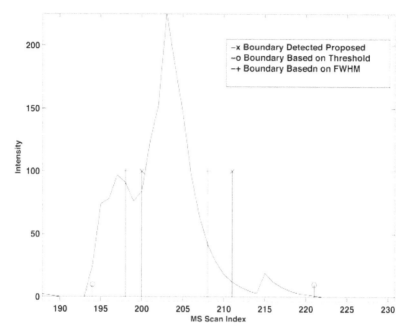

FIGURE 3: Comparison of different boundary detection methods. Comparison of C12 and C13 peaks reveals interference at C12 in scans 194 - 199.

quantify the peptide at other charge state and isotope positions. At a given "local m/z value" of low SNR, quantification consists of three processing steps: 1. Obtain a local MRC signal by averaging all MS peaks (using MRC) within the detected LC peak boundaries to optimally suppress noise; 2. Compare the translated peptide template with the local MRC signal for interference detection and removal. This step also provides an estimation of the scaling factor for the local MRC signal in reference to the peptide template, which can be multiplied to the total ion count of the template to derive the total ion count of the peptide at the local m/z value.

Local MRC signal are derived using weights proportional to the LC peak intensities obtained at the LC peak detection stage. The details of other processing steps are described below.

13.2.9 INTERFERENCE DETECTION AND REMOVAL

The input to this processing step includes the local MRC signal and the translated peptide template, whose correlation is calculated using the R-square statistic [15]. If the correlation is greater than 0.9, then it is considered that the interference signal does not exists. Otherwise, the local MRC signal is considered as interference corrupted, and we have to perform interference removal within its MS window.

We model a local MRC signal as the superposition of the translated peptide template (scaled by a) and an interference signal which is modeled as an order 1 polynomial. The interference removal problem is equivalent to the accurate estimation of the scale factor a and the polynomial parameters.

When assuming Gaussian noise, the Maximum Likelihood estimation of these parameters is equivalent to their least-square-estimation (LSE). Note that the correlation between the interference and the peptide template signal must be minimized to yield a good estimate of a. Otherwise, the estimated interference signal could contain partial template signal. Consequently, besides finding the LSE of parameters, the second objective is to find parameters that minimize the correlation between the template and the interference signal. In addition, there is the constraint that both the template and the interference signal should be positive at all m/z values. These requirements lead us to formulate a constrained multiple objective optimization problem. We utilize the Quadratic Programming algorithm [17] to numerically search for the solution of model parameters. The selection of polynomial order is based on the Bayesian Information Criteria (BIC) [18]. See [Additional file 1] for details.

Figure 4 shows an example of interference removal. The peptide template in Figure 4 is extracted at a high SNR. The local MRC signal is derived at a lower SNR. Due to interference, the local MRC signal deviates from the peptide template significantly. We employ the proposed interference removal method to estimate the interference and peptide signal. When performing quantification, the interference signal is not counted towards the total ion count.

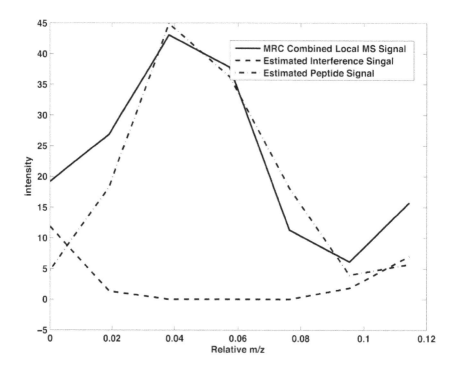

FIGURE 4: Interference Removal. The MS template is extracted at a high SNR. The local MRC MS signal is derived at a lower SNR. The local MRC MS signal deviates from the extracted MS template significantly.

Note that there exist various peak identification algorithms [19,20] that are specially designed to deal with the problem of overlapping peptide peaks. These algorithms are generally exponentially complex with the number of overlapping peaks considered. In this paper, the focus is on accurate quantification after peptide identification. Thus, the problem is simplified to only extract signals for the peptide of interest. The knowledge of overlapping peptides can help in improving quantification accuracy, but since peak identification algorithms may or may not provide such information, we uniformly treat overlapping signals as interference. The MRC process also has the effect of suppressing interfering signals since higher weights are given to tall MS peaks of the peptide of interest but not interfering peaks. This treatment also limits the computational complexity, which is linear to the number of peptides to be quantified.

13.2.10 QUANTIFICATION BASED ON LOCAL MRC SIGNAL

At the end of interference removal, the local MRC signal is cleaned of interference and the scale factor a is also derived. It is easy to show that the total ion count Cs of all MS peaks within the LC peak interval and the total ion count of the local MRC MS signal Cm has the relationship

$$C_s = C_m * \frac{1}{\sum_t w(t)^2} \tag{1}$$

where w(t) are normalizing weights used for MRC. Thus if the total ion count of the peptide template is C_t, the total ion count of the LC peak C_p can be estimated as:

$$C_p = C_t * a * \frac{1}{\sum_t w(t)^2}$$

where $C_t * a = Cm$ is the estimated total ion count of the local MRC signal.

In Figure 5, we show an example of the effect of noise reduction by MRC. At a lower SNR position, the peptide signal in an individual scan is very noisy (signal in dashed line). In contrast, the local MRC combined signal has a much higher SNR, and it is very close to the peptide template.

13.2.11 DATA COLLECTION AND PROCESSING

We developed our algorithm based on an LC-MS dataset collected from a tryptic digest of horse myoglobin at a concentration of 600 fmol (unless noted, all illustrations in this paper are generated based on this dataset). For

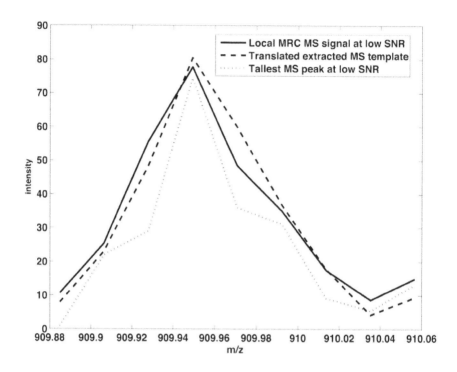

FIGURE 5: The Translated template at a lower isotope position. The peptide signal in an individual scan is very noisy (signal in dashed line). In contrast, the MRC combined signal has a much higher SNR, and it is very close to the extracted MS template.

reference, we also obtained an LC-MS/MS dataset for peptide sequence information at 100 fmol. LC-MS/MS was performed with a splitless nanoLC-2 D pump (Eksigent), a 50 μm-i.d. column packed with 10 cm of 5 μm-o.d. C18 particles, and a linear ion trap tandem mass spectrometer (LTQ-XLS; ThermoFisher). The top 7 most abundant eluting ions were fragmented by (data-dependent) collision-induced dissociation (CID). The LC gradient was 2 to 98% 0.1% formic acid/acetonitrile in 60 min (60-120 min) at 400 nL/min. Tandem mass spectra were extracted by Mascot Distiller version 2.3.1. Charge-state-deconvolution and deisotoping were not performed. All MS/MS samples were analyzed using Mascot (Matrix Science, London, UK; version 2.3.2). Mascot was set up to search the Swiss-Prot database assuming the digestion enzyme trypsin. Mascot was searched with a fragment ion mass tolerance of 0.80 Da and a parent ion tolerance of 2.0 Da. Oxidation of methionine and iodoacetamide derivative of cysteine were specified in Mascot as variable modifications. LC-MS was performed with a splitless nanoLC-2 D pump (Eksigent), a 50 μm-i.d. column packed with 10 cm of 5 micro-o.d. C18 particles, and a time-of-flight mass spectrometer (MicrOTOF; Bruker Daltonics). The LC gradient was 2 to 98% 0.1%formic acid/acetonitrile in 60 min (60-120 min) at 400 nL/min. Mascot search correctly linked 13 peptides observed in the sample to horse myoglobin with an 80% sequence coverage.

For algorithm verification, we downloaded a QTOF dataset from the repository of Seattle Proteome Center at http://regis-web.systemsbiology. net/PublicDatasets/. The repository was created for testing various algorithms. It contains LC-MS/MS datasets of an 18 protein digest. For details of data collection please refer to [21]. There are multiple LC-MS/MS datasets collected on various instruments within the repository. We downloaded datasets related to protein mixture 4 of the 18 protein mix. Among which, from a total of 21 runs on LTQ-FT, QStar and QTOF, we compiled a list of 784 LC-MS/MS-identified peptides for the same protein mixture. These peptides were all identified with a PeptideProphet™[22] score greater than 0:9. We also performed LC-MS peak detection using msInspect on one of the QTOF datasets QT 20060925_mix4_23.mzxml (mix4_23) that identified 1952 peptides. Subsequently we quantify these peptides by MRCQuant. MsInspect was selected because it is the most representative LC-MS peptide identification and quantification algorithm and has been shown to outperform other peak detection algorithms [23]. It applies a conservative noise threshold initially. Subsequently, MS scans are centroided; XICs are smoothed; LC peak length filter is applied; and LC peaks that appear and disappear together are pooled and treated as signals registered by identical peptides at different isotope positions and charge states. Subsequently, peptides are identified by comparing their theoretically predicted isotope patterns and measured isotope patterns using Kullback-Leibler(KL) distance. Other popular software packages such as ASAPRatio [8] differ slightly in the details, but the main procedure, MS peak detection in each MS scan followed by quantification based on XICs, is similar to that of msInspect. Among these software packages, msInspect provides relative quantification accuracy measurements in the form of KL distance, which enables us to compare performances. Other software packages do not provide this measurements, therefore, relative quantification accuracy cannot be accessed.

When using the msInspect software package (Build 599) to process mix4_23 dataset, we tried to optimize the number of peptides being reported. We selected the "walksmooth" option when running the command "findPeptides", and we set msInspect parameters "minpeaks" to 2 and "maxkl" to 10. The "walksmooth" option greatly improves the number of features as well as the KL scores reported. A total of 1952 features were reported. In comparison, if the default settings of msInspect are used, 933 features were reported with worse KL scores.

The peptides reported by msInspect were further processed by MRC-Quant. We used extracted templates at high SNRs as reference MS templates. We rejected some msInspect reported features either because: their reported msInspect KL scores are negative, or our algorithm determines that the LC peaks reported by msInspect cannot be found. The latter case could be caused by inaccurate mass reporting by msInspect. When the mass is reported inaccurately, the reference template and the local MS signal would deviate from each other significantly, and our algorithm rejects LC peaks when the R statistic between the reference template and the local MS signal is less than 0.4. Correcting the incorrectly reported mass is a peptide identification problem which is beyond the scope of this paper. This results in a peptide list of length 964 with accurately reported mass values.

13.2.12 RELATIVE QUANTIFICATION ACCURACY EVALUATION

To perform relative quantification accuracy evaluation, we need to introduce an appropriate metric. The ideal way to evaluate relative quantification accuracy is to compare the measured ratios of natural isotopes to that of theoretically predicted ones. However, none of the software packages report abundance levels at different isotope positions directly. MsInspect reports KL scores which can be used to access relative quantification accuracy indirectly. Given measured natural isotope ratios [p(1), p(2), ...] and theoretically predicted ones [q(1), q(2), ...], (When sequence information is available, natural isotope ratios can be calculated exactly. Otherwise, at a given mass, they can be estimated from its mass [24]), the KL score is evaluated using the following formula:

$$KL(p||q) = \sum_i p(i) * \log \frac{p(i)}{q(i)}$$

(2)

If two sets of isotope ratios entirely agree with one another, then their KL score equals to zero. Otherwise, a KL score is always positive, and the larger it is, the bigger the difference between the two sets of isotope ratios.

Different KL scores indicate different levels of quantification accuracy, and it is possible to compare the performance of different algorithms by the reported number of peptides at different KL score thresholds. For example, we can claim that algorithm one is better than algorithm two, if algorithm one reports more peptides with KL scores less than a threshold.

Obviously, we cannot set the KL threshold to infinity, and now the question becomes what could constitute an "acceptable range of KL thresholds". We know that given a KL score, there always exist the probability that it is the divergence between an arbitrarily generated and an authentic isotope distribution. The higher the KL score is, the higher the probability. If the KL score of a reported peptide is high, it is very probable that the real peptide signal does not exist, and the reported isotope distribution is generated based on observations of random noise. This probability is defined as the False-Detection-Rate (FDR), which can be converted from a KL score in reference to a KL null distribution (the distribution of KL scores between authentic peptide and arbitrarily generated isotope ratios). Obviously, when the FDR is high, it is not meaningful to compare the reported number of peptides between two algorithms anymore, since a significant portion of reported peptides should have been falsely detected. In this paper, we adopt a cutoff FDR of 12%, and we compare the number of reported peptides at different FDRs less than 0.12.

Given a KL score reported by an algorithm, to convert it to FDR, the p-value of the KL score is first generated based on the KL null distribution.

Subsequently, the FDR is estimated using the method described in [25] based on the p-value. The Matlab function, mafdr($\cdot$), is used to estimate the FDRs from the p-values.

The null distribution on KL score is generated by calculating the KL scores between arbitrarily generated isotope distributions with authentic ones. Without observations, an arbitrary distribution on isotopes is generated by drawing maxiso random numbers uniformly distributed on 0[1], and then these numbers are normalized to form a distribution. We generate authentic theoretical isotope distributions by randomly drawing mass values from the peptide list reported by msInspect, and then for these mass values, we calculate their theoretical isotope ratios using the method in [24].

13.3 RESULTS AND DISCUSSION

We applied MRCQuant to both peptide lists identified by msInspect and LC-MS/MS. The performance of MRCQuant is measured by the number of reported peptides at FDRs that are less then 0.12. Peptides reported with low FDRs/KLs are considered as accurately quantified ones. See Figure 6 for an illustration of the algorithm verification process. Note that the direct comparison of computing time between MRCQuant and msInspect is not possible because msInspect is a combined peak identification and quantification algorithm, while MRCQuant focuses on quantification only. The complexity of MRCQuant is linear in complexity, i.e. the processing time is linear to the number of peptides to be quantified. On a Dell T7500 workstation, the processing time for the msInspect list was below half an hour.

13.3.1 PERFORMANCE COMPARISON BETWEEN MRCQUANT AND MSINSPECT

We first compared the performance of MRCQuant to that of msInspect based on the msInspect reported peptide list. In Figure 7, we plot the number of reported peptides at different FDRs by MRCQuant and msInspect. From this figure, we can see that MRCQuant reports more accurately quantified

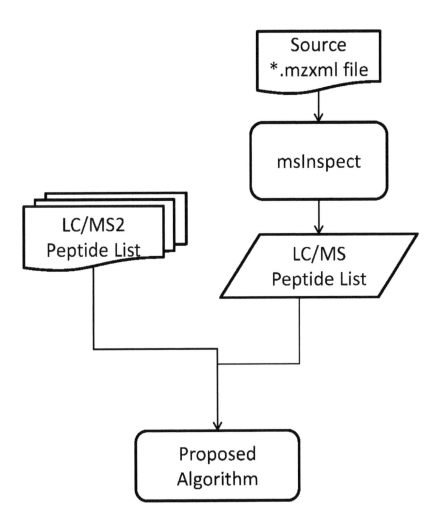

FIGURE 6: Verification process.

peptides than msInspect at low FDRs. We used reference and peptide templates extracted from LC-MS data for these calculations.

We also compared the performance of msInspect and MRCQuant based on LC-MS/MS-identified peptides. However, when allowing a 10 ppm tolerance, there are only 31 LC-MS/MS-identified peptides that overlap with msInspect-reported peptides. In other words, most peptides compiled from multiple LC-MS/MS runs were not reported by msInspect. With such a small number of overlaps, we could not perform a meaningful comparison. In contrast to the low detection rate of LC-MS/MS-identified peptides by msInspect, MRCQuant quantified 423 LC-MS/MS-identified peptides in total, among which, 203 have an FDR <= 0.1.

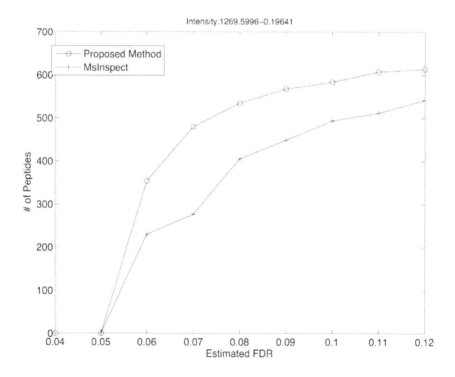

FIGURE 7: Number of detections v.s. estimated FDR of the proposed algorithm and msInspect. The proposed algorithm improves relative quantification accuracy greatly over msInspect on low FDR regions in the number of reported peptides.

13.3.2 PERFORMANCE AT DIFFERENT INTENSITY LEVELS

MRCQuant is mainly designed to correctly quantify peptides at low intensity levels where the effect of noise is the most detrimental. To evaluate the performance at different intensity levels, we sorted peptides according to their peak intensities reported by msInspect. Then, we divided peptides into 4 different groups according to their intensity levels, and we plotted the performance curves (the number of peptides v.s. FDR) as shown in Figure 8. MRCQuant clearly provides similar performance to msInspect in the high intensity region (33-1269); however, MRCQuant provides better performance over msInspect in lower intensity regions. Note that there are more peptides in low intensity regions (> 600) than in the high intensity

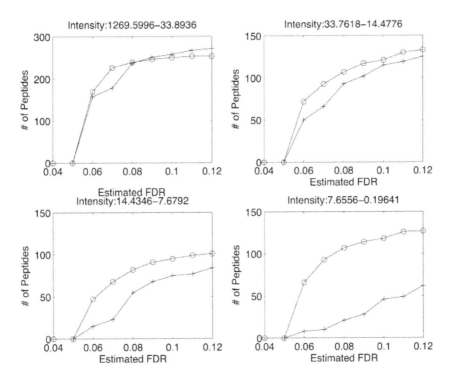

FIGURE 8: Performance at different intensity levels. As the intensity level lowers, the proposed algorithm provides better and better performance over msInspect in the number of reported peptides.

region (300). Thus, MRCQuant has a much better performance on the low intensity regions where most peptides can be found.

13.3.3 EFFECT OF USING DIFFERENT TEMPLATES

MRCQuant can be configured to use extracted or theoretically predicted reference templates for LC peak detection, and it can also be configured to use locally extracted peptide or reference templates for quantification. Thus, there are four possible ways of employing MRCQuant: (a). Use the extracted reference template for LC peak detection and use the extracted peptide template for quantification. (b). Use the theoretically predicted reference template for LC peak detection and use the extracted peptide template for quantification. (c). Use the extracted reference template for LC peak detection and quantification. (d). Use the theoretically predicted reference template for LC peak detection and quantification. We tested these four cases on the LC-MS/MS-identified peptide list. The performances are reported in Figure 9. The selection of templates greatly affects quantification performance. Case (a) uses the most accurate templates possible in both LC peak detection and quantification, and the result is the best with significantly higher number of reported peptides on the low FDR region. The comparison between case (a) and case (b) reveals the effect of mass drift on quantification accuracy. In case (a), the extracted reference templates are used, the mass drift in a specific LC-MS run is automatically addressed, and thus LC peak detection is more accurate. In case (b), the theoretically predicted MS reference template was not adjusted for mass drift and the resulted LC peak detection result is poor. Comparing case (a) and (c), we can see quantification accuracy degradation caused by not using extracted peptide templates. Slight variations in local signal peak shapes affect quantification accuracy significantly. We also compared the performance of using different templates based on msInspect generated peptide list. Again, more peptides are reported with low FRDs in case (a) than in other cases, which confirms the importance of using extracted templates.

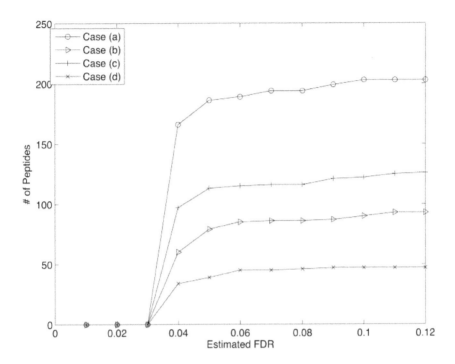

FIGURE 9: Performance with different template selections. (LC Peak Detection, Quantification): a. (Extracted Reference, Extracted Peptide); b. (Theoretical Reference, Extracted Peptide); c. (Extracted Reference, Extracted Reference); d.(Theoretical Reference, Theoretical Reference). The selection of templates greatly affects the quantification performance.

13.4 CONCLUSIONS

In this paper, we describe a new algorithm called "MRCQuant" for LC-MS relative quantification of "bottom-up" proteomics data based on extracted MS templates. Reference- and peptide- MS templates are extracted from scans with relatively high SNRs using MRC, a process that optimally suppresses noise.

Subsequently, these templates are used for detecting LC peak boundaries, detecting interference, and removing noise at lower SNRs. MRCQuant performs automatic mass drift correction by utilizing extracted MS templates which capture mass deviation from theoretical mass values. These techniques address major deficiencies in previous LC-MS quantification algorithms effectively. We demonstrate significant improvement in relative quantification accuracy with a larger number of detected peptides at low FDRs compared to msInspect. We expect that MRCQuant can be integrated with various LC-MS processing software to improve the overall performance. For example, MRCQuant can be readily modified and applied in label and label-free proteomic experiments for quantitative analysis. The proposed algorithm can also be incorporated in LC-MS peak detection algorithms that use isotope ratios.

REFERENCES

1. Aebersold R, Mann M: Mass spectrometry-based proteomics. Nature 2003, 422(6928):198-207.
2. Mueller L, Brusniak M, Mani D, Aebersold R: An assessment of software solutions for the analysis of mass spectrometry based quantitative proteomics data. Journal of proteome research 2008, 7(01):51-61.
3. Bantscheff M, Schirle M, Sweetman G, Rick J, Kuster B: Quantitative mass spectrometry in proteomics: a critical review. Analytical and bioanalytical chemistry 2007, 389(4):1017-1031.
4. Wang G, Wu W, Pisitkun T, Hoffert J, Knepper M, Shen R: Automated quantification tool for high-throughput proteomics using stable isotope labeling and LC-MSn. Analytical chemistry 2006, 78(16):5752.
5. Mann B, Madera M, Sheng Q, Tang H, Mechref Y, Novotny M: ProteinQuant Suite: a bundle of automated software tools for label-free quantitative proteomics. Rapid Communications in Mass Spectrometry 2008, 22(23):3823-3834.
6. Bellew M, Coram M, Fitzgibbon M, Igra M, Randolph T, Wang P, May D, Eng J, Fang R, Lin C, et al.: A suite of algorithms for the comprehensive analysis of complex protein mixtures using high-resolution LC-MS. Bioinformatics 2006, 22(15):1902.
7. Mueller L, Rinner O, Schmidt A, Letarte S, Bodenmiller B, Brusniak M, Vitek O, Aebersold R, Muller M: SuperHirn-a novel tool for high resolution LC-MS-based peptide/protein profiling. Proteomics 2007, 7(19):3470-80.
8. Li X, Zhang H, Ranish J, Aebersold R: Automated Statistical Analysis of Protein Abundance Ratios from Data Generated by Stable-Isotope Dilution and Tandem Mass Spectrometry. ANALYTICAL CHEMISTRY-WASHINGTON DC 2003, 75(23):6648-6657.

9. Leptos K, Sarracino D, Jaffe J, Krastins B, Church G: MapQuant: Open-source software for large-scale protein quantification. Proteomics 2006, 6(6):1770-1782.

10. Cox J, Mann M: MaxQuant enables high peptide identification rates, individualized ppb-range mass accuracies and proteome-wide protein quantification. Nature biotechnology 2008, 26(12):1367-1372.

11. Ong S, Mann M: A practical recipe for stable isotope labeling by amino acids in cell culture (SILAC). Nature protocols 2007, 1(6):2650-2660.

12. Du P, Stolovitzky G, Horvatovich P, Bischoff R, Lim J, Suits F: A noise model for mass spectrometry based proteomics. Bioinformatics 2008, 24(8):1070.

13. Shin H, Koomen J, Baggerly K, Markey M: Towards a noise model of MALDI TOF spectra. American Association for Cancer Research (AACR) advances in proteomics in cancer research 2004.

14. Goldsmith A: Wireless communications. Cambridge Univ Pr; 2005.

15. Draper N, Smith H: Applied Regression Analysis. Volume ch. 10. 3rd edition. Wiley-Interscience, New York; 1998.

16. Bayne C, Smith D: A new method for estimating isotopic ratios from pulse-counting mass spectrometric data. International Journal of Mass Spectrometry and Ion Processes 1984, 59(3):315-323.

17. Fletcher R: Practical Methods of Optimization: Vol. 2: Constrained Optimization. JOHN WILEY & SONS, INC., ONE WILEY DR., SOMERSET, N. J. 08873, 1981, 224 1981.

18. Liddle A: Information criteria for astrophysical model selection. Monthly Notices of the Royal Astronomical Society: Letters 2007, 377:L74-L78.

19. Renard B, Kirchner M, Steen H, Steen J, Hamprecht F: NITPICK: peak identification for mass spectrometry data. BMC bioinformatics 2008, 9:355.

20. Wang Y, Zhou X, Wang H, Li K, Yao L, Wong S: Reversible jump MCMC approach for peak identification for stroke SELDI mass spectrometry using mixture model. Bioinformatics 2008, 24(13):i407.

21. Klimek J, Eddes J, Hohmann L, Jackson J, Peterson A, Letarte S, Gafken P, Katz J, Mallick P, Lee H, et al.: The standard protein mix database: A diverse dataset to assist in the production of improved peptide and protein identification software tools. Journal of proteome research 2008, 7:96.

22. Keller A, Nesvizhskii A, Kolker E, Aebersold R: Empirical statistical model to estimate the accuracy of peptide identifications made by MS/MS and database search. Anal Chem 2002, 74(20):5383-5392.

23. Zhang J, Gonzalez E, Hestilow T, Haskins W, Huang Y: Review of Peak Detection Algorithms in Liquid-Chromatography-Mass Spectrometry. Current Genomics 2009, 10(6):388.

24. Valkenborg D, Assam P, Thomas G, Krols L, Kas K, Burzykowski T: Using a Poisson approximation to predict the isotopic distribution of sulphur-containing peptides in a peptide-centric proteomic approach. Rapid Commun Mass Spectrom 2007, 21(20):3387-91.

25. Benjamini Y, Hochberg Y: Controlling the false discovery rate: a practical and powerful approach to multiple testing. Journal of the Royal Statistical Society. Series B (Methodological) 1995, 57:289-300.

There are several supplemental files that are not available in this version of the article. To view this additional information, please use the citation information cited on the first page of this chapter.

AUTHOR NOTES

CHAPTER 1

Acknowledgments
The authors would like to thank Matt Hudson for critical review of the manuscript and Alvaro Hernandez and the High-Throughput Sequencing and Genotyping Unit in the W.M. Keck Center for Comparative and Functional Genomics at the University of Illinois at Urbana-Champaign for carrying out the library preparation and RNA sequencing. This work was funded by the National Soybean Research Laboratory's Soybean Disease Biotechnology Center.

CHAPTER 2

Competing Interests
The authors declare that they have no competing interests.

Acknowledgments
F.H. and E.E. are supported by National Science Foundation grants 0513612, 0731455, 0729049 and 0916676, and NIH grants K25-HL080079 and U01-DA024417. This research was supported in part by the University of California, Los Angeles subcontract of contract N01-ES-45530 from the National Toxicology Program and National Institute of Environmental Health Sciences to Perlegen Sciences.

This article has been published as part of *BMC Bioinformatics* Volume 12 Supplement 6, 2011: Proceedings of the First Annual RECOMB Satellite Workshop on Massively Parallel Sequencing (RECOMB-seq). The full contents of the supplement are available online at http://www.biomedcentral.com/1471-2105/12?issue=S6.

CHAPTER 3

Authors' Contributions
BL wrote the RSEM software, co-developed the methodology and experiments, carried out the computational experiments, and helped to draft the manuscript. CD co-developed the methodology and experiments, and wrote the manuscript. All authors read and approved the final manuscript.

Acknowledgments and Funding

We thank Victor Ruotti, Ron Stewart, Angela Elwell, and Jennifer Bolin for feedback on the software and for valuable discussions regarding RNA-Seq protocols. We also thank the reviewers of this manuscript for their constructive comments. BL was partially funded by Dr. James Thomson's MacArthur Professorship and by Morgridge Institute for Research support for Computation and Informatics in Biology and Medicine. CD was partially supported by NIH grant 1R01HG005232-01A1.

CHAPTER 4

Competing Interests
The authors declare that they have no competing interests.

Authors' Contributions
All the authors participated in the design of the study and wrote the manuscript together. All authors read and approved the final manuscript.

Acknowledgments

This work was supported by the National Institutes of Health [RR07801]; and the Civilian Research & Development Foundation [Grant Assistant Program RUB1-1578]; and Russian Foundation for Basic Research [09-04-01590-a, 10-01-00627-a, and 11-04-01162-a]. We thank JUC "Chromas", St.Petersburg State University, Russia, and Center for Developmental Genetics, Stony Brook University, USA, for sharing the confocal microscopes.

CHAPTER 5

Authors' Contributions

MAA developed the programs to perform the simulations with the synthetic dataset and the analysis of the real datasets. DG, EG and VS developed the database to retrieve the data from the Johnson experiment. PCS and APM developed the algorithm to predict the internal dimension of the factorization and performed the simulations using previous algorithms. RP and LMM provided biological insight to the project since its inception and selected the genes and their structures for the analysis. AR conceived the idea, supervised the project and developed the algorithm to predict the internal dimension of the factorization and performed the simulations using previous algorithms. All of the authors participated in the redaction, read and approved the final manuscript.

Acknowledgments

We gratefully acknowledge Dr Wang (Affymetrix) for providing the results of her experiments. This project was partially funded through the 'UTE project CIMA' and an FMMA grant to LMM. APM was partially funded by Spanish grants PR27/05-13964-BSCH, CAM-P2006/Gen-0166, TIN2005-05619 and by Spanish Ramón y Cajal program. We also thank Francesc Subirada (Oryzon Genomics) for his help in this study and the reviewers for their comments on the manuscript which have improved its readability and clarity significantly.

CHAPTER 6

Authors' Contributions

HY came up with the main frameworks of the methods, participated in the computational testing and drafted the manuscript. BHL was in charge of the computational coding and testing, and helped drafting the manuscript. ZQY participated in the method design and helped drafting the manuscript. CL supervised the statistical parts of the methods and modified the manuscript. YYL and YXL conceived of the study, and participated in its design and coordination and modified the manuscript. All authors read and approved the final manuscript.

Links

The Gene Expression Omnibus database http://www.ncbi.nlm.nih.gov/geo/
The DCGL package http://cran.r-project.org/web/packages/DCGL/index.html

Acknowledgments and Funding

We thank Dr. Christian Herder from German Diabetes Center at Heinrich Heine University Duesseldorf and Dr. Harald Grallert from Institute of Epidemiology, Helmholtz Zentrum München for they compiling the list of 52 T2D-associated genes. We also thank Prof. Michael Brent from Washington University in St. Louis for helpful discussions.

This work was supported by the National Natural Science Foundation of China (31000380, 30800641, 60970050), National Key Technologies R&D Program (2008BAI64B01, 2009AA022710, 2011CB910200), Shanghai Pujiang Program (09PJ1407900), and platform program of Chinese Academy of Sciences (KSCX2-EW-R-04).

CHAPTER 7

Competing Interests

The authors declare that they have no competing interests.

Authors' Contributions

MvdL conceived the project and designed the algorithm. HW implemented the algorithm, designed the simulation studies, and collected and analyzed the data. All authors participated in drafting the manuscript.

Acknowledgments

The authors want to thank Cathy Tugulus for sharing her codes and her helpful comments on this work. The authors also thank the reviewers for their precious appraisal of the earlier version of this manuscript. This work was by NIH R01 AI074345. The authors declare no conflicts of interest.

CHAPTER 8

Competing Interests

The authors declare that they have no competing interests.

Authors' Contributions

MM, MS, JCR, AH, ML designed the study, DZ, ML supervised the genotyping, LN, YL, JCR, MM, PS were responsible for clinical data collection, RK, EG performed the statistical analysis, EG wrote the manuscript and all authors contributed to and approved the final draft.

Acknowledgments

We are indebted to all patients whose participation made this study possible as well as to their treating physicians for the great cooperation. We thank the Centre National de Génotypage (CNG) and especially Mark Lathrop and Diana Zelenika for conducting the genotyping study and providing the control data. We also thank Alexis Sidoroff (Austria; data acquisition), Alexander Hellmer (Germany; data management), Konrad Bork, Uwe-Frithjof Haustein, Dieter Vieluf (Germany; clinical case review/expert committee), Davide Zenoni (Italy; data acquisition) and Jan Nico Bouwes Bavinck (The Netherlands; data acquisition) for their support in the study.

The RegiSCAR-study was funded by grants from the European Commission (QLRT-2002-01738), GIS-Institut des Maladies Rares and INSERM (4CH09G) in France, and by a consortium of pharmaceutical companies (Bayer Vital, Boehringer-Ingelheim, Cephalon, GlaxoSmithKline, MSD Sharp and Dohme, Merck, Novartis, Pfizer, Roche, Sanofi-Aventis, Servier). Maja Mockenhaupt received the Else Kröner Memorial Stipendium for support of clinical research through Else Kröner-Fresenius-Foundation. The Centre National de Génotypage (CNG), Paris, France, provided the funding for the genotyping.

CHAPTER 9

Competing Interests

The authors declare that they have no competing interests.

Authors' Contributions

D.H., N.Z., B.P., E.E. and E.H. developed the method. D.H., N.Z., B.P. performed the experiments. D.H., E.E. and E.H. wrote the manuscript.

Acknowledgments

D.H. and E.E. are supported by National Science Foundation grants 0513612, 0731455, 0729049 and 0916676, and NIH grants K25-HL080079 and U01-DA024417. H. is a faculty fellow of the Edmond J. Safra Bioinformatics program at Tel-Aviv University. E.H. and N.Z. were supported by the Israel Science Foundation grant no. 04514831. N.Z. was also supported by NIH Fellowship 5T32ES007142-27. B.P. was supported by NIH grant RC1 GM091332. This research was supported in part by the University of California, Los Angeles subcontract of contract N01-ES-45530 from the National Toxicology Program and National Institute of Environmental Health Sciences to Perlegen Sciences.

This article has been published as part of *BMC Bioinformatics* Volume 12 Supplement 6, 2011: Proceedings of the First Annual RECOMB Satellite Workshop on Massively Parallel Sequencing (RECOMB-seq). The full contents of the supplement are available online at http://www.biomedcentral.com/1471-2105/12?issue=S6.

CHAPTER 10

Authors' Contributions

XJ conceived the study, developed the DDAG model and the BNMBL score, conducted the experiments, and drafted the manuscript. RN identified the BN scores that were evaluated, performed the statistical analysis, and conceived and wrote Additional file 1. MB critically revised the manuscript for intellectual content concerning genetics. SV conceived the notion that we need not represent the relationships among SNPs, and critically revised the entire content of the manuscript. All authors read and approved the final manuscript.

Acknowledgments

The research reported here was funded in part by grant 1K99LM010822-01 from the National Library of Medicine.

CHAPTER 11

Competing Interests
The authors declare that they have no competing interests.

Authors' Contributions
PEM, ML, FG and DAT designed the study and directed its implementation. GA, TOM and AD carried out statistical analyses. MG and WC were responsible for data collection and database management. GA drafted the article that was further reviewed by PEM, FG and DAT. All authors read and approved the final manuscript.

Acknowledgments
The French-Canadian FVL family study was supported by grants from the Canadian Institutes of Health Research (MOP86466) and by the Heart and Stroke Foundation of Canada (T6484). The MARTHA studies were supported by a grant from the Program Hospitalier de la Recherche Clinique. G.A hold an "INSERM Poste d'accueil" position and T.O.M was supported by a grant from the Fondation pour la Recherche Médicale. F.G and P.W. hold Canada Research Chairs. A France-Canada Research Fund 2008 provided opportunities for face-to-face meetings of lead collaborators.

CHAPTER 12

Competing Interests
The authors declare that they have no competing interests.

Acknowledgments
The authors wish to thank Dr. Leslie T. Cooper of the Mayo Clinic for his permission to use the giant cell myocarditis data set; Dr. Manish Kohli of the Mayo Clinic for his permission to use the prostate cancer data set; Dr. LeeAnn Higgins of the University of Minnesota for providing step by step instructions for exporting data. The authors' work was supported by the Kemper Foundation, the University of Minnesota Biomedical In-

formatics and Computational Biology Program, United States National Cancer Institute CA15083 (Mayo Clinic Cancer Center), United States National Institutes of Health Grant CA 136393 (Mayo Clinic SPORE in Ovarian Cancer) from the National Cancer Institute. In addition, the prostate cancer study was supported by National Institutes of Health Grant 1R21CA133536-01A1 and the GCM study was supported by Grant Number 1 UL1 RR024150-01 from the National Center for Research Resources (NCRR), a component of the National Institutes of Health (NIH), and the NIH Roadmap for Medical Research. The contents of this publication are solely the responsibility of the authors and do not necessarily represent the official view of NCRR, NCI or NIH.

This article has been published as part of *BMC Bioinformatics* Volume 13 Supplement 16, 2012: Statistical mass spectrometry-based proteomics. The full contents of the supplement are available online at http://www. biomedcentral.com/bmcbioinformatics/supplements/13/S16.

CHAPTER 13

Authors' Contributions
JZ conceived, developed, and implemented the algorithm. She also prepared the initial manuscript. WH performed the LC-MS/MS experiments and revised the manuscript. KP advised and revised the manuscript. All authors have read and approved the final manuscript.

Availability and Requirements
Relevant data and source Matlab scripts are available at project home page: http://compgenomics.utsa.edu/MRCquant.html webcite

Acknowledgments
This work is supported by San Antonio Life Sciences Institute Research Enhancement, and a grant from National Institute of Health (NIH 2G12RR013646-11). The authors thank the RCMI Proteomics and Protein Biomarkers Cores at UTSA for assistance with experiment design, sample preparation, data collection, results interpretation, and manuscript preparation. We thank the Computational Biology Initiative (UTSA/UTHSC-SA) for providing access and training to the analysis software used. Lastly, the authors gratefully acknowledge the support of the Cancer Therapy and

Research Center (CTRC) at the University of Texas Health Science Center San Antonio, an NCI-designated Cancer Center. (NIH P30CA54174). This work is also partially supported by grants from the Virginia G. Piper Charitable Trust and the Flinn Foundation of Arizona.

INDEX